ORGANIC SYNTHESES

ORGANIC SYNTHESES

AN ANNUAL PUBLICATION OF SATISFACTORY
METHODS FOR THE PREPARATION
OF ORGANIC CHEMICALS

VOLUME 78

2002

A JOHN WILEY & SONS, INC., PUBLICATION

This book is printed on acid-free paper. ∞

Copyright © 2002 by Organic Syntheses, Inc.

Published by John Wiley & Sons, Inc., New York.

All rights reserved. Published simultaneously in Canada.

For ordering and customer service, call 1-800-CALL-WILEY.

Library of Congress Catalog Card Number: 21-17747
ISBN 0-471-23580-6

Printed in the United States of America

10 9 8 7 6 5 4 3 2 1

ORGANIC SYNTHESES

VOLUME	EDITOR-IN-CHIEF	PAGES
I*	†ROGER ADAMS	84
II*	†JAMES BRYANT CONANT	100
III*	†HANS THACHER CLARKE	105
IV*	†OLIVER KAMM	89
V*	†CARL SHIPP MARVEL	110
VI*	†HENRY GILMAN	120
VII*	†FRANK C. WHITMORE	105
VIII*	†ROGER ADAMS	139
IX*	†JAMES BRYANT CONANT	108
Collective Vol. I	A revised edition of Annual Volumes I–IX †HENRY GILMAN, *Editor-in-Chief* 2nd Edition revised by †A. H. BLATT	580
X*	†HANS THACHER CLARKE	119
XI*	†CARL SHIPP MARVEL	106
XII*	†FRANK C. WHITMORE	96
XIII*	†WALLACE H. CAROTHERS	119
XIV*	†WILLIAM W. HARTMAN	100
XV*	†CARL R. NOLLER	104
XVI*	†JOHN R. JOHNSON	104
XVII*	†L. F. FIESER	112
XVIII*	†REYNOLD C. FUSON	103
XIX*	†JOHN R. JOHNSON	105
Collective Vol. II	A revised edition of Annual Volumes X–XIX †A. H. BLATT, *Editor-in-Chief*	654
20*	†CHARLES F. H. ALLEN	113
21*	†NATHAN L. DRAKE	120
22*	†LEE IRVIN SMITH	114
23*	†LEE IRVIN SMITH	124
24*	†NATHAN L. DRAKE	119
25*	†WERNER E. BACHMANN	120
26*	†HOMER ADKINS	124
27*	†R. L. SHRINER	121
28*	†H. R. SNYDER	121
29*	†CLIFF S. HAMILTON	119
Collective Vol. III	A revised edition of Annual Volumes 20–29 †E. C. HORNING, *Editor-in-Chief*	890
30*	†ARTHUR C. COPE	115
31*	†R. S. SCHREIBER	122

Out of print.
† *Deceased.*

VOLUME	EDITOR-IN-CHIEF	PAGES
32*	RICHARD T. ARNOLD	119
33*	†CHARLES C. PRICE	115
34*	†WILLIAM S. JOHNSON	121
35*	†T. L. CAIRNS	122
36*	N. J. LEONARD	120
37*	JAMES CASON	109
38*	†JOHN C. SHEEHAN	120
39*	†MAX TISHLER	114
Collective Vol. IV	A revised edition of Annual Volumes 30–39 NORMAN RABJOHN, *Editor-in-Chief*	1036
40*	†MELVIN S. NEWMAN	114
41*	JOHN D. ROBERTS	118
42*	VIRGIL BOEKELHEIDE	118
43*	B. C. McKUSICK	124
44*	†WILLIAM E. PARHAM	131
45*	†WILLIAM G. DAUBEN	118
46*	E. J. COREY	146
47*	†WILLIAM D. EMMONS	140
48*	†PETER YATES	164
49*	KENNETH B. WIBERG	124
Collective Vol. V	A revised edition of Annual Volumes 40–49 HENRY E. BAUMGARTEN, *Editor-in-Chief*	1234
Cumulative Indices to Collective Volumes, I, II, III, IV, V †RALPH L. AND †RACHEL H. SHRINER, *Editors*		
50*	RONALD BRESLOW	136
51*	†RICHARD E. BENSON	209
52*	HERBERT O. HOUSE	192
53*	ARNOLD BROSSI	193
54*	ROBERT E. IRELAND	155
55*	SATORU MASAMUNE	150
56*	†GEORGE H. BÜCHI	144
57*	CARL R. JOHNSON	135
58*	†WILLIAM A. SHEPPARD	216
59*	ROBERT M. COATES	267
Collective Vol. VI	A revised edition of Annual Volumes 50–59 WAYLAND E. NOLAND, *Editor-in-Chief*	1208
60*	ORVILLE L. CHAPMAN	140
61*	†ROBERT V. STEVENS	165

Out of print.
†*Deceased.*

VOLUME	EDITOR-IN-CHIEF	PAGES
62	MARTIN F. SEMMELHACK	269
63	GABRIEL SAUCY	291
64	ANDREW S. KENDE	308

Collective Vol. VII — A revised edition of Annual Volumes 60–64 — 602
JEREMIAH P. FREEMAN, *Editor-in-Chief*

65	EDWIN VEDEJS	278
66	CLAYTON H. HEATHCOCK	265
67	BRUCE E. SMART	289
68	JAMES D. WHITE	318
69	LEO A. PAQUETTE	328

Reaction Guide to Collective Volumes I–VII and Annual Volumes 65–68
DENNIS C. LIOTTA AND MARK VOLMER, *Editors* — 854

Collective Vol. VIII — A revised edition of Annual Volumes 65–69 — 696
JEREMIAH P. FREEMAN, *Editor-in-Chief*

Cumulative Indices to Collective Volumes, I, II, III, IV, V, VI, VII, VIII
JEREMIAH P. FREEMAN, *Editor-in Chief*

70	ALBERT I. MEYERS	305
71	LARRY E. OVERMAN	285
72	DAVID L. COFFEN	333
73	ROBERT K. BOECKMAN, JR.	352
74	ICHIRO SHINKAI	341

Collective Vol. IX — A revised edition of Annual Volumes 70–74 — 840
JEREMIAH P. FREEMAN, *Editor-in-Chief*

75	AMOS B. SMITH III	257
76	STEPHEN F. MARTIN	340
77	DAVID J. HART	312
78	WILLIAM R. ROUSH	326

Collective Volumes, Collective Indices to Collective Volumes I–VIII, Volumes 75–76, and Reaction Guide are available from John Wiley & Sons, Inc.

*Out of print.
†Deceased.

NOTICE

With Volume 62, the Editors of *Organic Syntheses* began a new presentation and distribution policy to shorten the time between submission and appearance of an accepted procedure. The soft cover edition of this volume is produced by a rapid and inexpensive process, and is sent at no charge to members of the Organic Divisions of the American and French Chemical Society, The Perkin Division of the Royal Society of Chemistry, and The Society of Synthetic Organic Chemistry, Japan. The soft cover edition is intended as the personal copy of the owner and is not for library use. A hard cover edition is published by John Wiley & Sons, Inc. in the traditional format, and differs in content primarily in the inclusion of an index. The hard cover edition is intended primarily for library collections and is available for purchase through the publisher. Annual Volumes 70–74 have been incorporated into a new five-year version of the collective volumes of *Organic Syntheses* which has appeared as *Collective Volume Nine* in the traditional hard cover format. It is available for purchase from the publishers. The Editors hope that the new *Collective Volume* series, appearing twice as frequently as the previous decennial volumes, will provide a permanent and timely edition of the procedures for personal and institutional libraries. The Editors welcome comments and suggestions from users concerning the new editions.

Organic Syntheses has joined the age of electronic publication with the release of its website, www.orgsyn.org. This site is available free of charge to all chemists and contains all of the nine Collective as well as Annual Volumes and Indices.

To create the *Organic Syntheses* web site, the Board of Directors of *OS* formed a collaboration with CambridgeSoft Corporation (Cambridge, MA), producers of ChemOffice and ChemDraw, and DataTrace Publishing Company (Towson, MD), publishers of *ChemTracts*. *OS* fully funded this extensive effort. All of the information in the *OS* collective Volumes, Annual Volumes, and Indices were digitized, mapped, and converted to XML documents by DataTrace. CambridgeSoft developed the website incorporating the databases linking text and chemical structures using their proprietary *ChemOffice Webserver* software. Reaction diagrams are stored in a ChemFinder database to facilitate structure-based searching.

The *OS* website goes far beyond the scrop of the printed version and is fully searchable using a variety of techniques. Using the free *ChemDraw* plugin* for Netscape Navigator or Microsoft Internet Explorer, chemists can draw structural queries directly on the web page and combine structural or reaction transformation queries with full-text and bibliographic search terms, such as chemical name, reagents, molecular formula, apparatus, reagents, or even a hazard or warning phrase. The preparations are categorized into nearly 300 specific reaction types, allowing search by category.

*Because of browser incompatibility, at this time Macintosh users are limited to using Netscape, versions 4.5–4.73. Please read the hardware and software requirements and follow the instructions for plugin installation prior to attempting searches at the website.

NOMENCLATURE

Both common and systematic names of compounds are used throughout this volume, depending on which the Editor-in-Chief felt was more appropriate. The *Chemical Abstracts* indexing name for each title compound, if it differs from the title name, is given as a subtitle. Systematic *Chemical Abstracts* nomenclature, used in both the recent Collective Indexes for the title compound and a selection of other compounds mentioned in the procedure, is provided in an appendix at the end of each preparation. Registry numbers, which are useful in computer searching and identification, are also provided in these appendixes. Whenever two names are concurrently in use and one name is the correct *Chemical Abstracts* name, that name is preferred.

SUBMISSION OF PREPARATIONS

Organic Syntheses welcomes and encourages submission of experimental procedures which lead to compounds of wide interest or which illustrate important new developments in methodology. The Editorial Board will consider proposals in outline format as shown below, and will request full experimental details for those proposals which are of sufficient interest. Submissions which are longer than three steps from commercial sources or from existing *Organic Syntheses* procedures will be accepted only in unusual circumstances.

Organic Syntheses Proposal Format

1) Authors
2) Title
3) Literature reference or enclose preprint if available
4) Proposed sequence
5) Best current alternative(s)
6) a. Proposed scale, final product:
   b. Overall yield:
   c. Method of isolation and purification:
   d. Purity of product (%):
   e. How determined?

7) Any unusual apparatus or experimental technique?
8) Any hazards?
9) Source of starting material?
10) Utility of method or usefulness of product

Submit to: Dr. Jeremiah P. Freeman, Secretary
            Department of Chemistry
            University of Notre Dame
            Notre Dame, IN 46556

Proposals will be evaluated in outline form, again after submission of full experimental details and discussion, and, finally by checking experimental procedures. A form that details the preparation of a complete procedure (Notice to Submitters) may be obtained from the Secretary.

Additions, corrections, and improvements to the preparations previously published are welcomed; these should be directed to the Secretary. However, checking of such improvements will only be undertaken when new methodology is involved. Substantially improved procedures have been included in the Collective Volumes in place of a previously published procedure.

ACKNOWLEDGMENT

Organic Syntheses wishes to acknowledge the contributions of Discovery Partners Intl., Hoffmann-La Roche, Inc. and Merck & Co. to the success of this enterprise through their support, in the form of time and expenses, of members of the Boards of Directors and Editors.

HANDLING HAZARDOUS CHEMICALS
A Brief Introduction

General Reference: *Prudent Practices in the Laboratory*; National Academy Press; Washington, DC, 1995.

Physical Hazards

Fire. Avoid open flames by use of electric heaters. Limit the quantity of flammable liquids stored in the laboratory. Motors should be of the nonsparking induction type.

Explosion. Use shielding when working with explosive classes such as acetylides, azides, ozonides, and peroxides. Peroxidizable substances such as ethers and alkenes, when stored for a long time, should be tested for peroxides before use. Only sparkless "flammable storage" refrigerators should be used in laboratories.

Electric Shock. Use 3-prong grounded electrical equipment if possible.

Chemical Hazards

Because all chemicals are toxic under some conditions, and relatively few have been thoroughly tested, it is good strategy to minimize exposure to all chemicals. In practice this means having a good, properly installed hood; checking its performance periodically; using it properly; carrying out most operations in the hood; protecting the eyes; and, since many chemicals can penetrate the skin, avoiding skin contact by use of gloves and other protective clothing.

a. Acute Effects. These effects occur soon after exposure. The effects include burn, inflammation, allergic responses, damage to the eyes, lungs, or nervous system (e.g., dizziness), and unconsciousness or death (as from overexposure to HCN). The effect and its cause are usually obvious and so are the methods to prevent it. They generally arise from inhalation or skin contact, so should not be a problem if one follows the admonition "work in a hood and keep chemicals off your hands". Ingestion is a rare route, being generally the result of eating in the laboratory or not washing hands before eating.

b. Chronic Effects. These effects occur after a long period of exposure or after a long latency period and may show up in any of numerous organs. Of the chronic effects of chemicals, cancer has received the most attention lately. Several dozen chemicals have been demonstrated to be carcinogenic in man and hundreds to be carcinogenic to animals. Although there is no simple correlation between carcinogenicity in animals and in man, there is little doubt that a significant proportion of the chemicals used in laboratories have some potential for carcinogenicity in man. For this and other reasons, chemists should employ good practices.

The key to safe handling of chemicals is a good, properly installed hood, and the referenced book devotes many pages to hoods and ventilation. It recommends that in a laboratory where people spend much of their time working with chemicals there should be a hood for each two people, and each should have at least 2.5 linear feet (0.75 meter) of working space at it. Hoods are more than just devices to keep undesirable vapors from the laboratory atmosphere. When closed they provide a protective barrier between chemists and chemical operations, and they are a good containment device for spills. Portable shields can be a useful supplement to hoods, or can be an alternative for hazards of limited severity, e.g., for small-scale operations with oxidizing or explosive chemicals.

Specialized equipment can minimize exposure to the hazards of laboratory operations. Impact resistant safety glasses are basic equipment and should be worn at all times. They may be supplemented by face shields or goggles for particular operations, such as pouring corrosive liquids. Because skin contact with chemicals can lead to skin irritation or sensitization or, through absorption, to effects on internal organs, protective gloves are often needed.

Laboratories should have fire extinguishers and safety showers. Respirators should be available for emergencies. Emergency equipment should be kept in a central location and must be inspected periodically.

MSDS (Materials Safety Data Sheets) sheets are available from the suppliers of commercially available reagents, solvents, and other chemical materials; anyone performing an experiment should check these data sheets before initiating an experiment to learn of any specific hazards associated with the chemicals being used in that experiment.

DISPOSAL OF CHEMICAL WASTE

General Reference: *Prudent Practices in the Laboratory*, National Academy Press, Washington, D.C. 1995

Effluents from synthetic organic chemistry fall into the following categories:

1. **Gases**

   1a. Gaseous materials either used or generated in an organic reaction.
   1b. Solvent vapors generated in reactions swept with an inert gas and during solvent stripping operations.
   1c. Vapors from volatile reagents, intermediates and products.

2. **Liquids**

   2a. Waste solvents and solvent solutions of organic solids (see item 3b).
   2b. Aqueous layers from reaction work-up containing volatile organic solvents.
   2c. Aqueous waste containing non-volatile organic materials.
   2d. Aqueous waste containing inorganic materials.

3. **Solids**

   3a. Metal salts and other inorganic materials.
   3b. Organic residues (tars) and other unwanted organic materials.
   3c. Used silica gel, charcoal, filter aids, spent catalysts and the like.

The operation of industrial scale synthetic organic chemistry in an environmentally acceptable manner* requires that all these effluent categories be dealt with properly. In small scale operations in a research or academic set-

---

*An environmentally acceptable manner may be defined as being both in compliance with all relevant state and federal environmental regulations *and* in accord with the common sense and good judgement of an environmentally aware professional.

ting, provision should be made for dealing with the more environmentally offensive categories.

1a. Gaseous materials that are toxic or noxious, e.g., halogens, hydrogen halides, hydrogen sulfide, ammonia, hydrogen cyanide, phosphine, nitrogen oxides, metal carbonyls, and the like.

1c. Vapors from noxious volatile organic compounds, e.g., mercaptans, sulfides, volatile amines, acrolein, acrylates, and the like.

2a. All waste solvents and solvent solutions of organic waste.

2c. Aqueous waste containing dissolved organic material known to be toxic.

2d. Aqueous waste containing dissolved inorganic material known to be toxic, particularly compounds of metals such as arsenic, beryllium, chromium, lead, manganese, mercury, nickel, and selenium.

 3. All types of solid chemical waste.

Statutory procedures for waste and effluent management take precedence over any other methods. However, for operations in which compliance with statutory regulations is exempt or inapplicable because of scale or other circumstances, the following suggestions may be helpful.

Gases

Noxious gases and vapors from volatile compounds are best dealt with at the point of generation by "scrubbing" the effluent gas. The gas being swept from a reaction set-up is led through tubing to a (large!) trap to prevent suckback and on into a sintered glass gas dispersion tube immersed in the scrubbing fluid. A bleach container can be conveniently used as a vessel for the scrubbing fluid. The nature of the effluent determines which of four common fluids should be used: dilute sulfuric acid, dilute alkali or sodium carbonate solution, laundry bleach when an oxidizing scrubber is needed, and sodium thiosulfate solution or diluted alkaline sodium borohydride when a reducing scrubber is needed. Ice should be added if an exotherm is anticipated.

Larger scale operations may require the use of a pH meter or starch/iodide test paper to ensure that the scrubbing capacity is not being exceeded.

When the operation is complete, the contents of the scrubber should be handled as aqueous waste, as outlined in the "Liquids" section that follows. In many instances, this will require neutralization, followed by concentration to a minimum volume, or concentration to dryness before disposal as concentrated liquid or solid chemical waste.

Liquids

Every laboratory should be equipped with a waste solvent container in which *all* waste organic solvents and solutions are collected. The contents of these containers should be periodically transferred to properly labeled waste solvent drums and arrangements made for contracted disposal in a regulated and licensed incineration facility.**

Aqueous waste containing dissolved toxic organic material should be decomposed *in situ*, when feasible, by adding acid, base, oxidant, or reductant. Otherwise, the material should be concentrated to a minimum volume and added to the contents of a waste solvent drum.

Aqueous waste containing dissolved toxic inorganic material should be evaporated to dryness and the residue handled as a solid chemical waste.

Solids

Soluble organic solid waste can usually be transferred into a waste solvent drum, provided near-term incineration of the contents is assured.

Inorganic solid wastes, particularly those containing toxic metals and toxic metal compounds, used Raney nickel, manganese dioxide, etc. should be placed in glass bottles or lined fiber drums, sealed, properly labeled, and arrangements made for disposal in a secure landfill.** Used mercury is particularly pernicious and small amounts should first be amalgamated with zinc or combined with excess sulfur to solidify the material.

Other types of solid laboratory waste including used silica gel and charcoal should also be packed, labeled, and sent for disposal in a secure landfill.

Special Note

Since local ordinances may vary widely from one locale to another, one should always check with appropriate authorities. Also, professional disposal services differ in their requirements for segregating and packaging waste.

**If arrangements for incineration of waste solvent and disposal of solid chemical waste by licensed contract disposal services are not in place, a list of providers of such services should be available from a state or local office of environmental protection.

PREFACE

This volume of *Organic Syntheses*, the 78th volume in this series, provides 29 carefully checked and edited experimental procedures that describe important synthetic methods, transformations, reagents, and synthetic building blocks or intermediates with demonstrated utility in organic synthesis. There is no central theme to this Volume. The procedures in this compilation fall into four general areas: (1) reagents and methods for catalytic asymmetric synthesis; (2) organometallic chemistry, ligands, and transformations of organometallic reagents; (3) synthetically useful reagents and intermediates; and (4) useful synthetic transformations.

This volume begins with two procedures in the area of catalytic asymmetric synthesis. The first procedure describes the synthesis of **(R)-2-DIPHENYLPHOSPHINO-2′-METHOXY-1,1′-BINAPHTHYL** (MOP), a chiral ligand that has proven very useful in palladium-catalyzed hydrosilylation of olefins and palladium-catalyzed reduction of allylic esters by formic acid. The next procedure describes the catalytic asymmetric synthesis of nitroaldols using a chiral **LANTHANUM-LITHIUM-BINOL COMPLEX,** illustrated by the synthesis of **(2S,3S)-2-NITRO-5-PHENYL-1,3-PENTANEDIOL.**

In the area of organometallic chemistry, we begin with a procedure for the palladium-catalyzed amination of aryl halides and aryl triflates, illustrated by syntheses of **N-HEXYL-2-METHYL-4-METHOXYANILINE** and **N-METHYL-N-(4-CHLOROPHENYL)-ANILINE.** The next procedure describes the synthesis of **1,2,3,4-TETRAHYDROCARBAZOLE** by the palladium-catalyzed annulation of ketones with O-iodoaniline. Next, a procedure for the synthesis of **2,7-DIMETHYLNAPHTHALENE** via the nickel-catalyzed coupling of aryl O-carbamates with Grignard reagents is presented. The fourth procedure in this section describes the synthesis of **5-METHYL-2,2′-BIPYRIDINE** by a Negishi cross-coupling reaction of a 2-pyridyl zinc reagent. The preparation of **2-FLUORO-4-METHOXYANILINE** illustrates the Ullmann methoxylation reaction of a 2,5-dimethylpyrrole-blocked aniline. The section on organometallic chemistry concludes with a procedure for the synthesis of **1,4,7,10-TETRAAZACYCLODODECANE** (cyclen), which has been used as a ligand for MRI contrast agents.

The third section of this Volume presents procedures for the synthesis of useful reagents and intermediates. First, an efficient synthesis of **4,4′-BIS(CHLOROMETHYL)-2,2′-BIPYRIDINE** is described. 2,2′-Bipyridine derivatives have been used in supramolecular assembly, in bioinorganic contexts, in polymeric materials and in studies of redox catalysis. This procedure is followed by the preparation of **N,N′-DI-BOC-N″-TRIFLYLGUANIDINE,** which is a useful reagent for the guanidinylationof amines in solution and in solid phase synthesis. A synthesis of the powerful sulfenylating agent **S-METHYL METHANETHIOSULFONATE** is described next, illustrating a general method for the synthesis thiosulfonate thioesters via zinc reduction of sulfonyl halides. **(PHENYL)[2-(TRIMETHYLSILYL)PHENYL]IODONIUM TRIFLATE** is offered as an efficient and mild benzyne precursor, as demonstrated by the synthesis of **1,4-EPOXY-1,4-DIHYDRONAPHTHALENE.** The subsequent procedure describing the synthesis of **4-HYDROXY[1-^{13}C]BENZOIC ACID,** an important intermediate for biosynthesis studies, is only the second procedure published in *Organic Syntheses* in the area of isotopically labeled compounds. Two useful chiral building blocks are described in the procedure entitled **(1′R)-(-)-2,4-O-ETHYLIDENE-D-ERYTHROSE AND ETHYL(E)-(-)-4,6-O-ETHYLIDENE-(4S,5R,1′R)-4,5,6-TRIHYDROXY-2-HEXENOATE,** both of which have been used as starting materials in a variety of contexts. The next procedure describes the preparation of **PENTA-1,2-DIEN-4-ONE (ACETYLALLENE),** a valuable starting material for synthesis of α,β-unsaturated γ-lactones and as a dienophile in Diels-Alder reactions. **BICYCLOPROPYLIDENE** possesses unique reactivity toward a wide range of electrophiles and nucleophilic carbenes. **(E)-1-DIMETHYLAMINO-3-tert-BUTYLDIMETHYL-SILOXY-1,3-BUTADIENE** is a highly reactive diene for Diels-Alder reactions, as described in an accompanying procedure for the synthesis of **4-HYDROXYMETHYL-2-CYCLOHEXEN-1-ONE** via the Diels-Alder reaction with methyl acrylate. Finally, this section concludes with the preparation of **DIETHYL[(PHENYLSULFONYL)METHYL]PHOSPHONATE,** a reagent that is very useful for synthesis of α,β-unsaturated phenyl sulfones.

The final set of procedures describe general methods for important synthetic transformations. First, a synthesis of **(-)-(E,S)-3-(BENZYLOXY)-1-BUTENYL PHENYL SULFONE** via the Horner-Wadsworth-Emmons olefination of **(-)-(S)-2-(BENZYLOXY)PROPANAL** with diethyl [(phenylsulfonyl)methyl]phosphonate is described. In the accompanying procedure, (-)-(E,S)-3-(benzyloxy)-1-butenyl phenyl sulfone is converted into **((+)-(1R,2S,3R)-TETRACARBONYL[(1-3η)-1-(PHENYLSULFONYL)-BUT-2-EN-1-YL]IRON(1+) TETRAFLUOROBORATE,** a chiral cationic π-allyl iron(tetracarbonyl) complex that undergoes substitution reactions with a wide range of nucleophilic reagents. The synthesis of **tert-BUTYL 3a-METHYL-5-OXO-2,3,3A,4,5,6-HEXAHYDROINDOLE-1-CARBOXYL-ATE** illustrates a general procedure for Diels-Alder reactions of 2-amido-substituted furans, while the preparation of **11-OXATRICYCLO[4.3.1.1^{2,5}]-UNDEC-3-EN-10-ONE** demonstrates the [4+3] cycloaddition of aminoallyl cations with 1,3-dienes. The synthesis of amino acid ester isocyanates is illustrated by the preparation of **METHYL(S)-2-ISOCYANATO-3-PHENYL-PROPANOATE.** In the area of oxidation reactions, the procedure for the synthesis of **trans-2-METHYL-2,3-DIPHENYLOXIRANE** illustrates the use of a dioxirane reagent generated in situ from **TETRAHYDROTHIO-PYRAN-4-ONE AND OXONE.** The next procedure describes modified conditions for the Hoffmann rearrangement, as illustrated by the synthesis of **METHYL N-(p-METHOXYPHENYL)CARBAMATE.** Use of **TRI-BUTYLSTANNANE** (Bu$_3$SnH) for Barton-McCombie radical deoxygenation of alcohols is detailed in the synthesis of **3-DEOXY-1,2:5,6-BIS-O-(1-METHYLETHYLIDENE)-α-D-RIBO-HEXOFURANOSE,** in which the tin hydride reagent is generated catalytically from bis(tributyltin)oxide and poly(methylhydrosiloxane) (PMHS). The synthesis of **2-(3-OXOBUTYL)-CYCLOPENTANONE-2-CARBOXYLIC ACID ETHYL ESTER** describes a mild, FeCl$_3$ catalyzed Michael reaction of cyclopentanone-2-carboxylic acid ethyl ester and methyl vinyl ketone. Finally, a procedure for synthesis of **1-OXO-2-CYCLOHEXENYL-2-CARBONITRILE** is presented, illustrating a mild method for cynthesis of electron-deficient alkenes which are valuable reactants for cycloaddition and conjugate addition reactions.

Many individuals have contributed to this Volume of *Organic Syntheses*, and I would like to take this opportunity to thank them. Most importantly, a debt of gratitude is owed the authors who submitted the procedures that are published here. Their dedication and commitment to excellence in research is reflected in the important procedures that comprise this Volume. Thanks are also due to the members of the Editorial Board, and especially their students and postdoctoral associates for their tireless efforts in checking, and in some cases improving these procedures, thereby verifying that they are reproducible as written. Finally, I would like to thank Professor Jeremiah P. Freeman, Secretary to the Board, and Dr. Theodora W. Greene, the Assistant Editor of *Organic Syntheses*. It has been a privilege to work with Jerry, Theo, and all members of the Board. Without the substantial efforts of these colleagues, it would not have been possible for Volume 78 to be completed in a timely manner.

WILLIAM R. ROUSH

Ann Arbor, Michigan

NORMAN RABJOHN
May 1, 1915–September 2, 2000

Norman Rabjohn was an organic chemist of great insight and ability, a highly respected teacher, and a wonderfully reliable friend. At the time of this death in Columbia, Missouri, at the age of 85, he had held the title of Professor Emeritus of Chemistry of the University of Missouri for 17 years. His major contributions to *Organic Syntheses* were in his roles as Secretary during the period 1949–1959, as Treasurer from 1976 to 1980, and as Editor of Collective Volume IV (1963) of this series.

Norman was born in Rochester, New York, attended public schools there and also the University of Rochester, from which he received his B.S. degree in Chemistry in 1937, Magna cum Laude, and was elected to Phi Beta Kappa and Sigma Xi. During the following year, he worked at Eastman Kodak Company. It will be remembered that Eastman was the main commercial source of laboratory organic chemicals in the first half of the 20th century. The skill that Norm developed in organic synthesis was put to good use during the time that he spent at the University of Illinois, where he earned his M.S. and Ph.D. degrees (1939 and 1942, respectively). With his mentor, Professor R. C. Fuson, he published seven papers plus one preparative submission to *Organic Syntheses*, all of which described derivatives of mesitylene, including the first of the stable enols. These stable enols or vinyl alcohols, as they are called alternatively, are still of structural, theoretical, and experimental interest. Norm remained at Illinois for two years as an Instructor in Organic Chemistry and a participant in the Rubber Reserve Program of the War Production Board under the direction of Professor C. S. Marvel. He impressed me, when we were together at Illinois, as a builder of apparatus, e.g., for his study of the vapor-phase chlorination of aliphatic ketones, while his contributions to the Rubber Program included emulsion polymerization at high temperature and structure determination of butadiene polymers by ozonolysis.

During the years 1944–1948, Norm continued research in polymer chemistry at the Goodyear Tire and Rubber Company, where he coauthored three patents with Paul J. Flory on the vulcanization and curing of rubber at low temperature, along with four papers on the dependence of elastic properties and tensile strength of vulcanized rubber of the degree of cross-linking. Dr. Flory was later to receive the Nobel Prize in Chemistry following academic appointments at Cornell and Stanford Universities. Norman Rabjohn left Goodyear in 1948 to become Associate Professor of Chemistry at the University of Missouri and was advanced to Professor four years later. He maintained a long, distinguished and balanced career in research, teaching, and service at Missouri.

His academic research was focused to provide training for the 50 Ph.D. students who benefitted from his tutelage. Work on the synthesis and reactions of azodicarboxylic esters that had been started at Goodyear was continued, β-dialkylaminoethyl esters of sterically hindered alkyl-substituted benzoic acids were made that had longer local anaesthetic action than Procaine, and azo-nitrogen analogs of unsaturated acids and their derivatives were studied for potential antineoplastic activity. There was general interest in synthesis, product distribution in reactions, and occurrence or non-occurrence of rearrangements. Particular emphasis was placed on the synthesis and physical and chemical properties of highly substituted compounds, including dissymmetric tetrasubstituted methanes and compounds containing adjacent quaternary carbons. Compounds within these categories undergo reactions and rearrangements that depend upon the bulk and conformations of the groups involved. In the final instance, Kolbe electrolysis was used in an imaginative way to effect coupling of the bulky acid and ketoacid precursors. In addition, a timely study provided the ^{13}C NMR spectra of all of the highly branched acyclic compounds that had resulted from earlier syntheses. Norm wrote a definitive chapter on selenium dioxide oxidation for *Organic Reactions* early in his career and then capped this off more than two decades later with another chapter on the same subject in the same series, bringing us complete information on the reagent and its effective usage.

At one time or another, Norman taught all of the courses offered in organic chemistry at the University of Missouri. He had the gift of making the subject matter interesting, with emphasis on practical applications. He served two terms as department chair and two terms on the faculty council, and he was a key member of many university committees. Missouri honored him for his many contributions with the Faculty-Alumni Award in 1976. His colleagues, family, and former students established a professorship in his name at the University of Missouri, Columbia, in 1994. The first Rabjohn Distinguished Professor of Organic Chemistry, Michael Harmata, was appointed in 1999.

Norm was a veteran member of the American Chemical Society and a very active one. A most significant service to that organization was his term as chair of the national publications committee. He was involved in many of the planning sessions when Chemical Abstracts was making the transition to computer-searchable data bases. He took a turn as a member of the editorial board of *Chemical Reviews* and maintained membership in the American Association for the Advancement of Science, Phi Lambda Upsilon, Alpha Chi Sigma, the Missouri Academy of Science, and the MU Jefferson Club.

Norm is survived by his wife of nearly 57 years, Dora Rabjohn of Columbia, and by his son, James Rabjohn, of Aptos, California. He is remembered fondly by all his colleagues, past and present, on the Board of Organic Syntheses, Inc.

NELSON J. LEONARD

November 7, 2000

CONTENTS

(R)-2-DIPHENYLPHOSPHINO-2'-METHOXY-1,1'-BINAPHTHYL.   1
Yasuhiro Uozumi, Motoi Kawatsura, and Tamio Hayashi

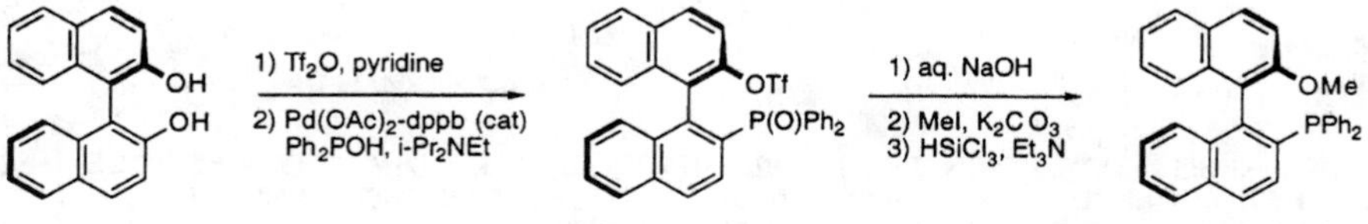

[(R)-MeO-MOP]

CATALYTIC ASYMMETRIC SYNTHESIS OF NITROALDOLS   14
USING A LANTHANUM-LITHIUM-BINOL COMPLEX:
(2S,3S)-2-NITRO-5-PHENYL-1,3-PENTANEDIOL.
Hiroaki Sasai, Shizue Watanabe, Takeyuki Suzuki, and Masakatsu Shibasaki

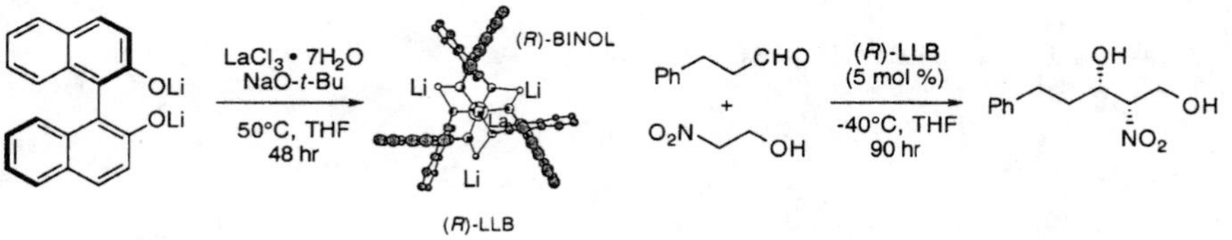

(R)-LLB

PALLADIUM-CATALYZED AMINATION OF ARYL HALIDES AND ARYL   23
TRIFLATES: N-HEXYL-2-METHYL-4-METHOXYANILINE AND N-METHYL-N-(4-
CHLOROPHENYL)ANILINE.
John P. Wolfe and Stephen L. Buchwald

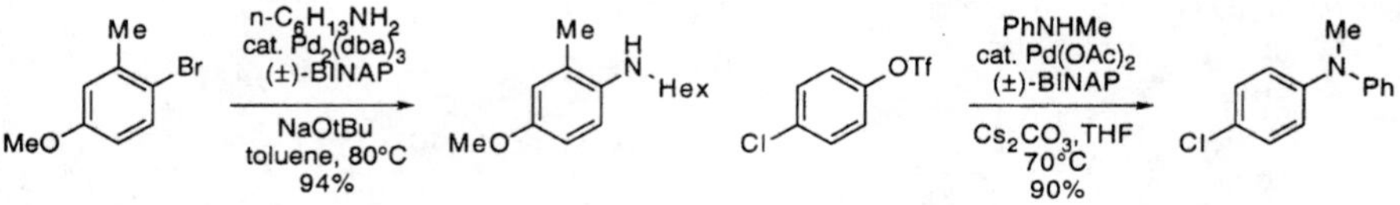

INDOLE SYNTHESIS BY Pd-CATALYZED ANNULATION OF KETONES   36
WITH o-IODOANILINE: 1,2,3,4-TETRAHYDROCARBAZOLE.
Cheng-yi Chen and Robert D. Larsen

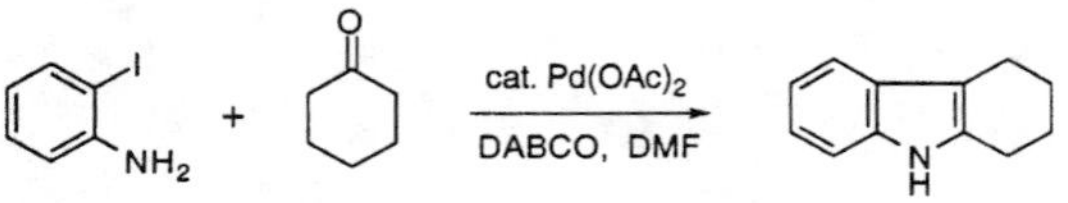

NICKEL-CATALYZED COUPLING OF ARYL O-CARBAMATES WITH GRIGNARD REAGENTS: 2,7-DIMETHYLNAPHTHALENE.

Carol Dallaire, Isabelle Kolber, and Marc Gingras

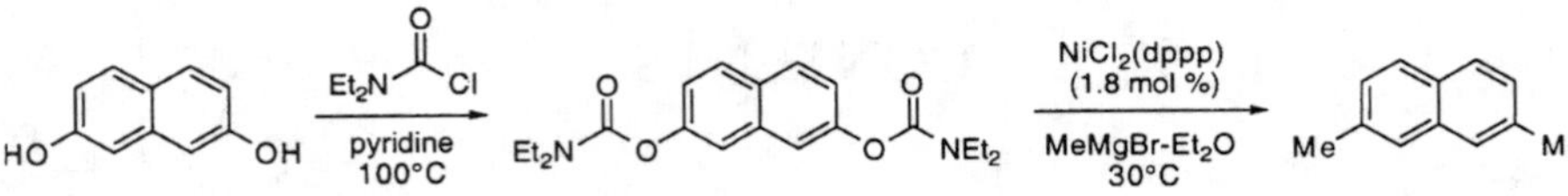

42

SYNTHESIS OF 4-, 5-, and 6-METHYL-2,2'-BIPYRIDINE BY A NEGISHI CROSS-COUPLING STRATEGY: 5-METHYL-2,2'-BIPYRIDINE.

Adam P. Smith, Scott A. Savage, J. Christopher Love, and Cassandra L. Fraser

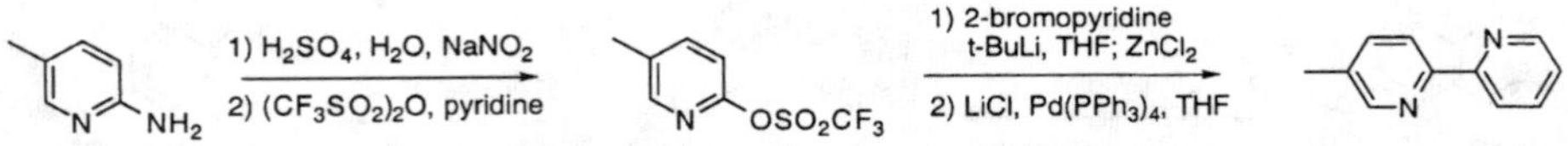

51

ULLMAN METHOXYLATION IN THE PRESENCE OF A 2,5-DIMETHYLPYRROLE-BLOCKED ANILINE: PREPARATION OF 2-FLUORO-4-METHOXYANILINE.

John A. Ragan, Brian P. Jones, Michael J. Castaldi, Paul D. Hill, and Teresa W. Makowski

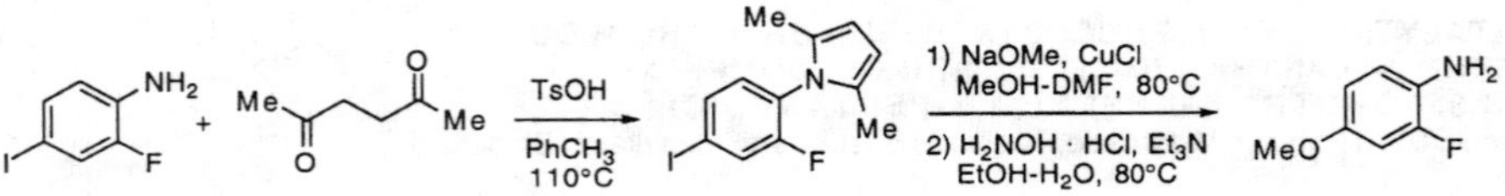

63

1,4,7,10-TETRAAZACYCLODODECANE.

David P. Reed and Gary R. Weisman

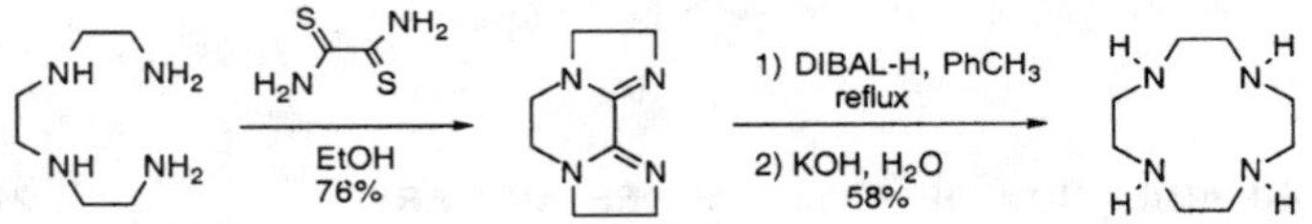

73

EFFICIENT SYNTHESIS OF HALOMETHYL-2,2'-BIPYRIDINES: 4,4'-BIS(CHLOROMETHYL)-2,2'-BIPYRIDINE.

Adam P. Smith, Jaydeep J. S. Lamba, and Cassandra L. Fraser

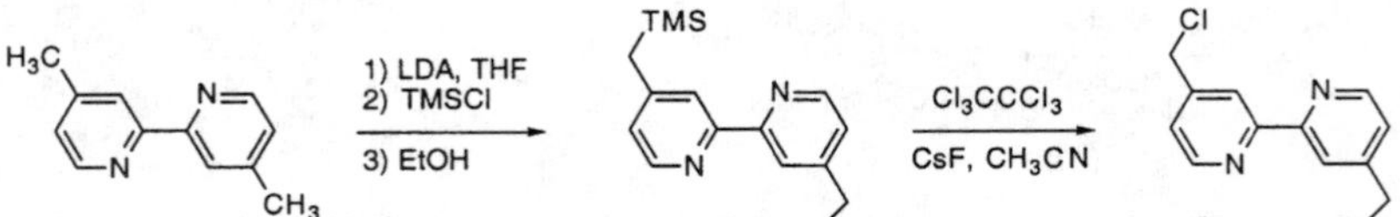

82

PREPARATION AND USE OF N,N'-DI-BOC-N"-TRIFLYLGUANIDINE.　91
Tracy J. Baker, Mika Tomioka, and Murray Goodman

REDUCTION OF SULFONYL HALIDES WITH ZINC POWDER:　99
S-METHYL METHANETHIOSULFONATE.
Fabrice Chemla and Philippe Karoyan

(PHENYL)[2-(TRIMETHYLSILYL)PHENYL]IODONIUM TRIFLATE.　104
AN EFFICIENT AND MILD BENZYNE PRECURSOR.
Tsugio Kitamura, Mitsuru Todaka, and Yuzo Fujiwara

4-HYDROXY[1-^{13}C]BENZOIC ACID.　113
Martin Lang, Susanne Lang-Fugmann, and Wolfgang Steglich

(1'R)-(-)-2,4-O-ETHYLIDENE-D-ERYTHROSE AND ETHYL　123
(E)-(-)-4,6-O-ETHYLIDENE-(4S,5R,1'R)-4,5,6-TRIHYDROXY-2-HEXENOATE.
M. Fengler-Veith, O. Schwardt, U. Kautz, B. Krämer, and V. Jäger

SYNTHESIS OF PENTA-1,2-DIEN-4-ONE (ACETYLALLENE).　135
Thierry Constantieux and Gérard Buono

BICYCLOPROPYLIDENE.

Armin de Meijere, Sergei I. Kozhushkov, and Thomas Späth

PREPARATION OF (E)-1-DIMETHYLAMINO-3-tert-BUTYLDIMETHYLSILOXY-1,3-BUTADIENE.

Sergey A. Kozmin, Shuwen He, and Viresh H. Rawal

[4+2] CYCLOADDITION OF 1-DIMETHYLAMINO-3-tert-BUTYLDIMETHYL-SILOXY-1,3-BUTADIENE WITH METHYL ACRYLATE: 4-HYDROXYMETHYL-2-CYCLOHEXEN-1-ONE.

Sergey A. Kozmin, Shuwen He, and Viresh H. Rawal

DIETHYL [(PHENYLSULFONYL)METHYL]PHOSPHONATE.

D. Enders, S. von Berg, and B. Jandeleit

SYNTHESIS OF (-)-(E,S)-3-(BENZYLOXY)-1-BUTENYL PHENYL SULFONE VIA A HORNER-WADSWORTH-EMMONS REACTION OF (-)-(S)-2-(BENZYLOXY)PROPANAL.

D. Enders, S. von Berg, and B. Jandeleit

(+)-(1R,2S,3R)-TETRACARBONYL[(1-3η)-1-(PHENYLSULFONYL)-BUT-2-EN-1-YL]IRON(1+) TETRAFLUOROBORATE.

D. Enders, B. Jandeleit, and S. von Berg

PREPARATION AND DIELS-ALDER REACTION OF A 2-AMIDO SUBSTITUTED FURAN: tert-BUTYL 3a-METHYL-5-OXO-2,3,3a,4,5,6-HEXAHYDROINDOLE-1-CARBOXYLATE. 202

Albert Padwa, Michael A. Brodney, and Stephen M. Lynch

[4 + 3] CYCLOADDITION OF AMINOALLYL CATIONS WITH 1,3-DIENES: 11-OXATRICYCLO[4.3.1.1^{2,5}]UNDEC-3-EN-10-ONE. 212

Jonghoon Oh, Chewki Ziani-Cherif, Jong-Ryoo Choi, and Jin Kun Cha

SYNTHESIS OF AMINO ACID ESTER ISOCYANATES: METHYL (S)-2-ISOCYANATO-3-PHENYLPROPANOATE. 220

James H. Tsai, Leo R. Takaoka, Noel A. Powell, and James S. Nowick

IN SITU CATALYTIC EPOXIDATION OF OLEFINS WITH TETRAHYDROTHIO-PYRAN-4-ONE AND OXONE: trans-2-METHYL-2,3-DIPHENYLOXIRANE. 225

Dan Yang, Yiu-Chung Yip, Guan-Sheng Jiao, and Man-Kin Wong

METHYL CARBAMATE FORMATION VIA MODIFIED HOFMANN REARRANGE-MENT REACTIONS: METHYL N-(p-METHOXYPHENYL)CARBAMATE. 234

Jeffrey W. Keillor and Xicai Huang

TRIBUTYLSTANNANE (Bu_3SnH)-CATALYZED BARTON-McCOMBIE DEOXYGENATION OF ALCOHOLS: 3-DEOXY-1,2:5,6-BIS-O-(1-METHYLETHYLIDENE)-α-D-RIBO-HEXOFURANOSE. 239
Jordi Tormo and Gregory C. Fu

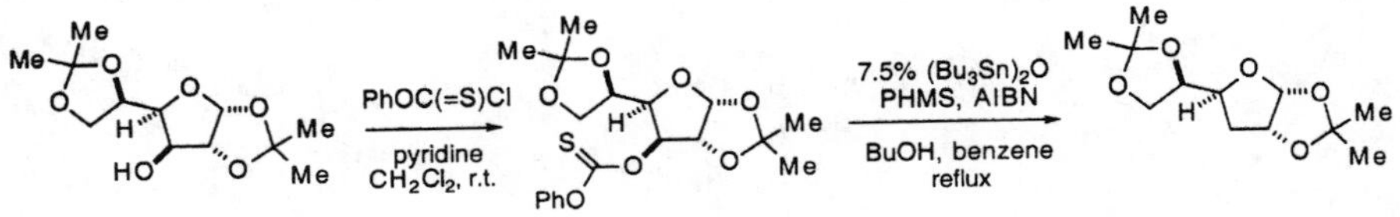

2-(3-OXOBUTYL)CYCLOPENTANONE-2-CARBOXYLIC ACID ETHYL ESTER. 249
Jens Christoffers

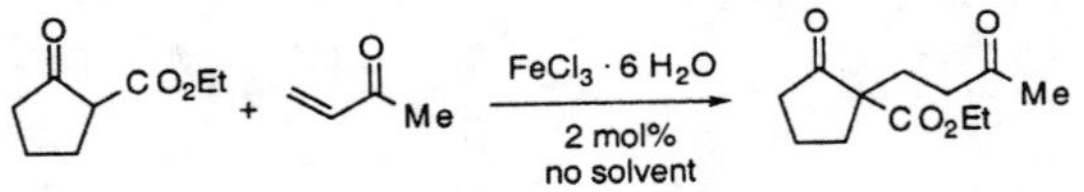

1-OXO-2-CYCLOHEXENYL-2-CARBONITRILE. 254
Fraser F. Fleming and Brian C. Shook

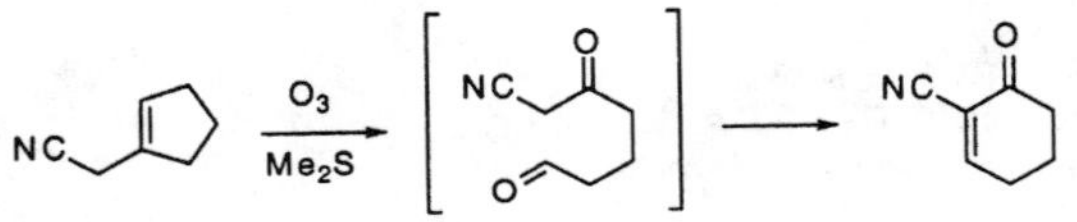

Unchecked Procedures 261

ORGANIC SYNTHESES

(R)-2-DIPHENYLPHOSPHINO-2'-METHOXY-1,1'-BINAPHTHYL

(Phosphine, (2'-methoxy[1,1'-binaphthalen]-2-yl)diphenyl-, (R)-)

A.

1 → Tf₂O, pyridine → 2

B. 2 → Pd(OAc)₂-dppb (cat), Ph₂POH, i-Pr₂NEt → 3

C. 3 → aq. NaOH → 4

D. 4 → MeI, K₂CO₃ → 5

E. 5 → HSiCl₃, Et₃N → 6 [(R)-MeO-MOP]

Submitted by Yasuhiro Uozumi, Motoi Kawatsura, and Tamio Hayashi.[1]
Checked by Sarge Salman and Louis S. Hegedus.

1. Procedure

Caution! All reactions should be conducted in a well-ventilated hood.

A. *(R)-2,2'-Bis(trifluoromethanesulfonyloxy)-1,1'-binaphthyl* (**2**). A dry, 200-mL, two-necked, round-bottomed flask is fitted with a magnetic stirring bar and a 30-mL pressure-equalizing addition funnel, and flushed with nitrogen gas. The flask is charged with 14.3 g (50.0 mmol) of (R)-(+)-1,1'-bi-2-naphthol (**1**) (Note 1), 12.0 mL (148 mmol) of pyridine (Note 2), and 100 mL of dichloromethane (Note 3), and the entire mixture is cooled to 0°C with an ice-water bath. Trifluoromethanesulfonic anhydride, (20.0 mL, 33.5 g, 119 mmol) (Note 2) is added dropwise over a period of 10 min to the stirred solution. After the mixture is stirred at 0°C for 6 hr, the reaction mixture is concentrated on a rotary evaporator. The residual brown oil is diluted with 200 mL of ethyl acetate and transferred to a 500-mL separatory funnel. The organic phase is washed with 5% hydrochloric acid (70 mL), saturated sodium bicarbonate (70 mL), and saturated sodium chloride (70 mL). The organic phase is dried over anhydrous sodium sulfate, and concentrated under reduced pressure on a rotary evaporator. The residue is chromatographed (10 x 20 cm column) on silica gel (700 g) (Note 4). The column is eluted with dichloromethane and the fractions are analyzed by TLC on silica gel (Note 5) using 30% dichloromethane-hexane as eluant. Fractions containing the product are combined and the solvent is evaporated on a rotary evaporator to give 26.3 g (96%) of **2** as a white powder (Note 6).

B. *(R)-(+)-2-Diphenylphosphinyl-2'-trifluoromethanesulfonyloxy-1,1'-binaphthyl* (**3**). A dry, 500-mL, Schlenk tube is fitted with a magnetic stirring bar and a rubber septum, and flushed with nitrogen gas. The flask is charged with 25.0 g (45.4 mmol) of

(R)-2,2'-bis(trifluoromethanesulfonyloxy)-1,1'-binaphthyl (**2**), 18.4 g (91.0 mmol) of diphenylphosphine oxide (Note 7), 1.02 g (4.54 mmol) of palladium acetate (Note 8), 1.94 g (4.55 mmol) of 1,4-bis(diphenylphosphino)butane (dppb) (Note 9), 23.4 g of diisopropylethylamine (181 mmol) (Note 10) and 200 mL of dimethyl sulfoxide (Note 11), and the entire mixture is heated with stirring at 100°C for 12 hr (Note 12). After the reaction mixture is cooled to room temperature, it is concentrated under reduced pressure (0.1-0.2 mm) on a rotary evaporator. The dark brown residue is diluted with 400 mL of ethyl acetate and transferred to a 1-L separatory funnel. The organic phase is washed successively with water (two 100-mL portions), 5% hydrochloric acid (100 mL), saturated sodium bicarbonate (100 mL), and saturated sodium chloride (100 mL), and the organic phase is dried over anhydrous sodium sulfate. After filtration the organic phase is concentrated under reduced pressure on a rotary evaporator, and the residue is chromatographed (14 x 30-cm column) on silica gel (ca. 1.5 kg) (Note 4). The column is eluted with 50% ethyl acetate-hexane and the fractions are analyzed by TLC on silica gel (Note 5) using the same eluant. Fractions containing the product are combined and the solvent is evaporated on a rotary evaporator to give 23.8 g (87%) of **3** as a white powder (Notes 13, 14).

C.   *(R)-(-)-2-Diphenylphosphinyl-2'-hydroxy-1,1'-binaphthyl* (**4**). A 100-mL round-bottomed flask containing a magnetic stirring bar is charged with 6.07 g (10.1 mmol) of (R)-(+)-2-diphenylphosphinyl-2'-trifluoromethanesulfonyloxy-1,1'-binaphthyl (**3**), 30 mL of 1,4-dioxane, and 14 mL of methanol. (Note 15) To the solution is added 14.1 mL of 3 N aqueous sodium hydroxide (NaOH) solution at ambient temperature. The reaction mixture is stirred for 12 hr then acidified (to pH 1) by the addition of a few drops of concd hydrochloric acid. The mixture is transferred to a separatory funnel and extracted twice with ethyl acetate (EtOAc). The organic phase is dried over magnesium sulfate (MgSO$_4$), filtered, and concentrated under reduced pressure to afford 6.19 g of **4** as a solid. (Note 16). This crude material is carried on to the next step without purification, assuming 100% yield.

D.   *(R)-(+)-2-Diphenylphosphinyl-2'-methoxy-1,1'-binaphthyl* (**5**).   A 250-mL round-bottomed flask is charged with crude **4**, 5.55 g (40.2 mmol) of potassium carbonate (K_2CO_3), and 66 mL of acetone. To this mixture is added 2.5 mL (40.2 mmol) of methyl iodide (MeI) (Note 17). The reaction mixture is refluxed for 3 hr. After the reaction is cooled to room temperature it is filtered through a Celite pad (Note 18), and the filter cake is washed with diethyl ether (Et_2O). The filtrate is concentrated under reduced pressure to give 6.88 g of **5** as a brown powder (Note 19). This crude material is carried on to the next step without purification, assuming 100% yield.

E.   *(R)-(+)-2-Diphenylphosphino-2'-methoxy-1,1'-binaphthyl* (**6**).   A 250-mL round-bottomed flask containing a magnetic stirring bar is charged with crude **5**, 7 mL (50 mmol) of triethylamine (Et_3N) (Note 20), and 84 mL of toluene. (Note 21). The mixture is cooled to 0°C then 4 mL (40 mmol) of trichlorosilane (Cl_3SiH) (Note 22) is added via syringe. The reaction is heated to 120°C and stirred for 5 hr. Upon cooling to ambient temperature the reaction is diluted with Et_2O and quenched with aqueous saturated sodium bicarbonate. The resulting suspension is filtered through a Celite pad, and the filter cake is washed with Et_2O. The organic phase is dried over $MgSO_4$ then concentrated under reduced pressure to afford 4.94 g of a yellow solid. This crude material is dissolved in a minimal amount of dichloromethane (CH_2Cl_2) and chromatographed on a 10 x 30-cm column containing 615 g of silica gel (SiO_2). The column is eluted with Et_2O to afford 4.08 g (8.71 mmol, 86% yield over last 3 steps) of **6** as an off-white powder (Note 23).

4

2. Notes

1. (R)-(+)-1,1'-Bi-2-naphthol (>99% op) was purchased from Aldrich Chemical Company, Inc., and used without further purification.

2. Pyridine and trifluoromethanesulfonic anhydride were purchased from Aldrich Chemical Company, Inc., and used without further purification.

3. Dichloromethane was purchased from Fisher Scientific, and distilled from calcium hydride before use.

4. Silica gel 60 230-400 mesh ASTM was used.

5. Merck silica gel 60F-254 plates were used.

6. Specific rotation value of **2**: $[\alpha]_D^{22}$ -143.5° (CHCl$_3$, c 1.24) [literature rotation for (S)-**2**;[2] $[\alpha]_D^{22}$ +142° (CHCl$_3$, c 1.035)]; ^{1}H NMR δ: 7.27 (d, 2 H, J = 8.7), 7.41 (t, 2 H, J = 7.8), 7.58 (t, 2 H, J = 7.8), 7.62 (d, 2 H, J = 9.6), 8.00 (d, 2 H, J = 8.3), 8.13 (d, 2 H, J = 9.2). If the material does not solidify, addition of equal amounts of diethyl ether and hexane, followed by reevaporation, should produce solid material.

7. Diphenylphosphine oxide is commercially available from Aldrich Chemical Company, Inc., and was recrystallized from 1:1 hexanes/ethyl acetate prior to use. The submitters have prepared and used this reagent. Preparation of diphenylphosphine oxide: To a solution of diethyl phosphite (65.0 mL, 505 mmol) in 250 mL of tetrahydrofuran (THF) is added sodium metal (11.5 g, 500 mg-atom) and the mixture is stirred under reflux for 20 hr. The resulting solution is added to 1.9 M solution (THF/Et$_2$O = 1/2) of phenylmagnesium bromide (580 mL, 1.10 mol) at 0°C and the mixture is refluxed for 6 hr. After the mixture is quenched with a small amount of water, it is diluted with ethyl acetate and washed once with 5% hydrochloric acid (HCl) and twice with water. The organic phase is dried over anhydrous sodium sulfate and concentrated under reduced pressure to give crude diphenylphosphine oxide. The

crude solid is purified by silica gel column chromatography (eluant: ethyl acetate) to give diphenylphosphine oxide (77.0 g, 75%).

8. Palladium acetate was purchased from Aldrich Chemical Company, Inc., and purified as follows: Palladium acetate is dissolved in hot benzene and filtered from insoluble material. After removal of the solvent, the residue is triturated with a small amount of diethyl ether to give brown powder that is collected by filtration, washed with diethyl ether and dried.

9. 1,4-Bis(diphenylphosphino)butane was purchased from Aldrich Chemical Company, Inc., and used without further purification.

10. Diisopropylethylamine was purchased from Aldrich Chemical Company, Inc., and used without further purification.

11. Dimethyl sulfoxide was purchased from Aldrich Chemical Company, Inc., and distilled from calcium hydride before use.

12. The use of 5 mol % of the catalyst [Pd(OAc)$_2$-dppb] also gave a high yield of **3**, but the submitters recommend the use of 10 mol % of the catalyst to ensure high chemical yield in the 12-hr reaction.

13. The physical properties of **3** are as follows: $[\alpha]_D^{20}$ +44.4° (CHCl$_3$, c 1.20), $[\alpha]_D^{22}$ +7.4° (CH$_2$Cl$_2$, c 1.40) [literature rotation for (R)-**3**;[3] $[\alpha]_D$ +6.29° (CH$_2$Cl$_2$, c 1.00)]; IR (KBr) cm^{-1} ν: 1410, 1205, 1140, and 945; ^{1}H NMR δ: 6.9-8.1 (m, 22 H, aromatic); ^{31}P NMR δ: 28.9 (s); EIMS m/z 603 (M+1), 454, 201 (base peak). Anal. Calcd for C$_{33}$H$_{22}$F$_3$O$_4$PS: C, 65.78; H, 3.68. Found: C, 65.67; H, 3.89.

14. Assignment of all peaks in the ^{13}C NMR is difficult because of ^{13}C-^{31}P coupling and the overlapping of peaks.

15. Methanol and 1,4-dioxane were purchased from Aldrich Chemical Company, Inc., and used without further purification.

16. An analytically pure sample is isolated by column chromatography on silica gel (see Note 4). The column is eluted with 50% ethyl acetate-hexane and the

6

fractions are analyzed by TLC on silica gel (Note 5) using the same eluent. Fractions containing the product are combined and the solvent is evaporated on a rotary evaporator to give **4** as a white powder. The physical properties of **4** are as follows (see Note 14): $[\alpha]_D^{20}$ -105° (CHCl$_3$, c 0.55), $[\alpha]_D^{20}$ -113° (CH$_2$Cl$_2$, c 1.00) [literature rotation for (R)-**4**;[3] $[\alpha]_D$ -108.3° (CH$_2$Cl$_2$, c 1.00)]; ^{1}H NMR δ: 6.35-8.10 (m, 22 H), 9.01 (br s, 1 H); ^{31}P NMR δ: 31.42 (s); EIMS m/z 470 (M+), 268 (base peak); HRMS calcd for C$_{32}$H$_{23}$PO$_2$ 470.1436, found 470. 1415. Anal. Calcd for C$_{32}$H$_{23}$O$_2$P: C, 81.69; H, 4.93. Found: C, 81.66; H, 4.96.

17. Methyl iodide, potassium carbonate, and acetone were purchased from Aldrich Chemical Company, Inc., and used without further purification.

18. Celite 535 (45 gals/sq.ft/hour), purchased from J.T. Baker, was used.

19. An analytically pure sample is isolated by column chromatography on silica gel (see Note 4). The column is eluted with ethyl acetate and the fractions are analyzed by TLC on silica gel (see Note 5) using the same eluent. Fractions containing the product are combined and the solvent is evaporated on a rotary evaporator to give **5** as a white powder. The physical properties of **5** are as follows (Note 14): $[\alpha]_D^{20}$ +121.5° (CHCl$_3$, c 1.30); ^{1}H NMR δ: 3.58 (s, 3 H), 6.75-8.01 (m, 22 H); ^{31}P NMR δ: 28.88(s); EIMS m/z 484 (M+), 453, 282 (base peak); HRMS calcd for C$_{33}$H$_{25}$O$_2$P 484.1592, found 484.1574. Anal. Calcd for C$_{33}$H$_{25}$O$_2$P: C, 81.80; H, 5.20. Found: C, 81.77; H, 5.38.

20. Triethylamine was purchased from Aldrich Chemical Company, Inc., and used without further purification.

21. Toluene was purchased from Aldrich Chemical Company, Inc., and distilled from calcium hydride before use.

22. Trichlorosilane was purchased from Aldrich Chemical Company, Inc., and used without further purification. While it is difficult to measure the volume of trichlorosilane used accurately because of its volatility, accurate measurement is not

7

essential in the reduction of **5**. The submitters found that a large excess of trichlorosilane does not interfere with reduction of phosphine oxide in Part E, and recommend the use of trichlorosilane (4 equiv or more to **5**) to complete the reduction in an appropriate reaction time.

23. Crystallization of the crude material from dichloromethane-hexane gave product **6** of 50% yield or lower. For efficiency of isolation, the submitters recommend purifying **6** by column chromatography. The physical properties of product **6** are as follows (see Note 14): $[\alpha]_D^{20}$ +95° (CHCl$_3$, c 0.27), $[\alpha]_D^{16}$ +75.7° (benzene, c 1.50) [literature rotation for (S)-**6**;[4] $[\alpha]_D^{16}$ -59.3° (benzene, c 1.0)]; mp 174-176°C (recrystallization from CH$_2$Cl$_2$/n-hexane); ^{1}H NMR δ: 3.34 (s, 3 H), 6.95-8.05 (m, 22 H); ^{31}P NMR δ: -12.74 (s); EIMS m/z 468 (M+), 437 (base peak); HRMS calcd for C$_{33}$H$_{25}$PO$_2$ 468.1643, found 468.1672. Anal. Calcd for C$_{33}$H$_{25}$OP: C, 84.60; H, 5.38. Found: C, 84.35; H, 5.44.

Waste Disposal Information

All toxic materials were disposed of in accordance with "Prudent Practices in the Laboratory"; National Academy Press; Washington, DC, 1995.

3. Discussion

Most of the chiral phosphine ligands prepared so far and used for catalytic asymmetric reactions are the bisphosphines,[5] which are expected to construct an effective chiral environment by bidentate coordination to metal; they have been demonstrated to be effective for several types of asymmetric reactions. On the other hand, there exist transition metal-catalyzed reactions where the bisphosphine-metal complexes cannot be used because of their low catalytic activity and/or low selectivity towards a desired reaction pathway. Therefore chiral monodentate phosphine ligands are required for the realization of new types of catalytic asymmetric reactions. Unfortunately, only a limited number of monodentate chiral phosphine ligands have been reported,[6] which with few exceptions are not so useful as bisphosphine ligands. Recently, the monodentate, optically active phosphine ligand, 2-diphenylphosphino-2'-methoxy-1,1'-binaphthyl (MeO-MOP), and its analogs[7] have been demonstrated to provide high enantioselectivity in palladium-catalyzed hydrosilylation of olefins[8] and palladium-catalyzed reduction of allylic esters by formic acid.[9] The procedures described here allow the convenient preparation of MOP and have advantages over previously published sequences.[4] MeO-MOP can be prepared in five steps from binaphthol without racemization and the overall yield is 72%. The key step in this process is the palladium-catalyzed monophosphinylation of 2,2'-bis(trifluoromethane-sulfonyloxy)-1,1'-binaphthyl, which was originally reported by Morgans and co-workers.[3] Under the slightly modified conditions, ditriflate (R)-**2** was efficiently converted into (R)-**3** (87% yield) without racemization. Hydrolysis of the remaining triflate with aqueous sodium hydroxide in 1,4-dioxane and methanol (2/1) gave (R)-**4**. The phenolic hydroxyl group of (R)-**4** was methylated by treatment with methyl iodide in the presence of potassium carbonate in acetone to give (R)-**5**. Reduction of phosphine oxide (R)-**5** was carried out with trichlorosilane and triethylamine[10] in toluene with heating to give the corresponding phosphine (R)-**6** (86% yield over last 3 steps).

A variety of MOP derivatives bearing various alkoxy or siloxy groups were readily prepared by changing the reagent used for the alkylation of **4**.[7] Furthermore, the presence of the triflate group[11] in compound **3** allows one to prepare a wide range of MOP derivatives functionalized at the 2'-position. Thus, the 2'-alkyl, carboxyl, cyano, aminomethyl groups, etc. were introduced into the 2'-position of MOP via transition metal-catalyzed cross-coupling, carbonylation, and cyanation reactions.[7,12]

Bis(substituted phenyl)phosphino groups were readily introduced into the binaphthyl by the palladium-catalyzed reaction with the corresponding diarylphosphine oxides. The same procedures used for the preparation of **3** were followed with di(p-methoxyphenyl)phosphine oxide and **2**. Subsequent hydrolysis, alkylation, and reduction processes gave 2-di(p-methoxyphenyl)phosphino-MOP.[7] The flexibility of the synthetic route allows fine tuning of the phosphine ligand by the introduction of several types of side chains and control of the steric and electronic effects of the phosphino group. Needless to say, the synthetic procedures shown here can be used for the preparation of MOPs having the (S)-absolute configuration by using (S)-binaphthol as a starting material. In addition, a MOP analog having the biphenanthryl skeleton (MOP-phen) was also prepared from optically active 4,4'-biphenanthrol through the same sequences mentioned above.[9b]

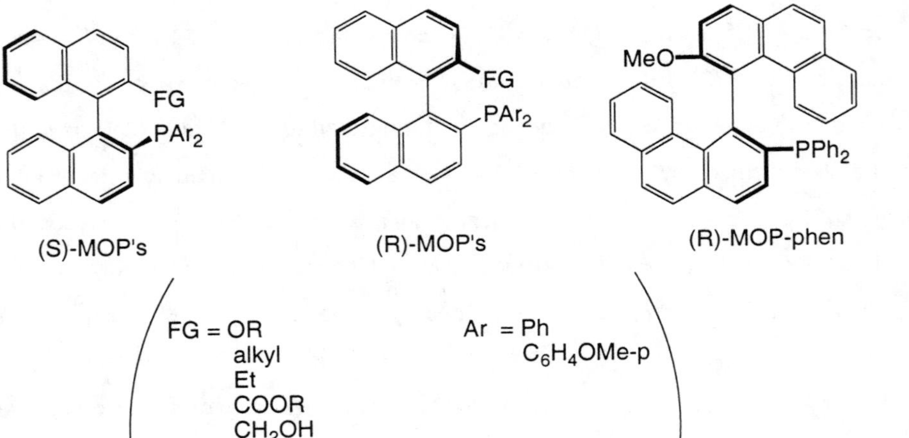

(S)-MOP's   (R)-MOP's   (R)-MOP-phen

FG = OR   Ar = Ph
     alkyl   C₆H₄OMe-p
     Et
     COOR
     CH₂OH
     CN
     CH₂NMe₂
     etc.
(R = H, alkyl, silyl, etc.)

10

1. Department of Chemistry, Faculty of Science, Kyoto University, Sakyo, Kyoto 606-8502, Japan

2. Vondenhof, M.; Mattay, J. *Tetrahedron Lett.* **1990**, *31*, 985.

3. Kurz, L.; Lee, G.; Morgans, Jr., D.; Waldyke, M. J.; Ward, T. *Tetrahedron Lett.* **1990**, *31*, 6321.

4. Hattori, T.; Shijo, M.; Kumagai, S.; Miyano, S. *Chem. Express* **1991**, *6*, 335.

5. For reviews: (a) Kagan, H. B. In "Asymmetric Synthesis"; Morrison, J. D. Ed.; Academic Press: London, 1985; Vol. 5, p. 1; (b) Kagan, H. B.; Sasaki, M. In "The Chemistry of Organophosphorus Compounds"; Hartley, F. R. Ed.; John Wiley and Sons: Chichester, 1990; Vol. 1, p. 51.

6. Examples of optically active monophosphine ligands so far reported: (a) (S)-(o-methoxyphenyl)cyclohexylmethylphosphine [(S)-CAMP]: Knowles, W. S.; Sabacky, M. J.; Vineyard, B. D. *J. Chem. Soc., Chem. Commun.* **1972**, 10; (b) neomentyldiphenylphosphine: Morrison, J. D.; Burnett, R. E.; Aguiar, A. M.; Morrow, C. J.; Phillips, C. *J. Am. Chem. Soc.* **1971**, *93*, 1301; (c) (S)-1-[(R)-2-(diphenylphosphino)ferrocenyl]ethyl methyl ether [(S)-(R)-PPFOMe]: Hayashi, T.; Mise, T.; Fukushima, M.; Kagotani, M.; Nagashima, N.; Hamada, Y.; Matsumoto, A.; Kawakami, S.; Konishi, M.; Yamamoto, K.; Kumada, M. *Bull. Chem. Soc. Jpn.* **1980**, *53*, 1138.

7. Uozumi, Y.; Tanahashi, A.; Lee, S.-Y.; Hayashi, T. *J. Org. Chem.* **1993**, *58*, 1945.

8. (a) Uozumi, Y.; Hayashi, T. *J. Am. Chem. Soc.* **1991**, *113*, 9887; (b) Uozumi, Y.; Lee, S.-Y.; Hayashi, T. *Tetrahedron Lett.* **1992**, *33*, 7185; (c) Uozumi, Y.; Hayashi, T. *Tetrahedron Lett.* **1993**, *34*, 2335; (d) Uozumi, Y.; Kitayama, K.; Hayashi, T. *Tetrahedron: Asymmetry* **1993**, *4*, 2419; (e) Uozumi, Y.; Kitayama, K.; Hayashi, T.; Yaragi, K.; Fukuyo, E. *Bull. Chem. Soc., Jpn.* **1995**, *68,* 713; (f)

Kitayama, K.; Uozumi, Y.; Hayashi, T. *J. Chem. Soc., Chem. Commun.* **1995**, 1533.

9. (a) Hayashi, T.; Iwamura, H.; Naito, M.; Matsumoto, Y.; Uozumi, Y.; Miki, M.; Yanagi, K. *J. Am. Chem. Soc.* **1994**, *116*, 775; (b) Hayashi, T.; Iwamura, H.; Uozumi, Y.; Matsumoto, Y.; Ozawa, F. *Synthesis* **1994**, 526; (c) Hayashi, T.; Iwamura, H.; Uozumi, Y. *Tetrahedron Lett.* **1994**, *35*, 4813.

10. Gilheany, D. G.; Mitchell, C. M. In "The Chemistry of Organophosphorus Compounds"; Hartley, F. R., Ed.; John Wiley and Sons: Chichester, 1990; Vol. 1, p.151 and references cited therein.

11. For reviews: (a) Stang, P. J.; Hanack, M.; Subramanian, L. R. *Synthesis* **1982**, 85; (b) Scott, W. J.; McMurry, J. E. *Acc. Chem. Res.* **1988**, *21*, 47; (c) Ritter, K. *Synthesis* **1993**, 735.

12. Uozumi, Y.; Suzuki, N.; Ogiwara, A.; Hayashi, T. *Tetrahedron* **1994**, *50*, 4293.

Appendix

Chemical Abstracts Nomenclature (Collective Index Number); (Registry Number)

(R)-2-Diphenylphosphino-2'-methoxy-1,1'-binaphthyl: Phosphine, (2'-methoxy[1,1'-binaphthalen]-2-yl)diphenyl-, (R)- (12); (145964-33-6)

(R)-2,2'-Bis(trifluoromethanesulfonyloxy)-1,1'-binaphthyl: Methanesulfonic acid, trifluoro-, [1,1'-binaphthalene]-2,2'-diyl ester, (R)- (12); (126613-06-7)

(R)-(+)-1,1'-Bi-2-naphthol: [1,1'-Binaphthalene]-2,2'-diol, (R)-(+)- (8);

[1,1'-Binaphthalene]-2,2'-diol, (R)- (9); (18531-94-7)

Pyridine (8,9); (110-86-1)

Trifluoromethanesulfonic anhydride: Methanesulfonic acid, trifluoro-, anhydride (8,9); (358-23-6)

(R)-(+)-2-Diphenylphosphinyl-2'-trifluoromethanesulfonyloxy-1,1'-binaphthyl:

Methanesulfonic acid, trifluoro-, 2'-(diphenylphosphinyl)[1,1'-binaphthalen]-2-yl ester, (R)- (12); (132532-04-8)

Diphenylphosphine oxide: Phosphine oxide, diphenyl- (8,9); (4559-70-0)

Palladium acetate: Acetic acid, palladium(2+) salt (8,9); (3375-31-3)

1,4-Bis(diphenylphosphino)butane (dppb): Phosphine, tetramethylenebis[diphenyl-(8); Phosphine, 1,4-butanediylbis[diphenyl- (9); (7688-25-7)

N,N-Diisopropylethylamine: Triethylamine, 1,1'-dimethyl- (8); 2-Propanamine, N-ethyl-N-(1-methylethyl)- (9); (7087-68-5)

Dimethyl sufoxide: Methyl sulfoxide (8); Methane, sulfinyl bis- (9); (67-68-5)

(R)-(-)-2-Diphenylphosphinyl-2'-hydroxy-1,1'-binaphthyl: [1,1'-Binaphthalene]-2-ol, 2'-(diphenylphosphinyl)-, (R)- (12); (132548-91-5)

1,4-Dioxane: CANCER SUSPECT AGENT: p-Dioxane (8); 1,4-Dioxane (9); (123-91-1)

(R)-(+)-2-Diphenylphosphinyl-2'-methoxy-1,1'-binaphthyl: Phosphine oxide, (2'-methoxy[1,1'-binaphthalen]-2-y1)diphenyl-, (R)- (13); (172897-73-3)

Methyl iodide: Methane, iodo- (8,9); (74-88-4)

Triethylamine (8); Ethanamine, N,N-diethyl- (9); (121-44-8)

Trichlorosilane: Silane, trichloro- (8,9); (10025-78-2)

Diethyl phosphite: Phosphonic acid, diethyl ester (8,9); (762-04-9)

Sodium (8,9); (7440-23-5)

Phenylmagnesium bromide: Magnesium, bromophenyl- (8,9); (100-58-3)

CATALYTIC ASYMMETRIC SYNTHESIS OF NITROALDOLS USING A LANTHANUM-LITHIUM-BINOL COMPLEX: (2S,3S)-2-NITRO-5-PHENYL-1,3-PENTANEDIOL

Submitted by Hiroaki Sasai,[1a] Shizue Watanabe, Takeyuki Suzuki, and Masakatsu Shibasaki.[1b]

Checked by Fabien Havas, Nabi A. Magomedov, and David J. Hart.

1. Procedure

A. Lanthanum-lithium-(R)-BINOL complex (LLB). A dry, 300-mL, three-necked flask equipped with a magnetic stirring bar, septum cap, and a rubber balloon filled with argon, is charged with (R)-2,2'-dihydroxy-1,1'-binaphthyl ((R)-BINOL; 6.49 g, 22.7 mmol, Note 1) and 119 mL of tetrahydrofuran (THF, Note 2) under an argon atmosphere. The system is placed in an ice-water bath and magnetic stirring is initiated. Via syringe, 28.4 mL (45.4 mmol) of a 1.60 M hexane solution of butyllithium (Note 3) is added to the cooled (R)-BINOL solution over 7 min and the pale yellow mixture is stirred for an additional 15 min. The cooling bath is removed and the THF solution of (R)-BINOL dilithium salt is allowed to reach room temperature.

A 500-mL, two-necked, round-bottomed flask equipped with a magnetic stirring bar, reflux condenser, and a septum cap is charged with 3.12 g of lanthanum trichloride heptahydrate (LaCl$_3$·7H$_2$O, 8.4 mmol, Note 4) and 100 mL of THF. The resulting suspension is sonicated for 30 min at room temperature (Note 5). To this suspension is added the above prepared solution of (R)-BINOL dilithium salt via syringe over 5 min with vigorous stirring (Note 6). After this mixture is stirred for 30 min at room temperature, a 0.52 M THF solution of sodium tert-butoxide (4.85 mL, 2.52 mmol, Note 7) is added via syringe over 5 min. The resulting suspension is stirred vigorously for 14 hr at room temperature and then stirred for 48 hr at 50°C. The reaction mixture is allowed to cool to room temperature without stirring, and the supernatant is used as a 0.03 M solution of lanthanum-lithium-(R)-BINOL catalyst [(R)-LLB].

B. (2S,3S)-2-Nitro-5-phenyl-1,3-pentanediol. A 500-mL, two-necked, round-bottomed flask equipped with a magnetic stirring bar, septum cap, and a rubber balloon filled with argon, is charged with 119 mL of THF and a 0.03 M THF solution of (R)-LLB catalyst (63.3 mL, 1.90 mmol) under an argon atmosphere. The system is cooled to -40°C and magnetic stirring is initiated (Note 8). The mixture is stirred for 30 min at -40°C, then 2-nitroethanol (3.00 mL, 41.8 mmol, Note 9) is added via syringe over 4 min. After 30 min of stirring at -40°C, 5.00 mL of 3-phenylpropanal (38.0 mmol, Note 10) is added via syringe over 5 min and the resulting solution is stirred for 90 hr. The reaction is monitored by TLC (Note 11). To the reaction mixture is added 150 mL of 1N hydrochloric acid (HCl). To this mixture is added 20 g of sodium chloride (NaCl) and the resulting mixture is transferred to a 2-L separatory funnel. The aqueous phase is extracted three times with ethyl acetate (400, 200, and 200 mL) and the combined organic phases are washed with 350 mL of aqueous saturated NaCl solution and dried over sodium sulfate. The solvent is removed with a rotary evaporator, and the resulting crude product is recrystallized from 1 : 1 hexane: ether (ca. 200 mL) to give 4.31g (50%) of analytically pure (2S,3S)-2-nitro-5-phenyl-1,3-pentanediol (98% ee, Notes 12-15).

2. Notes

1. (R)-BINOL was dried at 50°C for 2 hr under reduced pressure.

2. THF was freshly distilled from sodium benzophenone ketyl under an argon atmosphere.

3. BuLi was purchased from Kanto Chemical Company, Inc. and titrated prior to use.

4. Pulverized lanthanum chloride heptahydrate was purchased from Aldrich Chemical Company, Inc. (purity 99.9%), or Kanto Chemical Company, Inc. (purity 99.99%). The anhydrous salt is very hygroscopic and must be protected against moisture. It can, however, be handled very quickly in air without any special precaution.

5. The checkers used a Branson sonicator.

6. The transfer may also be made via cannula.

7. Sodium tert-butoxide was purchased from Aldrich Chemical Company, Inc.

8. When the checkers conducted the reaction at -30°C, a 45% yield of product with 50% ee was obtained.

9. 2-Nitroethanol was distilled under reduced pressure prior to use.

10. 3-Phenylpropanal was distilled under reduced pressure prior to use.

11. Acetone-hexane (1:2) or dichloromethane-methanol (20 : 1) is used as an eluant. Anisaldehyde is used as an indicator. Silica gel is used as the stationary phase.

12. The submitters indicate that purification of a small amount of crude 2-nitro-5-phenyl-1,3-pentanediol by reverse phase chromatography (Lobar LiChroprep RP-8, CH_3CN-H_2O 1:1) revealed that the yield, diastereoselectivity, and optical purity of the syn adduct is 79%, syn: anti = 11.5 : 1, and 89% ee, respectively.

13. The submitters obtained a second crop: 0.60 g (7% yield, 91% e.e.).

14. The enantiomeric excess was determined by HPLC analysis using DAICEL CHIRALPAK AD (hexane: i-PrOH 9:1, 0.7 mL/min, 254 nm). The retention times for the (2R,3R)-, (2S,3S)-, (2R,3S)- and (2S,3R)-derivatives are 22.6 min, 24 min, 17 min and 18 min, respectively. The analytical and spectral data of pure (2S,3S)-2-nitro-5-phenyl-1,3-pentanediol are as follows: $[\alpha]_D^{24}$ -17.7° (c 0.86, chloroform, 98% ee), mp 101-102°C; ^{1}H NMR δ: 1.83-1.95 (m, 2 H), 2.19 (br-t, 1 H, J = 6.2), 2.44 (br-d, 1 H, J = 7.5), 2.75 (m, 1 H), 2.90 (m, 1 H), 4.03-4.23 (m, 3 H), 4.57 (ddd, 1 H, J = 4.0, 5.2, 6.7), 7.17-7.35 (m, 5 H); ^{13}C NMR δ: 31.45 (CH), 35.15 (CH_2), 61.78 (CH_2), 69.53 (CH_2), 92.08 (CH), 126.33 (CH), 128.43 (CH), 128.64 (CH) 140.56 (C); IR (KBr) cm^{-1}: 3332, 1559, 1369; FABMS (glycerol) m/z 226 (M$^+$+1), 161, 91 (base peak). Anal. calcd for $C_{11}H_{15}NO_4$: C, 58.66; H, 6.71; N, 6.22. Found: C, 58.47; H, 6.77; N, 5.93. The checkers obtained a 53% yield of product, pure by combustion analysis, with 95% ee. One recrystallization of this material (85% recovery) gave product with greater than

98% ee. When the checkers used catalyst aged for 16 days at -20°C, product was obtained in 43% yield with a 95% ee.

15. The submitters recovered BINOL in quantiative yield without racemization. This material was reused after recrystallization from toluene.

Waste Disposal Information

All toxic materials were disposed of in accordance with "Prudent Practices in the Laboratory"; National Academy Press; Washington, DC, 1995.

3. Discussion

The lanthanum-lithium-BINOL (LLB) catalyst[2] can be prepared by three different methods that start with lanthanum trichloride,[3] lanthanum trichloride heptahydrate (procedure A),[4] or lanthanum tris(2-propoxide).[5,6] From a practical perspective, the most important method is procedure A, since this procedure is easy to perform, and the starting lanthanum trichloride heptahydrate is much cheaper than the corresponding anhydride or lanthanum tris(2-propoxide). Using methods similar to those for LLB, other rare earth-lithium-BINOL complexes can be prepared. For example, praseodymium-lithium-BINOL complex (PrLB) and neodymium-lithium-BINOL complex (NdLB) can be prepared from praseodymium tris(2-propoxide) and neodymium tris(2-propoxide), respectively. In some cases, these rare earth-lithium-BINOL complexes promote the nitroaldol reaction more efficiently than LLB to give nitroaldols in higher optical purities.[6,7] Furthermore, samarium-lithium-BINOL complex (SmLB), europium-lithium-BINOL complex (EuLB) and gadolinium-lithium-BINOL complex (GdLB) are effective in the asymmetric nitroaldol reaction of aromatic aldehydes with nitromethane.[6,7] The purity of rare earth tris(2-propoxide) may vary with supplier. The submitters have obtained their best results using rare earth tris(2-propoxide) purchased from Kojundo Chemical Laboratory Co., Ltd., Japan.

Concerning the source of the chiral catalysts, the introduction of trialkylsilylethynyl groups at the 6,6'-position of BINOL is effective for obtaining nitroaldols in higher enantiomeric excesses.[8] In the case of diastereo- and enantioselective nitroaldol reactions using nitroethane, nitropropane, and 2-nitroethanol, higher diastereomeric excesses are also observed using 6,6'-bis(triethylsilylethynyl)BINOL as a ligand.[8] Furthermore, the submitters have synthesized several therapeutically or biologically important compounds such as β-blockers,[7,9,10,11] norstatine derivative,[12] and threo-dihydrosphingosine.[8] Representative results are shown in Table I. To the best of the submitter's knowledge, rare earth-lithium-BINOL complexes are the only catalysts that efficiently promote asymmetric nitroaldol reactions.

1.   (a) Professor Sasai's present address: The Institute of Scientific and Industrial Research (ISIR), Osaka University, Mihogaoka, Ibaraki, Osaka 567-0047, Japan; (b) Graduate School of Pharmaceutical Sciences, The University of Tokyo, Hongo, Bunkyo-ku, Tokyo 113-0033, Japan.

2.   (R) or (S)-LLB catalyst can be purchased from Fluka Chemie AG as a 0.03M THF solution. See, tri-Lithium tris[(R)-1,1'-bi(2-naphtholato)]lanthanate solution, CAS: 161444-03-7, tri-Lithium tris[(S)-1,1'-bis(2-naphtholato)]lanthanate solution, CAS: 151736-98-0.

3.   Sasai, H.; Suzuki, T.; Arai, S.; Arai, T.; Shibasaki, M. *J. Am. Chem. Soc.* **1992**, *114*, 4418-4420.

4.   Sasai, H.; Suzuki, T.; Itoh, N.; Shibasaki, M. *Tetrahedron Lett.* **1993**, *34*, 851-854.

5. Sasai, H.; Suzuki, T.; Itoh, N.; Tanaka, K.; Date, T.; Okamura, K.; Shibasaki, M. *J. Am. Chem. Soc.* **1993**, *115*, 10372-10373.

6. Shibasaki, M.; Sasai, H. *Yuki Gosei Kagaku Kyokaishi* **1993**, *51*, 972-984.

7. Sasai, H.; Suzuki, T.; Itoh, N.; Arai, S.; Shibasaki, M. *Tetrahedron Lett.* **1993**, *34*, 2657-2660.

8. Sasai, H.; Tokunaga, T.; Watanabe, S.; Suzuki, T.; Itoh, N.; Shibasaki, M. *J. Org. Chem.* **1995**, *60*, 7388-7389.

9. Sasai, H.; Itoh, N.; Suzuki, T.; Shibasaki, M. *Tetrahedron Lett.* **1993**, *34*, 855-858.

10. Sasai, H.; Yamada, Y. M. A.; Suzuki, T.; Shibasaki, M. *Tetrahedron* **1994**, *50*, 12313-12318.

11. Sasai, H.; Suzuki, T.; Itoh, N.; Shibasaki, M. *Appl. Organomet. Chem., Lanthanoid Issue* **1995**, *9*, 421-426.

12. Sasai, H.; Kim, W.-S.; Suzuki, T.; Shibasaki, M.; Mitsuda, M.; Hasegawa, J.; Ohashi, T. *Tetrahedron Lett.* **1994**, *35*, 6123-6126.

Appendix

Chemical Abstracts Nomenclature (Collective Index Number); (Registry Number)

(R)-2,2'-Dihydroxy-1,1'-binaphthyl [(R)-BINOL]: [1,1'-Binaphthalene]-2,2'-diol, (R)- (+)-(8); [1,1'-Binaphthalene]-2,2'-diol, (R)- (9); (18531-94-7)

Butyllithium: Lithium, butyl- (8,9); (109-72-8)

Lanthanum trichloride heptahydrate: Aldrich: Lanthanum chloride heptahydrate (8); Lanthanum chloride, heptahydrate (9); (10025-84-0)

Sodium tert-butoxide: tert-Butyl alcohol, sodium salt (8); 2-Propanol, 2-methyl-, sodium salt (9); (865-48-5)

2-Nitroethanol: Ethanol, 2-nitro- (8,9); (625-48-9)

3-Phenylpropanal: Aldrich: Hydrocinnamaldehyde (8); Benzenepropanal (9);
(104-53-0)

TABLE I

ASYMMETRIC NITROALDOL REACTION USING LANTHANUM-LITHIUM-BINOL DERIVATIVE COMPLEX AS A CATALYST

Entry	Aldehyde	Nitro compound[a]	Conditions[b,c]	Nitroaldol	Yield (%)	syn / anti	ee (%) (of syn)
1	⅂⟍CHO	$MeNO_2$	(R)-LLB (3.3 mol%), -42 °C, 18 hr		80	—	85
2	Ph⟍⟍CHO	$MeNO_2$	(R)-LLB* (3.3 mol%), -40 °C, 91 hr		84	—	85
3	Ph⟍⟍CHO	⟍NO_2	(R)-LLB* (3.3 mol%), -20 °C, 75 hr		70	89 / 11	93
4	⬡⟍CHO	$MeNO_2$	(R)-LLB (3.3 mol%), -60 °C, 25 hr		73	—	93
5	(naphthyl-O-CH2-CHO)	$MeNO_2$	(R)-LLB (3.3 mol%), -50 °C, 60 hr		80	—	92
6	(MeO-...-O-CH2-CHO)	$MeNO_2$	(R)-LLB (5 mol%), -50 °C, 60 hr		90	—	94
7	(indolyl-O-CH2-CHO)	$MeNO_2$	(R)-LLB (10 mol%), -50 °C, 96 hr		76	—	92
8	⟍₁₄CHO	O_2N⟍⟍OH (3 mol eq)	(R)-LLB* (10 mol%), -40 °C, 163 hr		78	91 / 9	97
9	Ph⟍CHO / NPhth (96% ee)	$MeNO_2$	(R)-LLB (3.3 mol%), -40 °C, 72 hr	erythro : threo = >99 : 1	92% (erythro: 96% ee)		

a) Otherwise noted,10 mol eq of nitro compound was used. b) All reactions were carried out in THF. c) LLB* : lanthanum-lithium-(R)-6,6'-bis(triethylsilylethynyl)BINOL.

PALLADIUM-CATALYZED AMINATION OF ARYL HALIDES AND ARYL TRIFLATES: N-HEXYL-2-METHYL-4-METHOXYANILINE AND N-METHYL-N-(4-CHLOROPHENYL)ANILINE

(Benzenamine, 4-chloro-N-methyl-N-phenyl-)

A.

B.

Submitted by John P. Wolfe and Stephen L. Buchwald.[1]

Checked by Holly Norling and Louis S. Hegedus.

1. Procedure

A. N-Hexyl-2-methyl-4-methoxyaniline. A 250-mL, round-bottomed flask equipped with a magnetic stirbar and a rubber septum is flame-dried and allowed to cool to room temperature under an argon purge. The septum is removed and the flask is charged with tris(dibenzylideneacetone)dipalladium (0) (114 mg, 0.125 mmol, 0.5 mol% Pd) (Note 1), (±)-BINAP (233 mg, 0.375 mmol, 0.75 mol%) (Note 1), and sodium tert-butoxide (NaOtBu) (6.73 g, 70.0 mmol, 1.4 equiv) (Note 2). The septum is again placed on the flask, and the flask is purged with argon for 5 min. Toluene (50 mL) (Note 3) is added and stirring is started. The flask is charged with 4-bromo-3-

methylanisole (10.0 g, 50.0 mmol, 1.0 equiv) (Note 4), n-hexylamine (7.9 mL, 60.0 mmol, 1.2 equiv) (Note 5), and additional toluene (50 mL). The resulting dark red mixture is placed in an oil bath that is heated to 80°C with stirring until the aryl bromide has been completely consumed as judged by GC analysis (18-23 hr) (Note 6). The mixture is removed from the oil bath, allowed to cool to room temperature, then poured into a separatory funnel. The reaction flask is rinsed with ether (2 x 50 mL), brine (100 mL), deionized water (20 mL), and again with ether (50 mL). All rinses are added to the separatory funnel, the funnel is shaken, and the layers are separated. The aqueous layer is extracted with ether (50 mL), and the combined organic extracts are dried over anhydrous magnesium sulfate. The mixture is filtered; the magnesium sulfate is washed with ether (50 mL), filtered, and the organic solution is concentrated under reduced pressure to give the crude product as a brown oil. This oil is then distilled (bulb-to-bulb, bp 92°C at 0.001 mm) to afford 10.35 g (94%) of the desired product as a pale yellow oil (Note 7). A small amount of viscous material remains in the distillation flask following the distillation.

B. *N-Methyl-N-(4-chlorophenyl)aniline.* A 250-mL, three-necked, round-bottomed flask equipped with a reflux condenser, magnetic stirbar, one glass stopper, and rubber septa, covering the condenser and the remaining neck of the flask, is flame-dried and allowed to cool to room temperature under an argon purge. The septum is removed and the flask is charged with palladium acetate (337 mg, 1.5 mmol, 3.0 mol% Pd) (Note 1) and (±)-BINAP (1.4 g, 2.25 mmol, 4.5 mol%). The septum is again placed on the flask, and the flask is purged with argon for 5 min. Tetrahydrofuran (THF) (50 mL) (Note 8) is added and the mixture stirred at room temperature for 10 min until a peach-colored suspension forms. The flask is charged with 4-chlorophenyl trifluoromethanesulfonate (13.0 g, 50.0 mmol, 1.0 equiv) (Note 9), and N-methylaniline (6.5 mL, 60.0 mmol, 1.2 equiv) (Note 5). The septum is removed from the flask, and cesium carbonate (Cs_2CO_3) (22.8 g, 70.0 mmol, 1.4 equiv) (Note

24

10) is added under a flow of argon. The septum is again placed over the flask, and the flask is purged with argon for 30 seconds. Additional THF (50 mL) is added, the reaction mixture is immersed in a 70°C oil bath so that the level of the oil is even with the level of solvent in the flask and stirring is begun. The internal temperature of the reaction is monitored using a thermocouple and is found to be 60°C (±1°C) (Note 11). The mixture is stirred at this temperature until all the starting triflate has been consumed as judged by GC analysis (23-45 hr) (Notes 6, 12). The mixture is removed from the oil bath and allowed to cool to room temperature. Ether (100 mL) and hexanes (50 mL) are added to the flask and the solution formed is filtered through Celite. The flask is rinsed with ether (3 x 50 mL) and the rinses are filtered through Celite. The organic extracts are combined and concentrated under reduced pressure, and 1/1 (v/v) hexanes/ether (200 mL) is added. A yellow precipitate forms, and the mixture is filtered through a 1.5-inch deep plug of silica gel on a 3"-diameter type D fritted funnel. The flask is rinsed with 1/1 (v/v) hexanes/ether (2 x 100 mL), and the rinses are filtered through the plug of silica gel. The organic solution is concentrated under reduced pressure to give a light brown oil. The material is distilled (bulb to bulb, 0.002 mm). A low boiling fraction (bp 58-70°C) is collected; the distillation is stopped and the low boiling material is rinsed out of the receiver bulb. The distillation is resumed (bulb-to-bulb, bp 84°C at 0.002 mm) to give a yellow solid that melts at room temperature to give 9.72 g (90%) of the desired product as a pale yellow oil (Note 13).

2. Notes

1. $Pd_2(dba)_3$, palladium acetate, and (±)-BINAP were purchased from Strem Chemical Company and used without further purification. (The checkers recrystallized $Pd(OAc)_2$ from benzene prior to use. When it was used as obtained from the supplier, the reaction did not go to completion.)

2. Sodium tert-butoxide is purchased from Aldrich Chemical Company, Inc., and used without further purification. The bulk of this material is stored under nitrogen in a Vacuum Atmospheres Glovebox. Small portions (10-15 g) were removed from the glovebox in glass vials and weighed in the air. The checkers stored sodium tert-butoxide, purchased in 5-g bottles, in a desiccator and weighed it out in air.

3. Toluene is distilled under nitrogen from molten sodium.

4. 4-Bromo-3-methylanisole is purchased from Aldrich Chemical Company, Inc., and used without further purification.

5. All amines were purchased from Aldrich Chemical Company, Inc., and distilled under argon from calcium hydride (CaH_2).

6. Reaction times were considerably longer when run on a large scale than the previously published reaction times. This may be due to the lower internal temperature of the large scale reactions (for example, a reaction run on a 50-mmol scale in a 70°C oil bath was found to have an internal temperature of only 60°C). Longer reaction times in reactions that used cesium carbonate as a base may also be due in part to differences in stirring rate.

7. Analytical data for this compound are are as follows: ^{1}H NMR (300 MHz, $CDCl_3$) δ: 0.90 (t, 3 H, J = 6.6), 1.30-1.67 (m, 8 H), 2.13 (s, 3 H), 3.09 (t, 2 H, J = 7.2), 3.12 (s, br, 1 H), 3.74 (s, 3 H), 6.53-6.57 (m, 1 H), 6.69-6.71 (m, 2 H); ^{13}C NMR (75 MHz, $CDCl_3$) δ: 14.0, 17.6, 22.6, 26.9, 29.6, 31.6, 44.7, 55.6, 110.7, 111.5, 116.8, 123.5, 140.7, 151.4; IR (neat) cm[-1]: 3414, 2928, 1514, 1225, 1051. Anal. Calcd for $C_{14}H_{23}NO$: C, 75.97; H, 10.47. Found: C, 75.93; H, 10.45.

8. THF was distilled under argon from sodium benzophenone ketyl.

9. 4-Chlorophenyl trifluoromethanesulfonate[2a] is prepared according to the procedure of Stille[2b,c] as follows: A 500-mL, oven-dried, round-bottomed flask equipped with a Teflon stirbar and a rubber septum is allowed to cool to room temperature under an argon purge. The septum is removed and the flask was

charged with 4-chlorophenol (16.1 g, 125 mmol) (Note 14). The septum is again placed on the flask, the flask is purged with argon, and pyridine (125 mL) (Note 14) is added via syringe. The mixture is cooled to 0°C with stirring in an ice-water bath and trifluoromethanesulfonic anhydride (163 mmol) (Note 14) is added dropwise via syringe. After the addition is complete, the mixture is stirred at 0°C for 15 min, then allowed to warm to room temperature and stirred for 22 hr. The mixture is diluted with hexanes/ether (300 mL, 1/1 v/v) and transferred to a separatory funnel. The mixture is washed with aqueous hydrochloric acid (HCl) (1 M, 2 x 200 mL), and brine (200 mL), and the layers are separated. The organic layer is dried over anhydrous magnesium sulfate, filtered, and concentrated under reduced pressure. The crude material is Kugelrohr-distilled under vacuum (bp 90°C at water aspirator vacuum) to afford 31 g (95%) of the title compound as a colorless oil. This material is passed through a short plug of silica gel prior to use in palladium-catalyzed amination reactions.

10. Cesium carbonate was obtained from Chemetal and used without further purification. The bulk of this material was stored under nitrogen in a Vacuum Atmospheres Glovebox. Small portions (~50 g) were removed from the glovebox in glass bottles and weighed in the air. (The checkers stored the cesium carbonate in a desiccator under argon, and weighed it in air.)

11. The formation of a side product, N,N'-dimethyl-N,N'-diphenyl-1,4-phenylenediamine, was observed if the internal reaction temperature exceeded 65°C.

12. This reaction was run three times on a 50-mmol scale. Two of the three reactions were complete in <30 hr. The third reaction was run with a batch of triflate that had not been passed through silica as described above (Note 9); this reaction took 45 hr to proceed to completion. A second sample of the triflate was passed through a plug of silica; the reaction was repeated (on a 40-mmol scale) and proceeded to completion in 26 hr.

13. Analytical data for this compound are as follows: ^{1}H NMR (300 MHz), CDCl$_3$) δ: 3.28 (s, 3 H), 6.88-7.32 (m, 9 H); ^{13}C NMR (75 MHz, CDCl$_3$) δ: 40.3, 120.5, 121.4, 122.2, 125.5, 129.0, 129.3, 147.6, 148.6; IR (neat) cm^{-1}: 3061, 2943, 1601, 1452, 1156, 817; GC/MS (m/z) 219. Anal. Calcd for C$_{13}$H$_{12}$ClN: C, 71.87; H, 5.57. Found: C, 71.95; H, 5.81.

14. 4-Chlorophenol, trifluoromethanesulfonic anhydride, and anhydrous pyridine were purchased from Aldrich Chemical Company, Inc., and used without further purification.

Waste Disposal Information

All toxic materials were disposed of in accordance with "Prudent Practices in the Laboratory"; National Academy Press; Washington, DC, 1995.

3. Discussion

Aniline derivatives are frequently used in many areas of chemistry, including pharmaceuticals,[3a] agrochemicals,[3b] photography,[3c] pigments,[3d] xerography,[3e] and materials.[3f] Classic methods for the construction of aryl carbon-nitrogen bonds usually require harsh reaction conditions and/or activated substrates, and are often inefficient or unreliable.[4,5a]

The palladium-catalyzed amination of aryl halides has recently emerged as a powerful alternative to other methods for the synthesis of aryl amines.[5] This method allows for the cross-coupling of aryl halides and triflates with amines in the presence of a stoichiometric amount of a base and catalytic amounts of palladium complexes bearing tertiary phosphine ligands. As shown in Tables I and II, the Pd/BINAP catalyst is highly effective for reactions of aryl bromides and triflates with primary and cyclic

28

secondary amines. Use of the strong base NaOtBu allows for rapid reactions with low levels of the palladium catalyst (usually 0.05-0.5 mol % Pd) while a high degree of functional group tolerance is observed if Cs_2CO_3 is used as base.[5a,d,e,i] Catalytic aminations of aryl triflates are most effective with Cs_2CO_3;[5a,e] cleavage of the triflate moiety is observed in the presence of NaOtBu.[5a,e]

Catalysts based on bulky, electron-rich phosphine ligands have recently been developed that are effective for aminations of aryl chlorides, as well as aryl bromides and triflates.[5f,g,j] These catalysts also promote room temperature coupling reactions.[5f,j]

The examples described here demonstrate that a variety of aryl halides or triflates with differing substitution patterns, electronic properties, and functional groups can be coupled with primary and secondary amines in high yield. The reactions are only modestly air sensitive, require no special equipment or techniques, and are amenable to large scale synthesis.

1. Department of Chemistry, Massachusetts Institute of Technology, Cambridge, MA 02139.

2. (a) Niederprüm, H.; Voss, P.; Beyl, V. *Justus Liebigs Ann. Chem.* **1973**, 20-32; (b) Echavarren, A. M.; Stille, J. K. *J. Am. Chem. Soc.* **1987**, *109*, 5478-5486, (c) Stille, J. K.; Echavarren, A. M.; Williams, R. M.; Hendrix, J. A. *Org. Synth., Coll. Vol. IX* **1998**, 553.

3. (a) Negwer, M. "Organic-chemical Drugs and Their Synonyms: (An International Survey)", 7th ed.; Akademie Verlag GmbH: Berlin, 1994; (b) Montgomery, J. H. "Agrochemicals Desk Reference", 2nd ed.; CRC Press, 1997; (c) Loutfy, R. O.; Hsiao, C. K.; Kazmaier, P. M. *Photogr. Sci. Eng.* **1983**, *27*, 5-9; (d) "Pigment Handbook", 2nd ed.; John Wiley & Sons: New York, 1988; Vol. 1; (e) Schein, L. B. "Electrophotography and Development Physics", 2nd ed.,

Springer-Verlag: Berlin, 1992; (f) D'Aprano, G.; Leclerc, M.; Zotti, G.; Schiavon, G. *Chem. Mater.* **1995**, *7*, 33-42.

4. March, J. "Advanced Organic Chemistry: Reactions, Mechanisms and Structure", 4th ed.; Wiley, 1992.

5. (a) Wolfe, J. P.; Wagaw, S.; Marcoux, J.-F.; Buchwald, S. L. *Acc. Chem. Res.* **1998**, *31*, 805, and references cited therein; (b) Hartwig, J. F. *Angew. Chem., Int. Ed. Engl.* **1998**, *37*, 2046-2067 and references cited therein; (c) Wolfe, J. P.; Wagaw, S.; Buchwald, S. L. *J. Am. Chem. Soc.* **1996**, *118*, 7215-7216; (d) Wolfe, J. P.; Buchwald, S. L. *Tetrahedron Lett.* **1997**, *38*, 6359-6362; (e) Åhman, J.; Buchwald, S. L. *Tetrahedron Lett.* **1997**, *38*, 6363-6366; (f) Old, D. W.; Wolfe, J. P.; Buchwald, S. L. *J. Am. Chem. Soc.* **1998**, *120*, 9722-9723; (g) Hamahn, B. C.; Hartwig, J. F. *J. Am. Chem. Soc.* **1998**, *120*, 7369-7370; (h) Wolfe, J. P.; Buchwald, S. L. *J. Org. Chem.* **1996**, *61*, 1133-1135; (i) Wolfe, J. P.; Buchwald, S. L. *J. Org. Chem.* **2000**, *65*, 1144-1157; (j) Wolfe, J. P.; Tomori, H.; Sadighi, J. P.; Yin, J.; Buchwald, S. L. *J. Org. Chem.* **2000**, *65*, 1158-1174.

Appendix
Chemical Abstracts Nomenclature (Collective Index Number);
(Registry Number)

N-Methyl-N-(4-chlorophenyl)aniline: Benzenamine, 4-chloro-N-methyl-N-phenyl- (13); (174307-94-9)

Tris(dibenzylideneacetone)dipalladium (0): Palladium, tris(1,5-diphenyl-1,4-pentadien-3-one)di- (9); (52409-22-0)

rac-2,2'-Bis(diphenylphosphino)-1,1'-binaphthyl: rac-BINAP: Phosphine, [1,1'-binaphthalene]-2,2'-diylbis{diphenyl- (11); (98327-87-8)

4-Bromo-3-methylanisole: Anisole, 4-bromo-3-methyl- (8); Benzene, 1-bromo-4-methoxy-2-methyl- (9); (27060-75-9)

Hexylamine (8); 1-Hexanamine (9); (111-26-2)

Palladium acetate: Acetic acid, palladium(2+) salt (9); (3375-31-3)

4-Chlorophenyl trifluoromethanesulfonate: Methanesulfonic acid, trifluoro-, p-chlorophenyl ester (8); Methanesulfonic acid, trifluoro-, 4-chlorophenyl ester (9); (29540-84-9)

N-Methylaniline: Aniline, N-methyl- (8); Benzenamine, N-methyl- (9); (100-61-8)

Cesium carbonate: Carbonic acid, dicesium salt (9); (534-17-8)

4-Chlorophenol: TOXIC: Phenol, p-chloro- (8); Phenol, 4-chloro- (9); (106-48-9)

Pyridine (8,9); (110-86-1)

Trifluoromethanesulfonic anhydride: Methanesulfonic acid, trifluoro-, anhydride (8,9); (358-23-6)

TABLE I

BINAP/Pd-CATALYZED AMINATION OF ARYL BROMIDES USING NaOBu[a]

Entry	Halide	Amine	Product	Catalyst Loading (mol % Pd)	Reaction Time(hr)	Isolated Yield(%)
1			R=Hexyl	0.5	2	88
2		RNH₂	R=Bn	0.5	4	79
				0.05	7	79
3			R=Cyclic	0.5	18	83
4			R=t-Bu	4.0	25	76[b]
5		HexNH₂		0.5	<1	98[c]
				0.05	1.5	97
6		H₂NBn		0.5	2	81
7				0.5	17	90
8				2.0	18	79[b]
9		H₂NBn		0.5	18	87[b]

TABLE I (condt.)

BINAP/Pd-CATALYZED AMINATION OF ARYL BROMIDES USING NaOBu[a]

Entry	Halide	Amine	Product	Catalyst Loading (mol % Pd)	Reaction Time(hr)	Isolated Yield(%)
10				0.5 0.05	4 6	98[d] 94[d]
11				0.5	20	83
12				0.5	22	93[d]

a) Reaction conditions: 1.0 equiv ArBr, 1.1-1.2 equiv amine, 1.4 equiv NaOtBu, cat. Pd$_2$(dba)$_3$, cat. BINAP, toluene (2 mL/mmol halide), 80°C. (b) The reaction was conducted at 100°C. (c) Control experiments showed no formation of the desired product after 17 hr at 100°C in the absence of palladium. (d) No solvent was used in the reaction.

TABLE II

CATALYTIC AMINATION OF ARYL BROMIDES AND TRIFLATES USING Cs_2CO_3 [a]

Entry	Halide	Amine	Product	Catalyst [b]	Mol% Pd	Reaction Time(hr)	Yield %
1	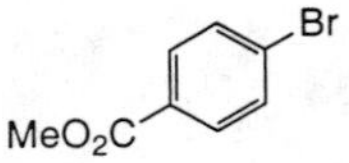 MeO_2C—C$_6$H$_4$—Br (para)	H_2NHex	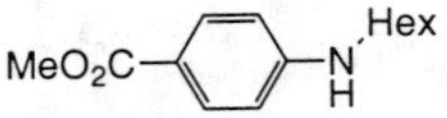 MeO_2C—C$_6$H$_4$—NH–Hex (para)	A	2	21	72
2	MeO_2C—C$_6$H$_4$—Br (para)	Me–NH–CH$_2$Ph	MeO_2C—C$_6$H$_4$—N(Me)–CH$_2$Ph (para)	B C	3 3	16 22	75 87[c]
3	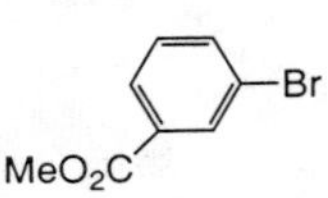 3-Br-C$_6$H$_4$—CO_2Me	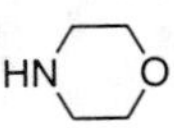 HN(morpholine)	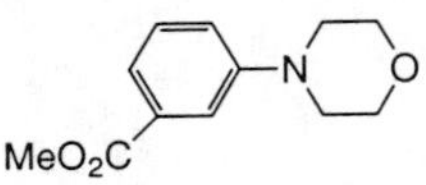 3-(morpholino)-C$_6$H$_4$—CO_2Me	B	1	20	86
4	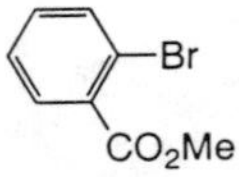 2-Br-C$_6$H$_4$—CO_2Me	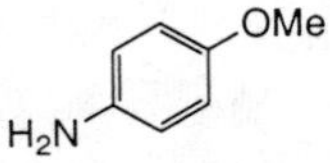 H_2N—C$_6$H$_4$—OMe (para)	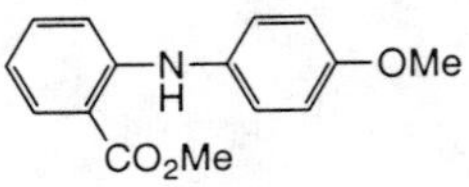 2-(CO_2Me)-C$_6$H$_4$—NH—C$_6$H$_4$—OMe	B	3	20	92
5	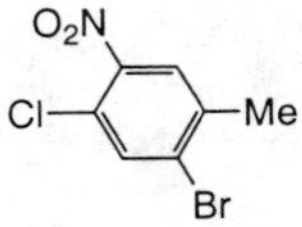 O_2N/Cl/Me/Br substituted benzene	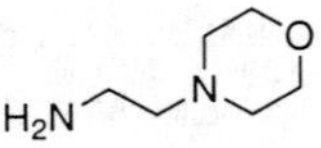 H_2NCH$_2$CH$_2$N(morpholine)	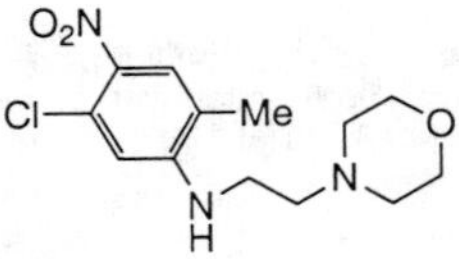 corresponding aniline product	A	2	16.5	75
6	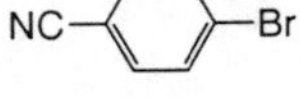 NC—C$_6$H$_4$—Br (para)	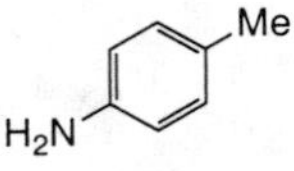 H_2N—C$_6$H$_4$—Me (para)	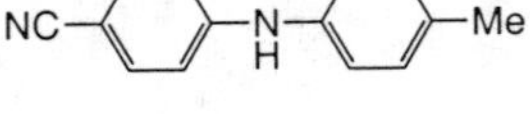 NC—C$_6$H$_4$—NH—C$_6$H$_4$—Me	A	1	26	80
7	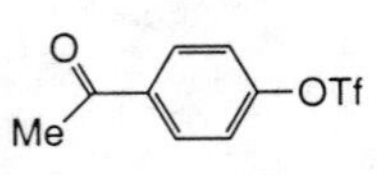 Me(C=O)—C$_6$H$_4$—OTf (para)	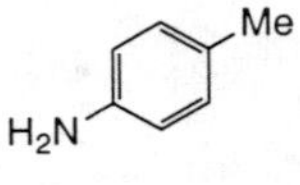 H_2N—C$_6$H$_4$—Me (para)	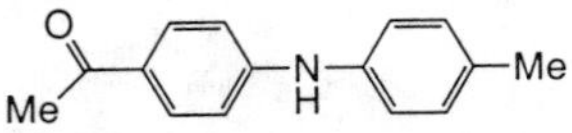 Me(C=O)—C$_6$H$_4$—NH—C$_6$H$_4$—Me	B	3	4	90[d]
8	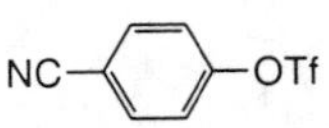 NC—C$_6$H$_4$—OTf (para)	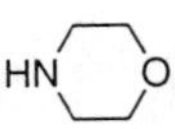 HN(morpholine)	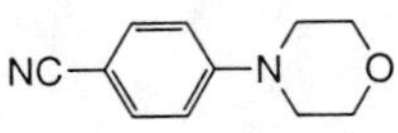 NC—C$_6$H$_4$—N(morpholine)	B	3	16	84[e]
9	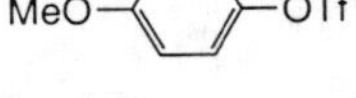 MeO—C$_6$H$_4$—OTf (para)	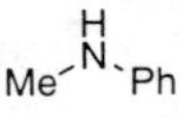 Me—NH—Ph	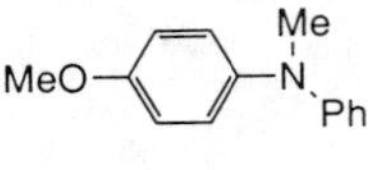 MeO—C$_6$H$_4$—N(Me)—Ph	B	3	16	88[e]

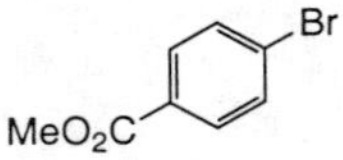

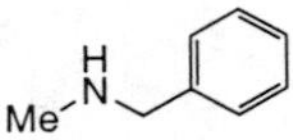

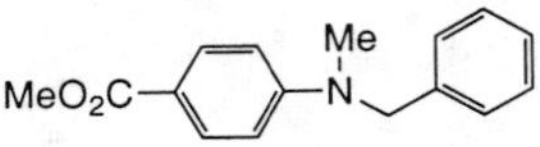

34

TABLE II (contd.)

CATALYTIC AMINATION OF ARYL BROMIDES AND TRIFLATES USING Cs_2CO_3[a]

Entry	Halide	Amine	Product	Catalyst[b]	Mol% Pd	Reaction Time(hr)	Yield %
10	Me—⟨Me⟩—OTf	BnNH$_2$	Me—⟨Me⟩—NH·Bn	B	5	5	90[f]
11	MeO$_2$C—⟨⟩—OTf	H$_2$NHex	MeO$_2$C—⟨⟩—N(H)Hex	B	3	5	87[f,g]
12	Ph(C=O)—⟨⟩—OTf	HN⟨pyrrolidine⟩	Ph(C=O)—⟨⟩—N⟨pyrrolidine⟩	B	3	6	92

(a) Reaction Conditions: 1.0 equiv halide, 1.2 equiv amine, 1.4 equi v Cs_2CO_3, cat. $Pd_2(dba)_3$ or $Pd(OAc)_2$, cat. BINAP (1.5 L/Pd), toluene (0.25 M), 100°C. (b) A=$Pd_2(dba)_3$/(S)-BINAP; B=$Pd(OAc)_2$/(S)-BINAP; C=$Pd(OAc)_2$/(±)-BINAP. (c) Base added last to the reaction mixture. (d) Reaction run in THF at 65°C. (e) Reaction run at 80°C. (f) 2.0 equiv of amine used in reaction. (g) Reaction run in 1,4-dioxane at 80°C.

INDOLE SYNTHESIS BY Pd-CATALYZED ANNULATION OF KETONES WITH o-IODOANILINE: 1,2,3,4-TETRAHYDROCARBAZOLE

(1H-Carbazole, 2,3,4,9-tetrahydro-)

Submitted by Cheng-yi Chen and Robert D. Larsen.[1]

Checked by Adam Charnley and Steven Wolff.

1. Procedure

To a 100-mL, two-necked flask, is added a mixture of cyclohexanone (5.9 g, 60 mmol), o-iodoaniline (4.4 g, 20 mmol), and 1,4-diazabicyclo[2.2.2]octane (DABCO) (6.7 g, 60 mmol) in N,N-dimethylformamide (DMF) (60 mL). The mixture is degassed three times via nitrogen/vacuum, followed by the addition of palladium acetate (Pd(OAc)$_2$) (2.24 mg, 0.1 mmol) (Note 1). The mixture is degassed twice and heated at 105°C for 3 hr or until completion of the reaction (Note 2). The reaction mixture is cooled to room temperature and partitioned between isopropyl acetate (150 mL) and water (50 mL). The organic layer is separated, washed with brine (50 mL), and concentrated under vacuum to dryness. The residue is chromatographed on 50 g of silica gel using 700 mL of ethyl acetate-heptane (1:6) as the eluent to give 2.22 g of 1,2,3,4-tetrahydrocarbazole (65%) as a pale brown solid (Note 3).

2. Notes

1. Both cyclohexanone and o-iodoaniline were purchased from Lancaster Synthesis and used directly in the reaction without further purification.

2. The reaction generally takes 3-5 hr to complete and is monitored by TLC (R_f = 0.50, SiO_2, eluted with EtOAc-heptane, 1:4).

3. The product is fully characterized: mp 116-118°C; IR (neat) cm^{-1}: 3401, 2928, 2848, 1470, 1305, 1235, 739; ^{1}H NMR (300 MHz, $CDCl_3$) δ: 1.86-1.99 (br m, 4 H), 2.74 (br t, 4 H, J = 6), 7.08-7.71 (br m, 2 H), 7.29 (m, 1 H), 7.49 (m, 1 H), 7.64 (br s, 1 H); ^{13}C NMR (75 MHz, $CDCl_3$) δ: 20.05, 22.20, 22.32, 22.42, 108.98, 109.61, 116.81, 118.12, 119.96, 126.82, 133.30, 134.66. Anal. Calcd for $C_{12}H_{13}N$: C, 84.17; H, 7.65; N, 8.18. Found: C, 82.87; H, 7.53; N, 7.84.

Waste Disposal Information

All toxic materials were disposed of in accordance with "Prudent Practices in the Laboratory"; National Academy Press; Washington, DC, 1995.

3. Discussion

The indole nucleus is a common and important feature of a variety of natural products and medicinal agents.[2] The traditional approach for preparing the indole nucleus is the Fischer indole reaction.[3] As this reaction has shortcomings, the palladium-catalyzed coupling of ortho-haloanilines is becoming an excellent alternative.[4] Recently, the submitters disclosed a new and efficient method for indole synthesis using a palladium-catalyzed annulation between o-iodoanilines and ketones (Scheme 1).[5]

Scheme 1

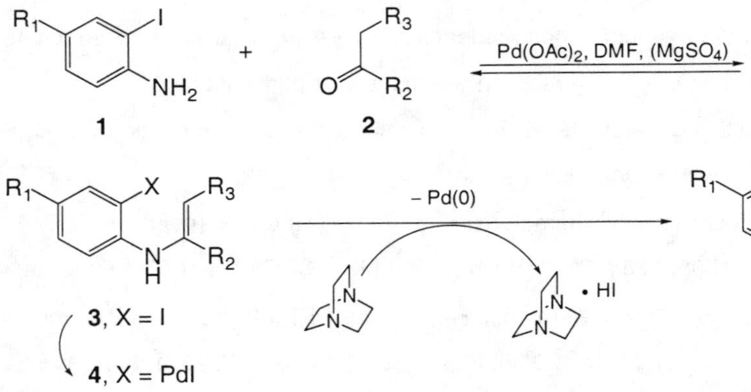

As illustrated in Chart 1, this reaction is applicable to a variety of o-iodoanilines and cyclic ketones to prepare the desired indoles in good yields. The coupling reaction is highly regioselective.[6] For example, condensation of o-iodoaniline 1, R=1-(1,2,4-triazolyl)methyl, with 2-methylcyclohexanone gave tetrahydrocarbazole 11 in 68% yield. Reaction of 3-methylcyclohexanone with o-iodoaniline formed tetrahydrocarbazole 12 predominantly. The reaction is also compatible with cyclopentanone and cycloheptanone (compounds 8 and 9). The reaction tolerates a variety of functional groups, especially the acid-sensitive ketal (10), carbamate (14), or the benzyl triazole[7] (8, 9 and 11). These compounds, which would be unstable under the conditions of the traditional Fischer indole reaction,[8] were conveniently synthesized using this method. The structurally interesting indole 15 was prepared from 3-quinuclidinone hydrochloride (1.0 equiv) in 55% yield. The interesting coupling of the indole nucleus onto a steroid was also achieved with 5α-cholestanone (1.0 equiv) affording 13 exclusively in 79% yield. Pyruvic acid and acetyl silane were also acceptable substrates, used to prepare indoles 16 and 17 in 82% and 64% yield respectively. Overall, the simple procedure, mild reaction conditions, and availability of the starting materials render this method a valuable addition to indole chemistry.

Chart 1

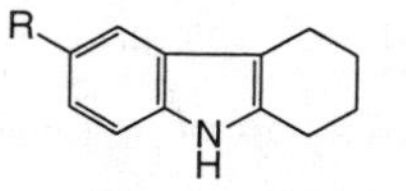

6, R = H, 65%
7, R = CN, 61%

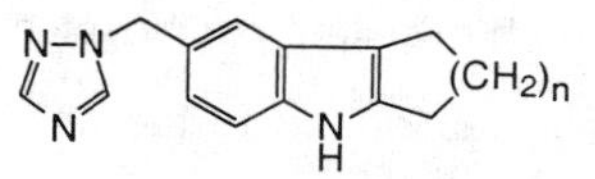

8, n = 0, 53%
9, n = 2, 72%

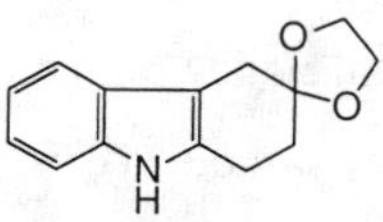

10 (55%)

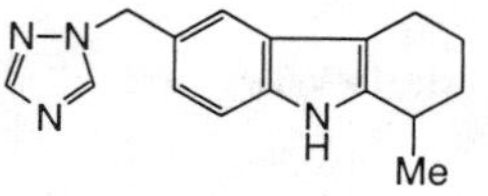

11 (68%)

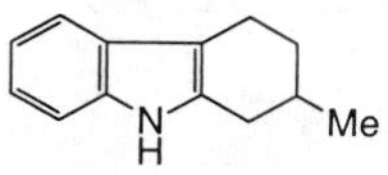

12 (65%)

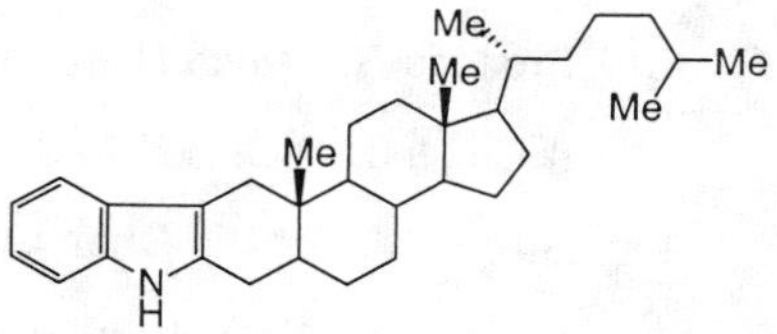

13 (79%)

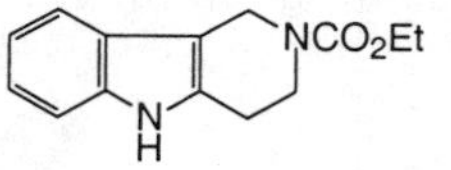

14 (78%)

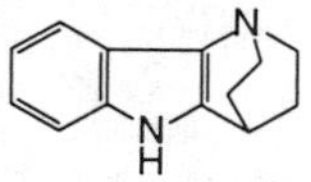

15 (55%)

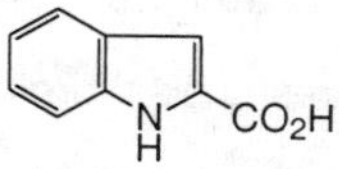

16 (82%)

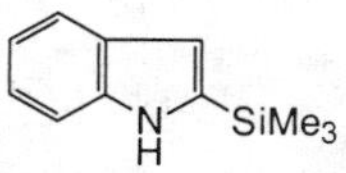

17 (64%)

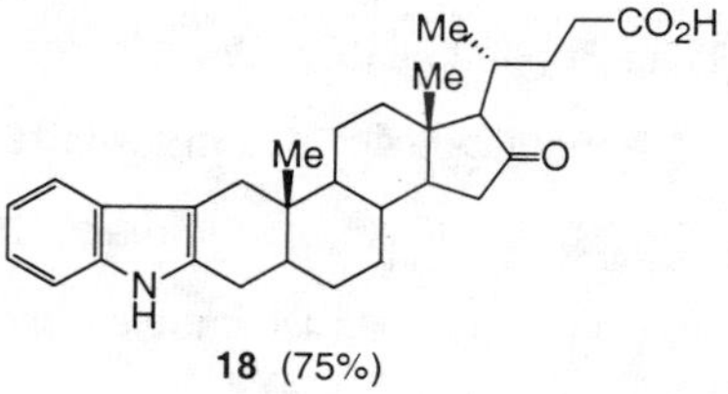

18 (75%)

The high regioselectivity of these reactions follows the same pattern as those of 2- and 3-substituted cyclohexanones when converted to enamines.[9] Apparently, $A^{(1,2)}$ and $A^{(1,3)}$ strain in the transition state controls the regiochemistry. The additive magnesium sulfate ($MgSO_4$), presumably acting as a dehydrating agent, was found to promote the annulation (for compounds **7, 8, 11, 13, 15** and **17**), indicating that the formation of the imine or enamine intermediate is critical to the reaction. The annulation of dehydrocholic acid and o-iodoaniline clearly demonstrated the high efficiency of this reaction as both excellent chemoselectivity and regioselectivity were observed (**18**). The coupling reaction led to the unique combination of indole and steroid moieties into one interesting molecule.

1. Process Research Department, Merck Research Laboratories, Division of Merck & Co., Inc., P.O. Box 2000, Rahway, NJ 07065.

2. Saxton, J. E. "The Chemistry of Heterocyclic Compounds"; John Wiley and Sons: New York, 1994; Vol. 25, Part IV.

3. For reviews on the Fischer indole reaction, see: Robinson, B. "The Fischer Indole Synthesis"; John Wiley and Sons: New York, 1982; Hughes, D. L. *Org. Prep. Proced. Int.* **1993**, *25*, 607; Gribble, G. W. *Contemp. Org. Synth.* **1994**, *1*, 145 and references cited therein.

4. (a) For a 'one-pot' synthesis of indoles under non-acidic conditions ($S_{RN}1$ reaction), see: Beugelmans, R.; Roussi, G. *J. Chem. Soc., Chem. Commun.* **1979**, 950; (b) Bard, R. R.; Bunnett, J. J. *J. Org. Chem.* **1980**, *45*, 1546; (c) Fukuyama, T.; Chen, X.; Peng, G. *J. Am. Chem. Soc.* **1994**, *116*, 3127; (d) Suzuki, H.; Thiruvikraman, S. V.; Osuka, A. *Synthesis* **1984**, 616 and references cited therein; (e) dAngelo, J.; Desmaele, D.; *Tetrahedron Lett.* **1990**, *31*, 879.

5. Chen, C.-y.; Lieberman, D. R.; Larsen, R. D.; Verhoeven, T. R.; Reider, P. J. *J. Org. Chem.* **1997**, *62*, 2676 and references cited therein.

6.   For a recent example of a regioselective Fischer indole reaction mediated by organoaluminum amides, see: Maruoka, K.; Oishi, M.; Yamamoto, H. *J. Org. Chem.* **1993**, *58*, 7638.

7.   Chen, C.-y; Lieberman, D. R.; Larsen, R. D.; Reamer, R. A.; Verhoeven, T. R.; Reider, P. J.; Cottrell, I. F.; Houghton, P. G. *Tetrahedron Lett.* **1994**, *35*, 6981.

8.   A Fischer indole reaction in pyridine has been reported: Welch, W. M. *Synthesis* **1977**, 645.

9.   Cook, A. G. "Enamines: Synthesis, Structure, and Reactions", 2nd ed.; Marcel Dekker, Inc.: New York, 1988.

Appendix
Chemical Abstracts Nomenclature (Collective Index Number); (Registry Number)

o-Iodoaniline: Aniline, o-iodo- (8); Benzenamine, 2-iodo- (9); (615-43-0)

1,2,3,4-Tetrahydrocarbazole: Carbazole, 1,2,3,4-tetrahydro- (8); 1H-Carbazole, 2,3,4,9-tetrahydro- (9); (942-01-8)

Cyclohexanone (8,9); (108-94-1)

1,4-Diazabicyclo[2.2.2]octane: DABCO (8,9); (280-57-9)

N,N-Dimethylformamide: CANCER SUSPECT AGENT: Formamide, N,N-dimethyl- (8,9); (68-12-2)

Palladium acetate: Acetic acid, palladium(2+) salt (8,9); (3375-31-3)

NICKEL-CATALYZED COUPLING OF ARYL O-CARBAMATES WITH GRIGNARD REAGENTS: 2,7-DIMETHYLNAPHTHALENE

(Naphthalene, 2,7-dimethyl-)

A.

B.

Submitted by Carol Dallaire,[1a] Isabelle Kolber,[1b] and Marc Gingras.[1c]

Checked by Mitsuru Kitamura and Koichi Narasaka.

1. Procedure

A. 2,7-Bis(diethylcarbamoyloxy)naphthalene.[2] Into a dry, 1-L, three-necked, round-bottomed flask, equipped with a mechanical stirrer and a condenser, are added 2,7-dihydroxynaphthalene (49.7 g, 0.310 mol, Note 1) and pyridine (700 mL, Note 2) under nitrogen. A dry, 200-mL pressure-equalizing dropping funnel fitted with a rubber septum is installed and charged with N,N-diethylcarbamoyl chloride (120 mL, 0.900 mol; Note 3). After the reaction flask is cooled in an ice bath for 30 min, N,N-diethylcarbamoyl chloride is added within 5 min to the vigorously stirred mixture. The ice bath is removed and the dark-brown solution is warmed to room temperature. The dropping funnel is removed under a stream of nitrogen and a thermometer is installed.

The solution is heated to 100°C (±5°C) for 2 days. TLC indicates a complete reaction (SiO$_2$, 2,7-dihydroxynaphthalene, R$_f$ = 0.46; acetone/hexane: 40/60). After the flask is cooled in an ice bath, hydrochloric acid (6 M, 250 mL) is poured into it over 10 min with vigorous stirring. A light-brown solid is formed. The brown mixture is poured into a 3-L Erlenmeyer flask and more hydrochloric acid (6 M, 350 mL) is added, followed by water (600 mL) in order to precipitate the compound further. The solid is filtered with a Büchner funnel and washed with water (500 mL). The crude product is dried under vacuum for several hours until a constant weight is obtained. The purity is found to be sufficient for the subsequent steps (111.0 g, 99% yield; mp 89.0-90.0°C, Note 4). However, a recrystallization is achieved by dissolving the crude compound in boiling 95% ethanol (300 mL), followed by the addition of water (200 mL). After the flask stands for 8 hr at room temperature, light-brown needles can be collected and washed with a solution of ethanol and water (50:50, 150 mL). The crystals are dried under vacuum to give pure 2,7-bis(N,N-diethylcarbamoyloxy)naphthalene (89.8 g, 0.250 mol, 81% yield; mp 89.6-90.5°C, Note 5). The filtrate is warmed, water (175 mL) is added, and the mixture is allowed to stand for 8 hr at room temperature to give additional crystals (14.0 g, 0.0391 mol, 13% yield; mp 89.0-90.0°C). The filtrate is cooled at 5°C to give a brown solid (5.7 g; mp 83.5-86.5°C), which is recrystallized from boiling hexanes (120 mL), to give needles of 2,7-bis(N,N-diethylcarbamoyloxy)naphthalene (3.22 g, 8.98 mmol; mp 88.0-89.5°C). The overall yield is 107 g, 0.299 mol, 97%.

  B. 2,7-Dimethylnaphthalene. A 2-L, three-necked, round-bottomed flask is equipped with a mechanical stirrer, reflux condenser, nitrogen inlet adapter and a 300-mL pressure-equalizing dropping funnel fitted with a rubber septum. All the glassware is oven-dried before assembly. Under a flow of nitrogen, the flask is charged with crystalline 2,7-bis(N,N-diethylcarbamoyloxy)naphthalene (70.3 g; 0.196 mol), the catalyst NiCl$_2$(dppp)$_2$ (1.90 g; 3.51 mmol, 1.8 mol % relative to 2,7-bis(N,N-diethylcarbamoyloxy)naphthalene, Note 6) and anhydrous diethyl ether (550 mL, Note

7). A red mixture is obtained. The dropping funnel is charged with an ethereal solution of methylmagnesium bromide (3 M in diethyl ether, 235 mL, 0.705 mol, Note 8), which is added dropwise over a period of 25 min. During the addition, the reaction mixture changes from red to pale brown and to green. The mixture is stirred at 30°C for 13 hr in order to complete the reaction (Note 9). TLC is used to follow the reaction [SiO_2, 2,7-bis(diethylcarbamoyloxy)naphthalene, R_f = 0.10; 2,7-dimethylnaphthalene, R_f = 0.74, (hexane/ethyl acetate: 80/20)]. The resulting dark brown mixture is cooled in an ice bath and the dropping funnel is charged with aqueous hydrochloric acid (6 M, 300 mL), which is slowly added to the reaction mixture over 25 min in order to maintain a gentle reflux. The aqueous layer is separated and extracted further with diethyl ether (50 mL). The combined organic layers are washed with aqueous hydrochloric acid (6 M, 3 x 100 mL), distilled water (150 mL), brine (200 mL), and dried over anhydrous sodium sulfate (25 g). After filtration and evaporation of the solvent, the compound is dried under vacuum to a constant weight, to afford a beige solid (30.2 g). The crude product is recrystallized from boiling 95% ethanol (350 mL) to give colorless crystals of 2,7-dimethylnaphthalene (22.1 g). Concentration of the mother liquors and another recrystallization, provides an additional amount of product (5.3 g). 2,7-Dimethylnaphthalene (overall: 27.4 g, 0.175 mol, 89%, mp 95-96°C, lit.[7]: mp 95-96°C, Note 10) is obtained as fluffy colorless leaflet crystals.

2. Notes

1. 2,7-Dihydroxynaphthalene (97%) was purchased from Aldrich Chemical Company, Inc., Acros Organics, or Tokyo Chemical Industry Co.

2. Laboratory grade pyridine was distilled from calcium hydride under nitrogen or used as received from Acros Organics (reagent grade, <0.1% water).

3. N,N-Diethylcarbamoyl chloride was used as received from Aldrich Chemical Company, Inc. (99%) or Tokyo Chemical Industry Co. (>95%). It was transferred to the dropping funnel via a syringe.

4. The purity was estimated from ^{1}H NMR (250 MHz, CDCl$_3$) and melting point.

5. The physical properties are as follows: R$_f$ = 0.65 (SiO$_2$, acetone/hexane = 40/60); ^{1}H NMR (250 MHz, CDCl$_3$) δ: 1.25 (broad s, 12 H), 3.45 (broad s, 8 H), 7.25 (dd, 2 H, J = 2.2, 8.7), 7.50 (d, 2 H, J = 2.2), 7.77 (d, 2 H, J = 8.9); ^{13}C NMR (62.9 MHz, CDCl$_3$) δ: 13.9, 14.7, 42.6, 118.5, 121.6, 129.3, 135.0, 150.3, 154.7; IR (CCl$_4$) cm^{-1}: 1724, 1155; MS (EI) m/e 358 (M$^{+ \cdot}$, 26%).

6. NiCl$_2$(dppp): [1,3-Bis(diphenylphosphino)propane]dichloronickel(II). **Important**: A loading of catalyst less than 1.8 mol % relative to the amount of 2,7-bis(N,N-diethylcarbamoyloxy)naphthalene gave erratic results. For example, 1.2-1.3 mol % sometimes gave an incomplete conversion, but additional catalyst (1.0 mol %) ensured completion of the reaction. This step was checked at least five times, each time with reproducible yields in the range of 86-89%. NiCl$_2$(dppp) was used as received from Acros Organics.

7. Diethyl ether was dried and purified by distillation under nitrogen from sodium and benzophenone, or was used as received from Kanto Chemical Co. (reagent grade, <0.005% water).

8. 3.0 M Methylmagnesium bromide solution in diethyl ether was purchased from Aldrich Chemical Company, Inc. and used without standardization. The Grignard reagent was transferred from the original bottle into the dropping funnel via a cannula under nitrogen.

9. In general, it was found that a dark brown color was indicative of a successful reaction. At 20°C, the reaction sometimes proceeds very slowly with poor reproducibility. The checkers recommended a temperature of ≥ 30°C.

10. The physical properties are as follows: R_f = 0.74 (SiO$_2$, hexane/ethyl acetate = 80/20). ^{1}H NMR (250 MHz, CDCl$_3$) δ: 2.49 (s, 6 H), 7.22 (d, 2 H, J = 8.0), 7.50 (s, 2 H), 7.65 (d, 2 H, J = 8.0); ^{13}C NMR (62.9 MHz, CDCl$_3$) δ: 21.7, 126.3, 127.2, 127.4, 130.0, 134.0, 135.4; MS (EI) m/e 156 (M$^{+\cdot}$, 100%).

Waste Disposal Information

All toxic materials were disposed of in accordance with "Prudent Practices in the Laboratory"; National Academy Press; Washington, DC, 1995.

3. Discussion

Nickel-catalyzed coupling of aryl O-carbamates and Grignard reagents is a promising methodology in synthetic organic chemistry.[3] Relatively cheap nickel catalysts are used in some efficient coupling procedures with a variety of easily made reagents such as vinyl, aryl and alkyl magnesium halides. The method avoids the classic use of expensive triflates in some palladium-catalyzed coupling procedures (for instance, the Stille coupling with costly and toxic organotins). Furthermore, palladium-catalyzed alkyl couplings are sometimes problematic because of β-hydride eliminations.

An application of the nickel catalysis is shown here in the formation of 2,7-disubstituted derivatives of naphthalene, which are less common in the library of commercial fine chemicals. For these reasons, the submitters developed some synthetic routes to 2,7-bis(diethylcarbamoyloxy)naphthalene, 2,7-dimethylnaphthalene[4] and 2,7-bis(bromomethyl)naphthalene that will facilitate access to a large family of 2,7-disubstituted naphthalenes.[5] The low cost of N,N-diethylcarbamoyl chloride, relative to triflic anhydride (for making aryl triflates), allows the formation of O-carbamates, and assures the incorporation of a wide variety of substituents with some relatively cheap nickel(II) catalysis and Grignard reagents.[3] Scheme 1 indicates a few possible uses of 2,7-dimethylnaphthalene from its Wohl-Ziegler dibromination with N-bromosuccinimide (NBS).[6] Recently, some effective procedures were published for making 2,7-dimethylnaphthalene and 2,7-bis(bromomethyl)naphthalene.[7] The best synthetic route required five steps to 2,7-bis(bromomethyl)naphthalene in an overall yield of ~34%, including the formation of a Grignard reagent, one separation of isomers and several unwanted by-products during two selective halogenation reactions. In addition, a major reactant was the relatively expensive 3,3-dimethoxy-2-butanone. The submitter's procedure is regiospecific at positions 2 and 7 on the naphthalene system and avoids separation of isomers. Furthermore, the dibromination of 2,7-dimethylnaphthalene could be accomplished by the Wohl-Ziegler reaction, which provided isolated yields equal to the photolytic procedure (64% in ref. 7), but no photolytic equipment was required. All the procedures described used simple purifications by recrystallization.

In spite of the poor availability of 2,7-bis(bromomethyl)naphthalene, it nevertheless has been used as an important convergent spacer and building block in supramolecular chemistry. For example, a bioinorganic model from a complexation with Cu(II), generated supramolecular cyclophanes by self-assembly, and encapsulated Lewis bases.[8] Helicoidal compounds, such as carbohelicenes, have recently been prepared from 2,7-bis(bromomethyl)naphthalene by Reetz[9] and Brunner.[7] Other syntheses are also known.[10] Because of the cost, the rather long synthetic sequences, and scarcity of these 2,7-disubstituted naphthalenes, the submitters believe that their procedures will encourage further uses of these synthons, either as supramolecular spacers with convergent functionalities or as important pharmaceutical intermediates. New receptors and helical structures could also be foreseen.

Scheme 1. A Few Synthetic Uses of 2,7-Dimethylnaphthalene and

2,7-Bis(bromomethyl)naphthalene

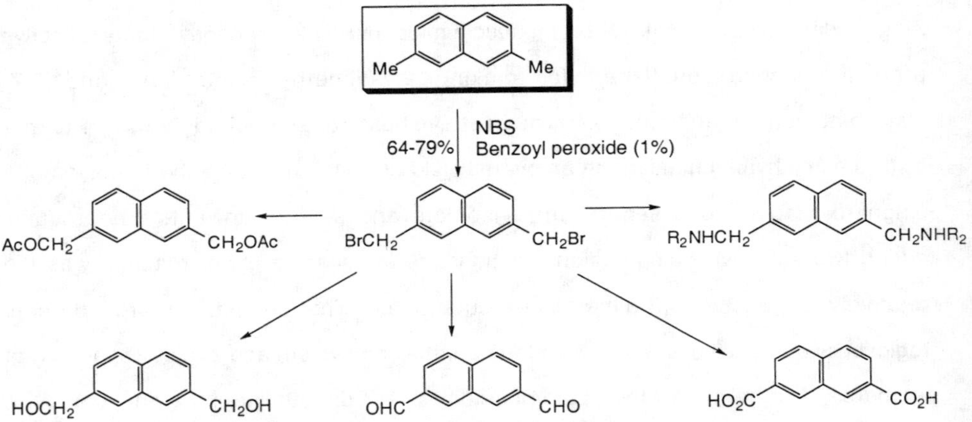

1.   (a) MDS Pharma Services, Laboratoire de Synthèse, 865, Boulevard Michèle-Bohec, Blainville, QC, Canada, J7C 5J6; (b) Université Libre de Bruxelles, CP 160/08, 50 Ave. F. D. Roosevelt, 1050 Brussels, Belgium; (c) Former Address: Laboratory of Supramolecular Chemistry and Catalysis, Organic Chemistry Division, CP 160-06, Université Libre de Bruxelles, 50 Ave. F. D. Roosevelt, 1050 Brussels, Belgium. Present address: Department of Chemistry, Faculty of Sciences, University of Nice-Sophia Antipolis, 28 Parc Valrose, 06108 Nice Cedex 2, France.

2.   Based in part from Kolber, I., Mémoire at Université Libre de Bruxelles, 1997.

3.   Sengupta, S.; Leite, M.; Soares Raslan, D.; Quesnelle, C.; Snieckus, V. J. Org. Chem. **1992**, 57, 4066-4068.

4.   Expensive 2,7-dimethylnaphthalene is available in small amounts from Aldrich Chemical Company, Inc. ($144 per gram).

5. Similar synthetic strategies to 2,7-disubstituted naphthalenes, but with different substitutents and Pd-reactions, have already been described: (a) Katz, T. J.; Liu, L.; Willmore, N. D.; Fox, J. M.; Rheingold, A. L.; Shi, S.; Nuckolls, C.; Rickman, B. H. *J. Am. Chem. Soc.* **1997**, *119*, 10054-10063; (b) Takeuchi, M.; Tuihiji, T.; Nishimura, J. *J. Org. Chem.* **1993**, *58*, 7388-7392.

6. (a) Jessup, P. J.; Reiss, J. A. *Aust. J. Chem.* **1976**, *29*, 173; (b) Baker, W.; Glocking, F.; McOmie, J. F. W. *J. Chem. Soc.* **1951**, 1118.

7. Terfort, A.; Görls, H.; Brunner, H. *Synthesis* **1997**, 79-86.

8. Maverick, A. W.; Buckingham, S. C.; Yao, Q.; Bradbury, J. R.; Stanley, G. G. *J. Am. Chem. Soc.* **1986**, *108*, 7430.

9. Reetz, M. T.; Beuttenmüller, E. W.; Goddard, R. *Tetrahedron Lett.* **1997**, *38*, 3211-3214.

10. (a) Yamamoto, K.; Ikeda, T.; Kitsuki, T.; Okamoto, Y.; Chikamatsu, H.; Nakazaki, M. *J. Chem. Soc., Perkin Trans. I* **1990**, 271-275; (b) Nakazaki, M.; Yamamoto, K.; Ikeda, T.; Kitsuki, T.; Okamoto, Y. *J. Chem. Soc., Chem. Comm.* **1983**, 787; (c) Katz, T. J.; Slusarek, W. *J. Am. Chem. Soc.* **1979**, *101*, 4259-4267; (d) Martin, R. H.; Marchant, M.-J.; Baes, M. *Helv. Chim. Acta* **1971**, *54*, 358-360.

Appendix

Chemical Abstracts Nomenclature (Collective Index Number); (Registry Number)

2,7-Dimethylnaphthalene: Naphthalene, 2,7-dimethyl- (8,9); (582-16-1)

2,7-Dihydroxynaphthalene: 2,7-Naphthalenediol (8,9); (582-17-2)

Diethylcarbamoyl chloride: CANCER SUSPECT AGENT: Carbamic chloride, diethyl- (8,9); (88-10-8)

[1,3-Bis(diphenylphosphino)propane]dichloronickel(II) CANCER SUSPECT AGENT: Nickel, dichloro[trimethylenebis[diphenylphosphine]]- (8); Nickel, dichloro[1,3-propanediylbis[diphenylphosphine]-PP']- (9); (15629-92-2)

Methylmagnesium bromide: Magnesium, bromomethyl- (8,9); (75-16-1)

SYNTHESIS OF 4-, 5-, and 6-METHYL-2,2'-BIPYRIDINE BY A NEGISHI CROSS-COUPLING STRATEGY: 5-METHYL-2,2'-BIPYRIDINE

(2,2'-Bipyridine, 5-methyl-)

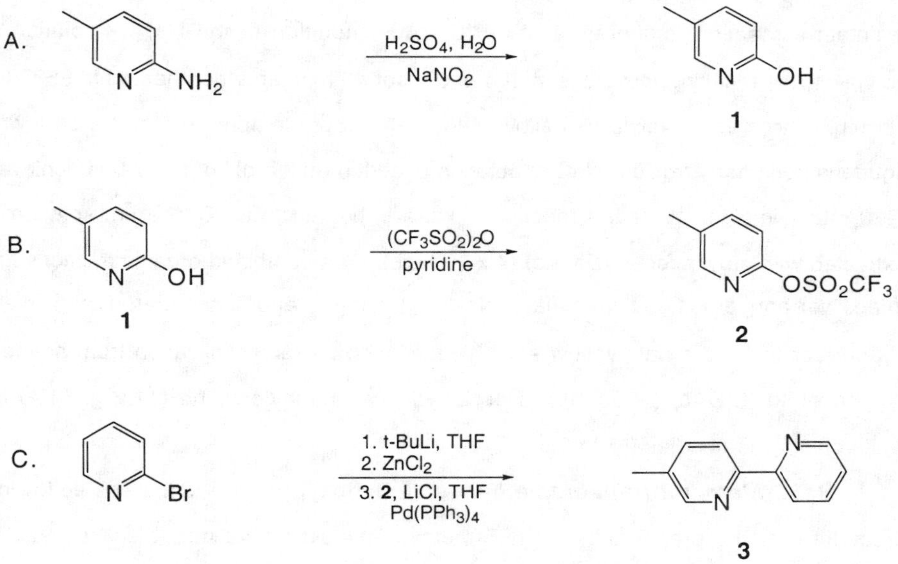

A.
N
NH2
H2SO4, H2O
NaNO2
N
OH
1
B.
N
OH
1
(CF3SO2)2O
pyridine
N
OSO2CF3
2
C.
N
Br
1. t-BuLi, THF
2. ZnCl2
3. 2, LiCl, THF
Pd(PPh3)4
N
N
3

Submitted by Adam P. Smith, Scott A. Savage, J. Christopher Love, and Cassandra L. Fraser.[1]

Checked by Erik Kuester and Louis S. Hegedus.

1. Procedure

A. 2-Hydroxy-5-methylpyridine (**1**) (Note 1). A 500-mL, two-necked, round-bottomed flask (Note 2) equipped with an internal thermometer and egg-shaped, Teflon-coated magnetic stirrer is charged with 150 mL of water (H_2O) and 40 g of concentrated sulfuric acid (H_2SO_4). This aqueous solution is cooled below 0°C by immersion in an acetone/ice bath, and 2-amino-5-methylpyridine (18.2 g, 168 mmol) is added (Note 3). The reaction mixture is treated with an aqueous solution of sodium nitrite ($NaNO_2$) (15.4 g, 223 mmol in 30 mL of H_2O) (Note 4) at a rate sufficient to maintain a reaction temperature of 0-5°C. After addition of the $NaNO_2$ solution is complete, the resulting mixture is stirred at 0°C for 45 min, and then heated to 95°C for 15 min. The reaction mixture is allowed to cool to room temperature and a 50% w/w aqueous sodium hydroxide (NaOH) solution is added until a pH of 6.5-7.0 is achieved (~30 mL) (Note 5). After the reaction mixture is heated to 60°C, the hot solution is extracted with ethyl acetate (EtOAc) (4 x 100 mL). The combined organic fractions are dried over anhydrous sodium sulfate (Na_2SO_4), filtered, and concentrated on a rotary evaporator to yield a pale-yellow solid. Purification by recrystallization from hot/cold ethyl acetate (EtOAc) (~300 mL) gives 2-hydroxy-5-methylpyridine (11.2 g, 61%) as white crystalline needles (Note 6).

   B. 5-Methyl-2-(trifluoromethanesulfonyl)oxypyridine (**2**). The following procedure for the preparation of the 5-methyl-2-pyridyl triflate may also be used to synthesize the 4- and 6-methyl derivatives. A 200-mL Schlenk flask (Note 2) containing a Teflon-coated, magnetic stirring bar and capped with a rubber septum is flushed with nitrogen. The flask is charged with 2-hydroxy-5-methylpyridine (**1**) (4.85 g, 44.4 mmol) and dry pyridine (140 mL) (Note 7). After the reactant dissolves, the flask is cooled to -12°C by immersion in an acetone/ice bath. Trifluoromethanesulfonic anhydride (15.1 g, 53.5 mmol) (Note 8) is added rapidly to the flask via syringe through the rubber septum. The solution is stirred at 0°C for 30 min and poured into a separatory funnel containing H_2O (150 mL). The mixture is extracted with dichloromethane (CH_2Cl_2) (3 x 100 mL) and the combined organic fractions are dried over anhydrous Na_2SO_4. Filtration and concentration on a rotary evaporator, followed by flash chromatography on 375 g of deactivated silica gel (Note 9) with 20%

EtOAc:80% hexanes, gives 9.89 g (92%) of 5-methyl-2-(trifluoromethane-sulfonyl)oxypyridine as a clear, colorless oil (Note 10).

C. 5-Methyl-2,2'-bipyridine (**3**). The 4- and 6-methyl-2,2'-bipyridines may also be prepared using the following procedure. A 500-mL, two-necked, round-bottomed flask (Note 2) with a Teflon-coated magnetic stirrer is placed in a dry ice/acetone bath (-78°C), then 80 mL of tetrahydrofuran (THF) (Note 11) and tert-butyllithium (tert-BuLi) (1.75 M in pentane, 52 mL, 91.0 mmol) (Note 12) are added to it, followed by dropwise addition of 2-bromopyridine (7.13 g, 4.3 mL, 45.1 mmol) (Note 13). The canary yellow THF solution becomes reddish-brown upon addition of the pyridyl bromide. After the solution is stirred at -78°C for 30 min, anhydrous zinc chloride ($ZnCl_2$) (13.3 g, 97.4 mmol) (Note 14) is added, and the reaction is stirred at 25°C for 2 hr. The 5-methyl-pyridyl triflate (**2**) (8.95 g, 37.1 mmol), lithium choride (LiCl) (3.18 g, 75.2 mmol) (Note 15), and tetrakis(triphenylphosphine) palladium ($Pd(PPh_3)_4$) (1.75 g, 1.5 mmol) (Note 16) are then added. The brownish-yellow reaction mixture is heated at reflux (Note 17) for 18 hr. After the solution is cooled, an aqueous solution of ethylenediaminetetraacetic acid (EDTA) (55 g, 148 mmol in 400 mL) (Note 18) is added and the pH is adjusted to ~8 with saturated aqueous sodium bicarbonate ($NaHCO_3$). The solution is stirred for 15 min then poured into a separatory funnel. The product is extracted with CH_2Cl_2 (3 x 200 mL). The combined organic fractions are dried over anhydrous Na_2SO_4, filtered, and concentrated using a rotary evaporator. Flash chromatography on 275 g of deactivated silica gel (Note 9) (20% EtOAc:80% hexanes) affords 5.94 g (94%) of 5-methyl-2,2'-bipyridine as a very pale yellow oil (Note 19).

2. Notes

1. This procedure is a modification of that reported by Adger and co-workers.[2] Both 2-hydroxy-4-methylpyridine and 2-hydroxy-6-methylpyridine can be obtained from Aldrich Chemical Company, Inc. However, it is more economical to prepare them in large quantities using this procedure from 2-amino-4-methylpyridine and 2-amino-6-methylpyridine, respectively.

2. Before use, all glassware, needles, and syringes were dried overnight in a 120°C oven.

3. 2-Amino-5-methylpyridine was purchased from Aldrich Chemical Company, Inc. and used as received. (Aldrich name: 2-Amino-5-picoline.)

4. Sodium nitrite was purchased from Aldrich Chemical Company, Inc. and used as obtained.

5. Sodium hydroxide pellets from Mallinckrodt Inc. were used as received.

6. The following characterization data was obtained: ^{1}H NMR (CDCl$_3$, 300 MHz) δ: 1.99 (s, 3 H), 6.43 (d, 1 H, J = 8.8), 7.06 (s, 1 H), 7.23 (dd, 1 H, J = 2.2, 9.5), 13.48 (s, 1 H); ^{13}C NMR (CDCl$_3$, 75 MHz) δ: 17.1, 116.2, 119.8, 132.5, 144.4, 164.9. Anal. Calcd for C$_6$H$_7$NO: C, 66.04; H, 6.46; N, 12.84. Found: C, 66.09; H, 6.31; N, 13.05.

7. Pyridine (99.9+% HPLC grade) was purchased from Aldrich Chemical Company, Inc., and used without further purification.

8. Trifluoromethanesulfonic anhydride, obtained from Aldrich Chemical Company, Inc., was used as received and weighed in a syringe inside a dry box. The checkers measured the anhydride volumetrically in a dry syringe, in a hood using a density of 1.68. Transfer in a dry box proved unnecessary.

9. Silica gel used for flash chromatography (particle size 0.035-0.075 mm) was obtained from VWR Scientific Products. Silica chromatography columns were deactivated by flushing with 10% triethylamine (Et$_3$N) in hexanes and then were washed with hexanes prior to use.

10. The product has the following properties: TLC R$_f$ = 0.54 (20% EtOAc:80% hexanes); ^{1}H NMR (CDCl$_3$, 300 MHz) δ: 2.37 (s, 3 H), 7.06 (d, 1 H, J = 8.1), 7.67 (dd, 1 H, J = 2.4, 8.5), 8.17 (s, 1 H); ^{13}C NMR (CDCl$_3$, 75 MHz) δ: 17.9, 114.9, 118.8 (q, J$_{CF}$ = 320.3), 134.7, 141.6, 148.8, 154.2. Anal. Calcd for C$_7$H$_6$F$_3$NO$_3$S: C, 34.86; H, 2.51; N, 5.81. Found: C, 34.99; H. 2.19; N, 5.70.

11. THF was dried and purified by passage through alumina solvent purification columns[3] or by distillation over sodium/benzophenone.

12. A 1.6 M solution of tert-BuLi in pentane was obtained from Aldrich Chemical Company, Inc. It is crucial to have at least 2 equiv of tert-BuLi for the lithium-halogen exchange. Depressed yields (25-60%) were obtained when less than 2 equiv were used. The tert-BuLi is titrated prior to its use in each reaction using the following procedure.[4] To a 50-mL Schlenk flask is added N-benzylbenzamide (274 mg, 1.3 mmol) (as received from Aldrich Chemical Company, Inc.) and THF (10 mL) (Note 11). The solution is cooled to -43°C (acetonitrile/dry ice) and tert-BuLi is added dropwise to the blue endpoint (color persists for >30 s). The molarity is calculated using a 1:1 stoichiometric ratio of N-benzylbenzamide to tert-BuLi (just greater than 1 equivalent of alkyllithium needed to reach the endpoint).

13. 2-Bromopyridine was purchased from Aldrich Chemical Company, Inc., and used as received.

14. Zinc chloride, obtained from Strem Chemicals Inc., was flame-dried to remove excess H$_2$O and stored in a dry box prior to use. Weighing out flame-dried zinc chloride on the bench rather than in a dry box resulted in reduced yields. When a

1$\underline{M}$ solution of the above flame-dried zinc chloride was prepared in THF and transferred by syringe, the published yields were obtained.

15. Granular lithium chloride from Mallinckrodt, Inc. was stored in a dry box prior to use. The checkers stored the LiCl in a desiccator before use.

16. The Pd(PPh$_3$)$_4$ catalyst can be purchased from Aldrich Chemical Company, Inc., or Strem Chemicals Inc. However, it was easily prepared using the procedure of Coulson[5] for the synthesis delineated here.

17. Pentane is removed by distillation (bp 36°C). A reflux temperature of 70-75° is required for the reaction to proceed to completion.

18. Ethylenediaminetetraacetic acid, disodium salt dihydrate, 99+% was obtained from Aldrich Chemical Company, Inc. and used as received. The EDTA mixture was heated gently to facilitate dissolution and was allowed to cool to room temperature prior to use.

19. The analytical data for 5-methyl-2,2'-bipyridine are as follows: TLC R$_f$ = 0.46 (20% EtOAc:80% hexanes); ^{1}H NMR (CDCl$_3$, 300 MHz) δ: 2.43 (s, 3 H), 7.35 (dd, 1 H, J = 4.6, 7.7), 7.71 (d, 1 H, J = 7.7), 7.87 (t, 1 H, J = 7.3), 8.39 (d, 1 H, J = 7.7), 8.48 (d, 1 H, J = 7.3), 8.55 (s, 1 H), 8.70 (d, 1 H, J = 4.6); ^{13}C NMR (CDCl$_3$, 75 MHz) δ: 17.7, 120.0, 120.2, 122.8, 132.8, 136.2, 136.8, 148.5, 149.1, 153.0, 155.7. Anal. Calcd for C$_{11}$H$_{10}$N$_2$: C, 77.62; H, 5.92; N, 16.46. Found: C, 77.66; H, 5.98; N, 16.37.

Waste Disposal Information

All toxic materials were disposed of in accordance with "Prudent Practices in the Laboratory"; National Academy Press; Washington, DC, 1995.

3. Discussion

As ligands for metal ions, 2,2'-bipyridines find wide application in chemistry. They have been used in studies of supramolecular assembly,[6] in bioinorganic contexts,[7] and in polymeric materials,[8] as well as in discrete small-molecule analogues.

Traditionally, methyl-2,2'-bipyridines (methyl bpys) have been prepared by the Kröhnke method, which involves reaction of pyridinium salts with α,β-unsaturated ketones followed by treatment with ammonium acetate to effect cyclization.[9] They have also been made by coupling pyridyllithium reagents with pyridyl sulfoxides,[10] by Ni and other metal-catalyzed cross-coupling reactions,[11] by the Ullman reaction[9] and by use of α-oxoketene dithioacetals among a variety of other routes.[12] Many methods lead to mixtures of isomers or they produce dimethyl byproducts. Nearly all of them afford products in moderate yields at best. The cross-coupling of a pyridyl zinc reagent and a pyridyl triflate in the presence of a catalytic amount of palladium by the Negishi method[13] as described here constitutes an efficient, large scale, high yield synthesis of 4-, 5-, and 6-methyl-2,2'-bipyridine. These methyl bpys are readily converted to bromomethyl and chloromethyl analogues,[14] which are valuable starting materials for further derivatization.[15] Moreover, the halomethyl bipyridines have been used as ligand initiators in controlled polymerizations.[16]

TABLE I

SYNTHESIS OF 2-PYRIDYL TRIFLATES

Product	R_1	R_2	R_3	Yield (%)
4-Methyl Triflate	CH_3	H	H	95
5-Methyl Triflate	H	CH_3	H	95
6-Methyl Triflate	H	H	CH_3	94

TABLE II

SYNTHESIS OF METHYL-2,2'-BIPYRIDINES

Product	R_1	R_2	R_3	Yield (%)
4-Methyl-2,2'-bipyridine	CH_3	H	H	96
5-Methyl-2,2'-bipyridine	H	CH_3	H	94
6-Methyl-2,2'-bipyridine	H	H	CH_3	93

1.   Department of Chemistry, University of Virginia, Charlottesville, VA 22904-4319.

2.   Adger, B. M.; Ayrey, P.; Bannister, R.; Forth, M. A.; Hajikarimian, Y.; Lewis, N. J.; O'Farrell, C.; Owens, N.; Shamji, A. *J. Chem. Soc., Perkin Trans. I* **1988**, 2791.

3.   Pangborn, A. B.; Giardello, M. A.; Grubbs, R. H.; Rosen, R. K.; Timmers, F. J. *Organometallics* **1996**, *15*, 1518.

4.   Burchat, A. F.; Chong, J. M.; Nielsen, N. *J. Organomet. Chem.* **1997**, *542*, 281.

5.   Coulson, D. R. *Inorg. Synth.* **1990**, *28*, 107.

6.   (a) Boulas, P. L.; Gómez-Kaifer, M.; Echegoyen, L. *Angew. Chem., Int. Ed. Engl.* **1998**, *37*, 216; (b) Mamula, O.; von Zelewsky, A.; Bernardinelli, G. *Angew. Chem., Int. Ed. Engl.* **1998**, *37*, 290.

7.   (a) Gray, H. B.; Winkler, J. R. *Annu. Rev. Biochem.* **1996**, *65*, 537; (b) Dandliker, P. J.; Holmlin, R. E.; Barton, J. K. *Science* **1997**, *275*, 1465.

8. For a recent review see: Matyjaszewski, K., Ed. "Controlled Radical Polymerizations" *ACS Symp. Ser.* **1998**, *685*, 2-30.

9. Kröhnke, F. *Synthesis* **1976**, 1.

10. (a) Uenishi, J.; Tanaka, T.; Nishiwaki, K.; Wakabayashi, S.; Oae, S.; Tsukube, H. *J. Org. Chem.* **1993**, *58*, 4382; (b) Kawai, T.; Furukawa, N.; Oae, S. *Tetrahedron Lett.* **1984**, *25*, 2549.

11. Tiecco, M.; Testaferri, L.; Tingoli, M.; Chianelli, D.; Montanucci, M. *Synthesis* **1984**, 736 and references therein.

12. Potts, K. T.; Winslow, P. A. *J. Org. Chem.* **1985**, *50*, 5405 and references therein.

13. (a) Negishi, E.-i.; King, A. O.; Okukado, N. *J. Org. Chem.* **1977**, *42*, 1821; (b) Negishi, E.-i.; Takahashi, T.; King, A. O. *Org. Synth., Coll. Vol. VII* **1993**, 430; (c) Larsen, M.; Jorgensen, M. *J. Org. Chem.* **1997**, *62*, 4171.

14. Savage, S. A.; Smith, A. P.; Fraser, C. L. *J. Org. Chem.* **1998**, *63*, 10048.

15. (a) Collins, J. E.; Lamba, J. J. S.; Love. J. C.; McAlvin, J. E.; Ng, C.; Peters, B. P.; Wu, X.; Fraser, C. L. *Inorg. Chem.* **1999**, *38*, 2020; (b) Smith, A. P.; Corbin, P. S.; Fraser, C. L. *Tetrahedron Lett.* **2000**, *41*, 2787.

16. (a) Collins, J. E.; Fraser, C. L. *Macromolecules* **1998**, *31*, 6715; (b) Wu, X.; Fraser, C. L. *Macromolecules* **2000**, *33*, 4053; (c) Fraser, C. L.; Smith, A. P.; Wu, X. *J. Am. Chem. Soc.* **2000**, *122*, 9026.

Appendix

Chemical Abstracts Nomenclature (Collective Index Number);
(Registry Number)

2-Hydroxy-5-methylpyridine: 2(1H)-Pyridinone, 5-methyl- (8,9); (1003-68-5)

2-Amino-5-methylpyridine: Aldrich Name: 2-Amino-5-picoline: HIGHLY TOXIC:

3-Picoline, 6-amino- (8); 2-Pyridinamine, 5-methyl- (9); (1603-41-4)

Sodium nitrite: Nitrous acid, sodium salt (8,9); (7632-00-0)

5-Methyl-2-(trifluoromethanesulfonyl)oxypyridine: Methanesulfonic acid, trifluoro-,

5-methyl-2-pyridinyl ester (13); (154447-03-7)

4-Methyl-2-pyridyl triflate: Methanesulfonic acid, trifluoro-, 4-methyl-2-pyridinyl ester

(13); (179260-78-7)

6-Methyl-2-pyridyl triflate: Methanesulfonic acid, trifluoro-, 6-methyl-2-pyridinyl ester

(13); (154447-04-8)

Trifluoromethanesulfonic anhydride: Methanesulfonic acid, trifluoro-, anhydride (8,9);

(358-23-6)

5-Methyl-2,2'-bipyridine: 2,2'-Bipyridine, 5-methyl- (9); (56100-20-0)

4-Methyl-2,2'-bipyridine: 2,2'-Bipyridine, 4-methyl- (9); (56100-19-7)

6-Methyl-2,2'-bipyridine: 2,2'-Bipyridine, 6-methyl- (9); (56100-22-2)

tert-Butyllithium: Lithium, tert-butyl- (8); Lithium, (1,1-dimethylethyl)- (9); (594-19-4)

2-Bromopyridine: HIGHLY TOXIC: Pyridine, 2-bromo- (8,9); (109-04-6)

Zinc chloride (8,9); (7646-85-7)

Lithium chloride (8,9); (7447-41-8)

Tetrakis(triphenylphosphine)palladium(0): Palladium, tetrakis(triphenylphosphine)-

(8); Palladium, tetrakis(triphenylphosphine)-, (T-4)- (9); (14221-01-3)

Ethylenediaminetetraacetic acid, disodium salt dihydrate: Acetic acid (ethylenedinitrilo)tetra-, disodium salt, dihydrate (8); Glycine, N,N'-1,2-ethanediylbis[N-(carboxymethyl)-, disodium salt, dihydrate (9); (6381-92-6)

2-Hydroxy-4-methylpyridine: 2(1H)-Pyridinone, 4-methyl- (9); (13466-41-6)

2-Hydroxy-6-methylpyridine: 2(1H)-Pyridinone, 6-methyl- (9); (3279-76-3)

2-Amino-4-methylpyridine: Aldrich Name: 2-Amino-4-picoline: HIGHLY TOXIC: 4-Picoline, 2-amino- (8); 2-Pyridinamine, 4-methyl- (9); (695-34-1)

2-Amino-6-methylpyridine: Aldrich Name: 2-Amino-6-picoline: HIGHLY TOXIC: 2-Picoline, 6-amino- (8); 2-Pyridinamine, 6-methyl- (9); (1824-81-3)

N-Benzylbenzamide: Benzamide, N-benzyl- (8); Benzamide, N-(phenylmethyl)- (9); (1485-70-7)

ULLMAN METHOXYLATION IN THE PRESENCE OF A 2,5-DIMETHYLPYRROLE-BLOCKED ANILINE: PREPARATION OF 2-FLUORO-4-METHOXYANILINE

(Benzenamine, 2-fluoro-4-methoxy-)

Submitted by John A. Ragan, Brian P. Jones, Michael J. Castaldi, Paul D. Hill, and Teresa W. Makowski. [1]

Checked by Samuel W. Ridenour and David J. Hart.

1. Procedure

A.   1-(2-Fluoro-4-iodophenyl)-2,5-dimethyl-1H-pyrrole. A 500-mL, single-necked, round-bottomed flask with a Teflon-coated magnetic stir bar is charged with 2-fluoro-4-iodoaniline (50.0 g, 211 mmol) (Note 1), p-toluenesulfonic acid (0.40 g, 2.1 mmol) (Note 2), 250 mL of toluene (Note 3), and acetonylacetone (29.7 mL, 253 mmol) (Note 4). A Dean-Stark trap is attached to the flask, and the solution is warmed to reflux for 1 hr (Note 5). After the solution is cooled to room temperature, it is transferred to a 500-mL separatory funnel, and washed with one 50-mL portion of aqueous saturated sodium bicarbonate ($NaHCO_3$), five 50-mL portions of water (Note 6), and one 50-mL portion of brine. The organic phase is dried over anhydrous magnesium sulfate ($MgSO_4$), filtered, and concentrated to provide a brown, free-flowing solid (67.5 g, 101.5% of theory) (Notes 7 and 8).

B.   1-(2-Fluoro-4-methoxyphenyl)-2,5-dimethyl-1H-pyrrole. A 500-mL, single-necked, round-bottomed flask equipped with a condenser and Teflon-coated magnetic stir bar is flame-dried under a positive pressure of nitrogen, and charged with 1-(2-fluoro-4-iodophenyl)-2,5-dimethyl-1H-pyrrole (66.6 g, 211 mmol), sodium methoxide (NaOMe) (34.2 g, 633 mmol) (Note 9), copper(I) chloride (3.13 g, 31.7 mmol) (Note 10), 230 mL of methanol (Note 11) and 70 mL of dimethylformamide (DMF) (Note 12). The resulting slurry is placed in an 80°C oil bath for 90 min (Note 13) so that a gentle reflux is maintained, then cooled to room temperature. The slurry is poured into a rapidly stirring mixture of 500 mL of isopropyl ether (Note 14), 220 mL of aq 5% ammonium chloride (NH_4Cl) (Note 15), and 350 mL of water in a 2-L Erlenmeyer flask, rinsing with several small portions of methanol. The resulting slurry is stirred overnight (Note 16). It is filtered through a 3"-pad of Celite in a coarse-frit glass funnel, and the filtrate is transferred to a 2-L separatory funnel. The phases are separated, and the aqueous phase is extracted with three 50-mL portions of isopropyl ether. The combined organic phases are washed with 200 mL of aqueous 10% NH_4OH (Note 17), filtered through a silica gel pad (Note 18), and concentrated to provide a brown, free-flowing solid (42.0 g, 92% crude yield) (Note 19). The crude product is recrystallized by dissolving it in 100 mL of hot hexanes and stirring overnight at room temperature, then collecting the resulting brown solid (34.7 g, 75% yield) (Note 20).

C. *2-Fluoro-4-methoxyaniline.* A 1-L, single-necked flask equipped with a reflux condenser and Teflon-coated magnetic stirring bar is charged with 1-(2-fluoro-4-methoxyphenyl)-2,5-dimethyl-1H-pyrrole (34.7 g, 158 mmol), hydroxylamine hydrochloride (110 g, 1.58 mol) (Note 21), triethylamine (44.0 mL, 316 mmol) (Note 22), 300 mL of 95% ethanol, and 150 mL of water. The resulting solution is warmed to reflux for 20 hr (Notes 23 and 24), then cooled to room temperature. The reaction is quenched by pouring into a rapidly stirred solution of 200-300 mL of ice-cold 1 N hydrochloric acid (HCl). This solution is washed with two 250-mL portions of isopropyl ether, the pH is adjusted to 9-10 by careful addition of 6 N sodium hydroxide (NaOH), and the resulting mixture is extracted with two 250-mL portions of isopropyl ether. The final organic phase is dried over $MgSO_4$, filtered, and concentrated to an oily, brown solid. This material is triturated with several portions of isopropyl ether with warming on a steam bath to dissolve as much product as possible, decanting away from an insoluble cream colored solid (Note 25). Concentration of these extracts provides a brown solid (18.1 g, 81% crude yield), which is recrystallized by dissolving in 18 mL of hot isopropyl ether, and slowly adding 80 mL of hexanes. After the product is placed in an ice bath for 1 hr, it is isolated as a brown solid (16.0 g, 72% yield), which is further purified by trituration at room temperature with 160 mL of water for 16 hr (Note 26): filtration and drying in a vacuum oven (with little if any heating) provide analytically pure product: 13.0 g, 58% yield (Note 27).

2. Notes

1. 2-Fluoro-4-iodoaniline was obtained from Aldrich Chemical Company, Inc., and used without further purification. It can also be prepared by iodination of 2-fluoroaniline.[2]

2. p-Toluenesulfonic acid monohydrate was obtained from Aldrich Chemical Company, Inc., and used without further purification.

3. A.C.S. reagent grade toluene was obtained from J. T. Baker and used as received.

4. Acetonylacetone (hexane-2,5-dione) was obtained from Aldrich Chemical Company, Inc., and used without further purification.

5. The reaction can be monitored by TLC (20:1 hexane-ethyl acetate, UV visualization, SM R_f = 0.11, product R_f = 0.56). The submitters indicate that the reaction can also be monitored by GC/MS (Hewlett-Packard 5890 GC/MS, HP-1 column (12 m x 0.2 mm x 0.33 μm), 1 mL/min flow rate, injector temp. 280°C, oven temp. 133°C for 0.1 min, then ramp 19°C/min to 310°C, hold for 1.65 min): SM R_f = 1.53 min, product R_f = 3.10 min.

6. Multiple aqueous washes assist in removing any excess acetonylacetone.

7. ^{1}H NMR indicates reasonably pure product, with trace amounts of toluene and acetonylacetone.

8. The product shows the following physical properties: mp 68-70°C; ^{1}H NMR (400 MHz, CDCl$_3$) δ: 2.03 (s, 6 H), 5.94 (s, 2 H), 7.02 (t, 1 H, J = 8), 7.64 (m, 2 H); ^{13}C NMR (CDCl$_3$) (8 of 9 lines observed) δ: 12.5, 92.8, 106.5, 126.1 (d, J = 23), 128.9, 131.9, 134.0 (d, J = 4), 158.1 (d, J = 254); MS (EI): m/z 268 (100); HRMS (FAB) calcd for C$_{12}$H$_{11}$NFI (M[+]) 315.9999, found 315.9995.

9. Sodium methoxide was obtained from Aldrich Chemical Company, Inc., and used without further purification. Out of a total of five separate batches of NaOMe (all

new, unopened bottles) used over a 24-month period, the submitters had one occasion where the reaction failed to proceed beyond 15-20% conversion. When a new bottle of NaOMe (different lot number) was used, the reaction worked as usual. The NaOMe that failed in the Ullman coupling was found to have limited solubility in methanol (MeOH) (bottles that worked displayed MeOH solubilities of >100 mg/mL), suggestive of contamination by significant quantities of NaOH, possibly from adventitious water introduced during re-packaging or manufacture.

10. Cuprous chloride (CuCl) was obtained from Aldrich Chemical Company, Inc., and used without further purification.

11. A.C.S. reagent grade methanol was obtained from J. T. Baker and used as received.

12. Sure-seal DMF was obtained from Aldrich Chemical Company, Inc., and used as received.

13. The submitters indicate that the reaction can be monitored by GC/MS (same conditions as Note 5): SM R_f = 3.10 min, product R_f = 2.50 min.

14. A.C.S. reagent grade isopropyl ether was obtained from J. T. Baker and used as received.

15. The ammonium chloride solution was prepared by dissolution of 50 g of NH_4Cl (Aldrich Chemical Company, Inc.) in 950 mL of distilled water.

16. The Erlenmeyer is loosely capped with a piece of aluminum foil or a cork stopper. The submitters have run this step, which assists in removal of copper salts, for as short as 16 hr to as long as 60 hr with no change in outcome.

17. The ammonium hydroxide was prepared by dilution of commercial 28-30% NH_4OH (Baker) with 9 volumes of distilled water.

18. Silica gel, 100 g of 230-400 mesh, was used in an 8-cm diameter, medium-frit glass funnel.

19. The crude product is quite pure by ^{1}H NMR, and can be carried directly into the deprotection if desired.

20. The product shows the following physical properties: mp 67-69°C; H NMR (400 MHz, CDCl$_3$) δ: 1.97 (s, 6 H), 3.82 (s, 3 H), 5.89 (s, 2 H), 6.73 (s, 1 H), 6.75 (d, 1 H, J = 8), 7.12 (t, 1 H, J = 8); ^{13}C NMR (CDCl$_3$) (9 of 10 lines observed) δ: 12.4, 55.7, 102.3 (d, J = 25), 105.6, 106.0, 109.9 (d, J = 3), 129.5, 130.7, 159.1 (d, J = 260); MS (EI): m/z 219 (100); HRMS (FAB) calcd for C$_{13}$H$_{14}$NFO (M+H) 220.1138, found 220.1127. Product color ranges from light brown to dark brown with no effect on physical and spectral properties.

21. Hydroxylamine hydrochloride was obtained from Fisher Scientific Company and used as received.

22. A.C.S. reagent grade triethylamine was obtained from J. T. Baker and used as received.

23. Because of the potentially explosive nature of hydroxylamine[3] the reaction should be kept behind a blast shield while heating.

24. The submitters indicate that the reaction can be monitored by GC/MS (same conditions as Note 5): SM R$_f$ = 2.50 min, product R$_f$ = 0.90 min.

25. ^{1}H NMR of this material shows just two singlets (δ 2.48 and 1.91), suggesting that it is derived from acetonylacetone (possibly the bis-oxime).

26. This water reslurry serves to remove trace residues of the δ 2.48 and 1.91 impurity referred to in Note 25.

27. The product shows the following physical properties (ref 4): mp 46.8-47.1°C; ^{1}H NMR (300 MHz, CDCl$_3$) δ: 3.34 (br s, 2 H), 3.72 (s, 3 H), 6.53 (m, 1 H), 6.61 (dd, 1 H, J = 3, 10), 6.72 (dd, 1 H, J = 9, 10); ^{13}C NMR (100 MHz, CDCl$_3$) δ: 56.2, 102.7, 110.2, 118.1, 128.1, 153.3, 152.4. Anal. Calcd for C$_7$H$_8$NFO: C, 59.57; H, 5.71; N, 9.92. Found: C, 59.62; H, 5.74; N, 9.99. The product color ranges from light brown to dark brown with no effect on physical or spectral properties.

Waste Disposal Information

All toxic materials were disposed of in accordance with "Prudent Practices in the Laboratory"; National Academy Press; Washington, DC, 1995.

3. Discussion

2-Fluoro-4-methoxyaniline has been previously prepared by nitration of 2-fluorophenol[5] followed by alkylation with dimethyl sulfate and reduction of the nitro group.[4] Nitration delivers a mixture of regioisomers that require chromatographic separation. The submitters recently required a practical, multi-gram synthesis of this compound, and concerns with the potential thermal hazards of a nitration reaction, poor nitration regioselectivity, and handling of dimethyl sulfate led them to investigate an alternative synthesis. The commercial availability of 2-fluoro-4-iodoaniline motivated them to investigate an Ullman coupling.[6] They investigated a variety of blocking groups for the aniline moiety (the unprotected aniline failed to couple under standard conditions), and found the 2,5-dimethylpyrrole blocking group to be uniquely suited for this purpose.[7] Several other substrates were also investigated, as summarized in the Table. Interestingly, electron-withdrawing groups on the aromatic ring led to significantly lower yields (e.g., entries 7 and 8). However, the regiochemistry between the 2,5-dimethylpyrrole substituent and the iodide was of little consequence to the yield of methoxylation.

1. Process Research and Development, Pfizer Central Research, Groton, CT 06340.

2. Krüger, G.; Keck, J.; Noll, K.; Pieper, H. *Arzneim.-Forsch.* **1984**, *34*, 1612.

3. Sandler, S. R. "Organic Functional Group Preparations", 2nd ed.; Academic Press: New York, 1986; Vol. 3, p. 490.

4. Norris, R. K.; Sternhell, S. *Aust. J. Chem.* **1972**, *25*, 2621.

5. Hodgson, H. H.; Nixon, J. *J. Chem. Soc.* **1928**, 1879.

6. Chiu, C. K.-F. "Review of Alkenyl and Aryl C-O Bond Forming Reactions"; Chiu, C. K.-F., Ed.; Pergamon: New York, 1995; Vol. 2; pp. 683-685.

7. Ragan, J. A.; Makowski, T. W.; Castaldi, M. J.; Hill, P. D. *Synthesis* **1998**, 1599.

Product	% Yield	Product	% Yield
	96		97
	95		0
	92		34
	94		

Appendix

Chemical Abstracts Nomenclature (Collective Index Number); (Registry Number)

2-Fluoro-4-methoxyaniline (8,9); (458-52-6)

1-(2-Fluoro-4-iodophenyl)-2,5-dimethyl-1H-pyrrole: 1H-Pyrrole, 1-(2-fluoro-4-iodophenyl)-2,5-dimethyl- (14); (217314-30-2)

2-Fluoro-4-iodoaniline: Aniline, 2-fluoro-4-iodo- (8,9); (29632-74-4)

p-Toluenesulfonic acid monohydrate (8); Benzenesulfonic acid, 4-methyl-, monohydrate (9); (6192-52-5)

Acetonylacetone: 2,5-Hexanedione (8,9); (110-13-4)

1-(2-Fluoro-4-methoxyphenyl)-2,5-dimethyl-1H-pyrrole: 1H-Pyrrole, 1-(2-fluoro-4-methoxyphenyl)-2,5-dimethyl- (14); (217314-31-3)

Sodium methoxide: Methanol, sodium salt (8,9); (124-41-4)

Copper(I) chloride: Copper chloride (8,9); (7758-89-6)

N,N-Dimethylformamide: CANCER SUSPECT AGENT: Formamide, N,N-dimethyl- (8,9); (68-12-2)

Hydroxylamine hydrochloride (8); Hydroxylamine, hydrochloride (9); (5470-11-1)

1,4,7,10-TETRAAZACYCLODODECANE

A.

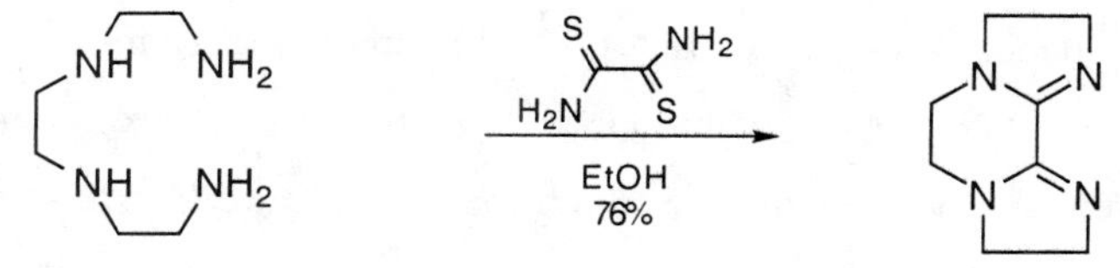

NH
NH2
NH
NH2
S
NH2
H2N
S
EtOH
76%
N
N
N
N

1

B.

1) 6 DIBAL-H, PhCH3
reflux 16 hr
2) KOH, H2O
58%

2

Submitted by David P. Reed and Gary R. Weisman.[1]

Checked by Maya Escobar and Stephen F. Martin.

1. Procedure

Caution: Hydrogen sulfide (H_2S) is generated in Part A of this procedure. The reaction and associated operations must be carried out with provision for H_2S trapping in an efficient hood.

A. 2,3,5,6,8,9-Hexahydrodiimidazo[1,2-a:2',1'-c]pyrazine (**1**). A 500-mL, three-necked, round-bottomed flask is equipped with a 125-mL pressure-equalizing addition funnel, a Teflon-coated magnetic stirring bar, a fritted-gas dispersion tube (initially closed) connected to a nitrogen manifold, and a reflux condenser fitted with a nitrogen (N_2) inlet tube connected to the nitrogen manifold. The nitrogen manifold exit line is routed through two fritted-gas washing bottles charged with 30% aqueous sodium

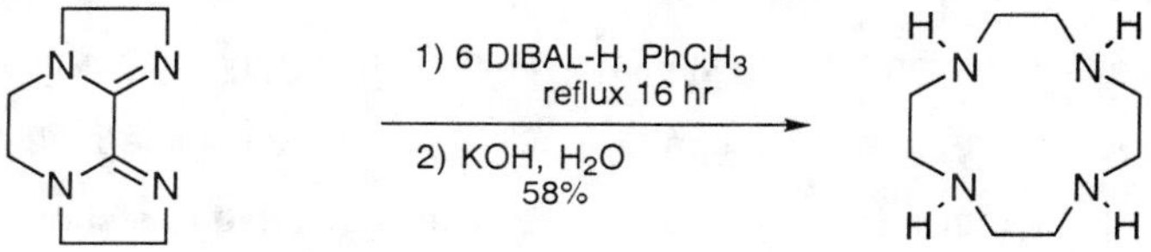

hydroxide (NaOH) in order to scrub H_2S evolved in the reaction (Note 1). The reaction flask is charged with 10.0 g (83.2 mmol) of dithiooxamide (Note 2) and 50 mL of absolute ethanol. A solution of 12.2 g (83.2 mmol) of triethylenetetramine (Note 3) in 50 mL of absolute ethanol is introduced into the reaction flask in one portion via the addition funnel. The magnetically stirred reaction mixture is heated at reflux for 4 hr under nitrogen with evolution of H_2S and ammonia (NH_3) (Note 4). The mixture is then cooled to room temperature and residual H_2S and NH_3 are purged from the reaction mixture for 3 hr by entrainment with nitrogen, which is bubbled through the submerged fritted-gas dispersion tube. Ethanol is removed by rotary evaporation, and the residue is dissolved in 150 mL of chloroform ($CHCl_3$). The insoluble material is removed by gravity filtration through a glass wool plug that is inserted in a short-stem glass funnel. $CHCl_3$ is then removed by rotary evaporation to give 14.0 g of crude product. This solid is taken up in 50 mL of boiling toluene, insoluble impurities are removed by filtration through a glass wool plug, and the flask and funnel are rinsed with a second 50-mL aliquot of boiling toluene (Note 5). The combined filtrates are concentrated to afford 13.4 g of light yellow crystalline product. Sublimation of this material (0.05 mm, 110°C) affords 10.4 g (76%) of pure (>99%) white product (Notes 6, 7, 8).

B. *1,4,7,10-Tetraazacyclododecane* (**2**). A 1-L, three-necked, round-bottomed flask charged with 8.96 g (54.6 mmol) of 2,3,5,6,8,9-hexahydrodiimidazo[1,2-a:2',1'-c]pyrazine is equipped with a reflux condenser fitted with nitrogen inlet tube, a 500-mL pressure-equalizing addition funnel, and a Teflon-coated magnetic stirring bar. The system is flushed with N_2 prior to cannulation of 218 mL (327 mmol) of 1.5 M diisobutylaluminum hydride (DIBAL-H) in toluene (Note 9) to the addition funnel. The reaction flask is cooled in an ice/water (H_2O) bath and the DIBAL-H solution is added to the reaction flask with stirring over 5 min. The reaction mixture is heated at reflux under nitrogen for 16 hr (Note 10). The reaction flask is again cooled in an ice/H_2O bath prior to the addition of 200 mL of toluene. Excess DIBAL-H is quenched by the

74

cautious dropwise addition of 20 mL of 3 M aqueous potassium hydroxide (KOH) solution. When gas evolution ceases, 350 mL of 3 M aqueous KOH is added in one portion and the two-phase mixture is transferred to a separatory funnel (Notes 11, 12). The phases are separated and chipped ice is added to the aqueous phase, which is further extracted with ice-cold $CHCl_3$ (12 x 150 mL). The combined organic extracts are dried over sodium sulfate (Na_2SO_4), filtered, and the solvents are removed by rotary evaporation to afford 6.19 g of white crystalline solid. Sublimation (0.05 mm, 90°C) of this material gives 5.44 g (58%) of product **2** (>98% purity by NMR; Notes 13, 14, 15).

2. Notes

1. The nitrogen manifold (Tygon tubing is suitable) is connected as follows, in this order: (a) nitrogen source, (b) T-connector to fritted-gas dispersion tube with shutoff valve or clamp, (c) shutoff valve or clamp (enables nitrogen to be routed through fritted-gas dispersion tube when closed and dispersion tube is opened), (d) T-connector to nitrogen inlet tube on reflux condenser, (e) safety flask, (f) gas washing bottle #1, (g) gas washing bottle #2, and (h) mineral oil exit bubbler (See Figure 1).

2. Dithiooxamide was purchased from Fluka Chemical Corp.

3. Triethylenetetramine was purchased from Aldrich Chemical Company, Inc., as a hydrate. Anhydrous triethylenetetramine must be used in this procedure. The anhydrous tetraamine was obtained by azeotropic distillation of water (Dean-Stark trap, 3 days) from a solution of 125 g of the commercial hydrate in 150 mL of toluene. Analysis by ^1H NMR verified the removal of water, and no further purification was necessary.

4. Dithiooxamide dissolved to give a homogeneous orange solution soon after the initiation of heating.

5. The hot filtration must be carried out quickly to avoid crystallization of product. This step can be omitted, but a second sublimation may then be necessary to obtain product of sufficient purity for reduction to cyclen.

6. Compound **1** has the following physical and spectroscopic properties: mp 148-150°C (lit[2] mp 150-151°C); ^{1}H NMR (CDCl$_3$, 360 MHz) δ: 3.26 (s, 4 H), 3.35 [apparent t (XX' of AA'XX'), 4 H, J_{appar} = 9.6], 3.86 [apparent t (AA' of AA'XX'), 4 H, J_{appar} = 9.6]; ^{13}C NMR (CDCl$_3$, 90.56 MHz) δ: 45.3, 52.1, 53.9, 155.4; IR (KBr) cm^{-1}: 1629 (C=N); MS (EI) 164.15 (M)$^+$; Anal. Calcd for C$_8$H$_{12}$N$_4$: C, 58.52; H, 7.37; N, 34.12. Found: C, 58.38; H, 7.55; N, 34.22.

7. Bisamidine **1** is hydrolyzed in water (in minutes to hours depending upon purity). While it is not necessary to handle **1** in a dry atmosphere, it is prudent to store it in a desiccator.

8. The checkers found that when the reaction was conducted on 1/2 scale, significantly lower yields (55 - 70% before sublimation) were obtained.

9. DIBAL-H in toluene (1.5 M) was purchased from Aldrich Chemical Company, Inc.

10. The submitters found that a small scale (0.4 g of **1**) reaction with 5 equivalents of DIBAL-H at reflux for 8 hr afforded product in 94% crude yield. However, these conditions gave incomplete reduction and a lower yield when the reaction was scaled up to 10 g of **1**. Therefore, the number of equivalents of DIBAL-H was increased to 6 and the reaction was run for 16 hr.

11. A small amount of solid remains undissolved, but this tends to be distributed in the aqueous phase, making filtration at this stage unnecessary.

12. Originally,[2] a NaF/H$_2$O workup was used. Soxhlet extraction of the solids generated in the work-up was required to obtain good yields of crude **2**. The present aqueous KOH work-up simplifies the procedure and gives comparable or better yields of crude **2**.

13. Compound **2** has the following physical and spectroscopic properties: mp 105-107°C; ^{1}H NMR (CDCl$_3$, 360 MHz) δ: 2.69 (s, 16 H), 2.16 (br s, 4 H); ^{13}C NMR (CDCl$_3$, 90.56 MHz) δ: 46.11. ^{1}H NMR relative integrations are consistent with anhydrous **2**. There has been much confusion in the literature concerning the mp of **2**. Stetter and Mayer[3] originally reported mp 35°C. Buøen, et al. reported mp 119-120°C.[4] Zhang and Busch subsequently reported mp 36-38°C.[5] Aldrich and Fluka list melting point ranges of 110-113°C (97%) and 105-110°C (≥97%) respectively in their catalogs. The submitter's mp range for **2** (calibrated thermometer) is lower than that reported in reference 4, but is consistent with the mp range of sublimed material (no detectable impurities by high S/N NMR) that they have prepared by the Richman-Atkins procedure[6] (mp 105-109°C).

14. The checkers have found it necessary to perform as many as twelve extractions with cold chloroform to obtain the product from the aqueous solution.

15. The submitters obtained yields as high as 88% at scale.

Waste Disposal Information

All toxic materials were disposed of in accordance with "Prudent Practices in the Laboratory"; National Academy Press; Washington, DC, 1995.

3. Discussion

The title compound, **2**,[3] ("cyclen") and its derivatives are important ligands,[7] some of which have biomedical applications[8] (for example, as ligand components of MRI contrast agents). Cyclen is commercially available, but expensive.[9]

This procedure is a modification of the method originally reported by Weisman and Reed.[2] In the first reaction of the two-step sequence (Step A), a two-carbon, permanent, covalently-bound template[10] is introduced by way of dithiooxamide to convert triethylenetetramine to tricyclic bisamidine **1**. Step A is analogous to the synthesis of 2,2'-bi-2-imidazoline reported by Forssell in 1891.[11] Step B is a double reductive ring expansion, which converts the two amidine (template) carbons of bisamidine **1** to a -CH$_2$CH$_2$- unit of **2**. The reaction is conceptually based upon Yamamoto and Maruoka's highly regioselective DIBAL-H reduction of bicyclic amidines to ring-expanded cyclic diamines.[12]

The advantages of this procedure are: (a) it is short and relatively efficient (44-68% overall yield), (b) it is atom-economic,[13] (c) starting materials are readily available, (d) purifications are simple, and (e) it permits preparation of moderate quantities of product with modest effort. The disadvantages are the production of hydrogen sulfide *(highly toxic)* in Step A and the required use of DIBAL-H, an active hydride reducing agent. However, the former can be efficiently trapped and the latter can be handled safely at the reported scale.

There are alternative methods for preparation of cyclen. Since the mid-1970's, the standard method for preparation of cyclen has been based upon the general Stetter-Richman-Atkins synthesis of macrocyclic polyamines,[6,14] a medium-dilution cyclization approach that uses tosyl protection of nitrogen. The cyclen syntheses developed by Richman and Atkins[6] (5 steps) and related modifications[2,15] (4 steps), while very reliable, are still labor-intensive sequences that suffer from atom economy and solvent requirement problems. These problems are largely overcome by the shorter approach documented here. Three additional syntheses of **2** have recently appeared in the literature.[16-18] The syntheses (each 3 steps) rely upon carbon templating for preorganization, subsequent cyclization, and final template removal. Such an approach may prove superior for large scale production of **2**, since active hydride reducing agents are avoided. However, the procedure reported here is very satisfactory for the laboratory-scale preparation of **2**.

1. Department of Chemistry, University of New Hampshire, Durham, NH 03824

2. (a) Weisman, G. R.; Reed, D. P. *J. Org. Chem.* **1996**, *61*, 5186; (b) *Correction: J. Org. Chem.* **1997**, *62*, 4548.

3. Stetter, H.; Mayer, K.-H. *Chem. Ber.* **1961**, *94*, 1410.

4. Buøen, S.; Dale, J.; Krane, J. *Acta Chem. Scand. Ser. B* **1984**, *B38*, 773.

5. Zhang, R.; Busch, D. H. *Inorg. Chem.* **1993**, *32*, 4920.

6. (a) Richman, J. E.; Atkins, T. J. *J. Am. Chem. Soc.* **1974**, *96*, 2268; (b) Atkins, T. J.; Richman, J. E.; Oettle, W. F. *Org. Synth., Coll. Vol. VI* **1988**, 652.

7. (a) Bradshaw, J. S.; Krakowiak, K. E.; Izatt, R. M. "Aza-Crown Macrocycles"; *The Chemistry of Heterocyclic Compounds*; Taylor, E. C., Series Ed.,Wiley: New York, 1993; Vol. 51; (b) Dietrich, B.; Viout, P.; Lehn, J.-M. "Macrocyclic Chemistry: Aspects of Organic and Inorganic Supramolecular Chemistry"; VCH: Weinheim, 1993.

8. (a) Parker, D. In "Crown Compounds"; Cooper, S. R., Ed.; VCH: New York, 1992; (b) Jurisson, S.; Berning, D.; Jia, W.; Ma, D. *Chem. Rev.* **1993**, *93*, 1137.

9. Major specialty chemical suppliers' cyclen (**2**) prices: $214-290 (U.S.) per gram (1997); $236-295 per gram (2000) except that one supplier lowered the price to $3 per gram!

10. Hoss, R.; Vögtle, F. *Angew. Chem., Int. Ed. Engl.* **1994**, *33*, 375.

11. Forssell, G. *Chem. Ber.* **1891**, *24*, 1846.

12. Yamamoto, H.; Maruoka, K. *J. Am. Chem. Soc.* **1981**, *103*, 4186.

13. Trost, B. M. *Science* **1991**, *254*, 1471.

14. Stetter, H.; Roos, E.-E. *Chem. Ber.* **1954**, *87*, 566.

15. (a) Vriesema, B. K.; Buter, J.; Kellogg, R. M. *J. Org. Chem.* **1984**, *49*, 110; (b) Chavez, F.; Sherry, A. D. *J. Org. Chem.* **1989**, *54*, 2990.

16. (a) Sandnes, R. W.; Vasilevskis, J.; Undheim, K.; Gacek, M. (Nycomed Imaging A/s, Norway) PCT Int. Appl. WO 96 28,432, 19 Sept 1996; *Chem. Abstr.* **1996**, *125:* 301031c; (b) Sandnes, R. W.; Gacek, M.; Undheim, K. *Acta Chem. Scand.* **1998**, *52*, 1402.

17. Athey, P. S.; Kiefer, G. E.; (Dow Chem. Co. USA): U.S. Patent 5,587,451, 24 Dec 1996; *Chem. Abstr.* **1997**, *126*, 144300r.

18. Hervé, G.; Bernard, H.; Le Bris, N.; Yaouanc, J.-J.; Handel, H.; Toupet, L. *Tetrahedron Lett.* **1998**, *39*, 6861.

Appendix

Chemical Abstracts Nomenclature (Collective Index Number); (Registry Number)

1,4,7,10-Tetraazacyclododecane (9); (294-90-6)

Hydrogen sulfide: HIGHLY TOXIC (8,9); (7783-06-4)

2,3,5,6,8,9-Hexahydrodiimidazo[1,2-a:2′,1′-c]pyrazine:

Diimidazo[1,2-a:2′,1′-c]pyrazine, 2,3,5,6,8,9-hexahydro- (13); (180588-23-2)

Dithiooxamide: Ethanedithioamide (9); (79-40-3)

Triethylenetetramine (8); 1,2-Ethanediamine, N,N′-bis(2-aminoethyl)- (9); (112-24-3)

Chloroform: HIGHLY TOXIC. CANCER SUSPECT AGENT: (8); Methane, trichloro- (9); (67-66-3)

Diisobutylaluminum hydride: Aluminum, hydrodiisobutyl- (8); Aluminum, hydrobis(2-methylpropyl)- (9); (1191-15-7)

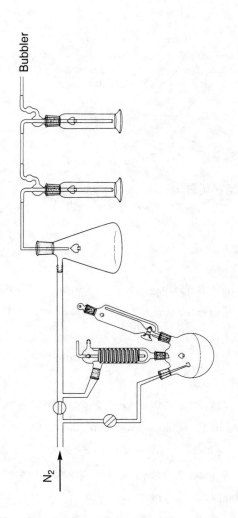

Figure 1

EFFICIENT SYNTHESIS OF HALOMETHYL-2,2'-BIPYRIDINES: 4,4'-BIS(CHLOROMETHYL)-2,2'-BIPYRIDINE

(2,2'-Bipyridine, 4,4'-bis(chloromethyl)-)

Submitted by Adam P. Smith, Jaydeep J. S. Lamba, and Cassandra L. Fraser.[1]

Checked by Motoki Yamane and Koichi Narasaka.

1. Procedure

A. 4,4'-Bis[(trimethylsilyl)methyl]-2,2'-bipyridine. A 500-mL, two-necked, round-bottomed flask (Note 1), equipped with a nitrogen inlet, magnetic stirrer, and rubber septum is charged with tetrahydrofuran (THF) (90 mL) (Note 2) and diisopropylamine (9.8 mL, 69.7 mmol) (Note 3). The reaction mixture is cooled to -78°C and a solution of butyllithium (n-BuLi) (1.7 M in hexanes, 36.0 mL, 61.4 mmol) (Note 4) is added. The solution is stirred at -78°C for 10 min, warmed to 0°C and stirred for 10 min, then cooled back to -78°C. A solution of 4,4'-dimethyl-2,2'-bipyridine (5.14 g, 27.9 mmol)

(Note 5) in THF (130 mL) (Note 2), prepared in a 250-mL, two-necked, round-bottomed flask under a nitrogen atmosphere, is added via cannula to the cold lithium diisopropylamide (LDA) solution. The resulting maroon-black reaction mixture is stirred at -78°C for 1 hr, then chlorotrimethylsilane (TMSCl) (8.85 mL, 69.7 mmol) (Note 6) is rapidly added via syringe. After the solution becomes pale blue-green (~10 sec after the TMSCl addition), the reaction is quenched by rapid addition of absolute ethanol (10 mL). (Note: the reaction should be quenched regardless of color change after a maximum of 15 seconds to avoid over silylation). The cold reaction mixture is poured into a separatory funnel (1 L) containing aqueous saturated sodium bicarbonate (NaHCO$_3$, ~200 mL) and allowed to warm to ~25°C. The product is extracted with dichloromethane (CH$_2$Cl$_2$, 3 x 300 mL); the combined organic fractions are shaken with brine (~200 mL) and dried over sodium sulfate (Na$_2$SO$_4$). Filtration and concentration on a rotary evaporator affords 8.85 g (97%) of 4,4'-bis[(trimethylsilyl)methy]-2,2'-bipyridine as a slightly off-white crystalline solid (Note 7).

B. *4,4'-Bis(chloromethyl)-2,2'-bipyridine.* Into a 500-mL, two-necked, round-bottomed flask (Note 1) equipped with a magnetic stirring bar are placed 5.22 g (15.9 mmol) of 4,4'-bis[(trimethylsilyl)methyl]-2,2'-bipyridine, 15.1 g (63.6 mmol) of hexachloroethane (Cl$_3$CCCl$_3$, Note 8) and 9.65 g (63.6 mmol) of cesium fluoride (CsF, Note 9) at 25°C under a nitrogen atmosphere. Acetonitrile (260 mL) (Note 10) is added and the heterogeneous reaction mixture is stirred at 60°C for ~3.5 hr (or until TLC indicates that all TMS starting material is consumed). After the mixture is cooled to 25°C, it is poured into a separatory funnel containing ethyl acetate (EtOAc) and water (H$_2$O, ~100 mL each). The product is extracted with EtOAc (3 x 100 mL); the combined organic fractions are shaken with brine (~200 mL) and dried over Na$_2$SO$_4$. Filtration and concentration on a rotary evaporator, followed by flash chromatography using deactivated silica gel (60% EtOAc: 40% hexanes) (Note 11), gives 3.67 g (91%) of the chloride as a white solid (Note 12).

2. Notes

1. Before use, all glassware, needles, and syringes were dried overnight in a 120°C oven.

2. THF was dried and purified by passage through alumina solvent purification columns[2] or by distillation over sodium/benzophenone.

3. Diisopropylamine was purchased from Aldrich Chemical Company, Inc., and distilled over calcium hydride (CaH_2) prior to use.

4. A 1.7 M solution of n-BuLi in hexanes was obtained from Aldrich Chemical Company, Inc. The n-BuLi is titrated prior to its use in each reaction using the following procedure.[3] To a 50-mL, round-bottomed flask (Note 1), equipped with nitrogen inlet and a magnetic stirrer is added N-benzylbenzamide (854 mg, 4.0 mmol) (as received from Aldrich Chemical Company, Inc.) and THF (40 mL) (Note 2). The solution is cooled to -42°C (acetonitrile/dry ice) and n-BuLi is added dropwise to the blue endpoint (color persists for >30 sec). The molarity is calculated using a 1:1 stoichiometric ratio of N-benzylbenzamide to n-BuLi. (Just greater than 1 equivalent of alkyllithium is needed to reach the endpoint).

5. 4,4'-Dimethyl-2,2'-bipyridine was obtained from GFS Chemicals, Inc. or Tokyo Chemical Industry Co. and used as received.

6. Chlorotrimethylsilane (TMSCl) was purchased from Aldrich Chemical Company, Inc., and used as obtained.

7. The following characterization data was obtained: mp 90-92°C; ^{1}H NMR (CDCl$_3$, 300 MHz) δ: 0.04 (s, 18 H), 2.21 (s, 4 H), 6.94 (d, 2 H, J = 5.01), 8.05 (br s, 2 H), 8.46 (d, 2 H, J = 5.00); ^{13}C NMR (CDCl$_3$, 75 MHz) δ: -2.2, 27.1, 120.4, 123.0, 148.3, 150.8, 155.5. Anal. Calcd for $C_{18}H_{28}N_2Si_2$: C, 65.79; H, 8.59; N, 8.53. Found: C, 65.78; H, 8.43; N, 8.76. It has been noted that desilylation occurs after standing in deuterochloroform (CDCl$_3$) overnight. The resulting methyl derivatives have also

been observed in certain purified TMS bipyridine samples when stored over time. Therefore, it is best to convert these intermediates to the corresponding halides in a timely fashion.

8. Hexachloroethane (Cl_3CCCl_3), obtained from Aldrich Chemical Company, Inc., was used as received.

9. Cesium fluoride was purchased from Acros Organics, Inc. or Soekawa Chemicals Co. and stored in a dry box prior to use.

10. Acetonitrile was distilled over CaH_2 and stored in a 500-mL Kontes flask prior to use.

11. Silica gel used for flash chromatography (particle size 0.035-0.075 mm) was obtained from VWR Scientific Products. Silica chromatography columns were deactivated by flushing with 10% triethylamine in hexanes and then were washed with hexanes prior to use.

12. Spectral properties are as follows: mp 98-100°C; ^1H NMR ($CDCl_3$, 300 MHz) δ: 4.63 (s, 4 H), 7.38 (dd, 2 H, J = 1.9, 5.0), 8.43 (s, 2 H), 8.70 (d, 2 H, J = 4.6); ^{13}C NMR ($CDCl_3$, 75 MHz) δ: 43.9, 120.1, 122.8, 146.7, 149.4, 155.8. Anal. Calcd for $C_{12}H_{10}Cl_2N_2$: C, 56.94; H, 3.98; N, 11.07. Found: C, 56.82; H, 4.04; N, 11.01.

Waste Disposal Information

All toxic materials were disposed of in accordance with "Prudent Practices in the Laboratory"; National Academy Press; Washington, DC, 1995.

3. Discussion

Halomethylbipyridines are typically synthesized either by radical halogenation[4] or from hydroxymethylbipyridine precursors.[5] Radical methods often give rise to mixtures of halogenated species that are difficult to separate with flash chromatography. A solution to this problem, involving the selective reduction of polyhalogenated by-products with diisobutylaluminum hydride (DIBAL-H), has resulted in slight improvements in overall yields.[6] While the synthesis of halomethyl compounds from hydroxymethyl precursors is more efficient than radical halogenation, such procedures involve many steps, each of which give intermediates in moderate to high yields.[5] Direct trapping of bpy $(CH_2Li)_n$ with electrophiles has proved unsuccessful for the generation of halide products.[5a] The quenching of LDA-generated carbanions with TMSCl prior to halogenation as described here constitutes an efficient, high yield synthesis of halomethyl bpys substituted at various positions around the ring system.[7,8]

Currently, 2,2'-bipyridine derivatives figure prominently in supramolecular assembly,[9] in bioinorganic contexts,[10] in studies of redox electrocatalysis[4a] and in polymeric materials.[11] Halomethyl bpys and their various metal complexes have also been used as initiators for controlled polymerizations of several different monomers including styrene and 2-alkyl-2-oxazolines.[12]

TABLE I

SYNTHESIS OF (TRIMETHYLSILYL)METHYL-2,2'-BIPYRIDINES

Methyl-2,2'-bipyridine $\xrightarrow[\text{(3) EtOH}]{\begin{array}{l}\text{(1) LDA, THF}\\\text{(2) TMSCl}\end{array}}$

Product	R_1	R_2	R_3	R_4	Yield (%)
4-(Trimethylsilyl)-methyl-2,2'-bipyridine	$TMSCH_2$	H	H	H	93
5-(Trimethylsilyl)-methyl-2,2'-bipyridine	H	$TMSCH_2$	H	H	99
6-(Trimethylsilyl)-methyl-2,2'-bipyridine	H	H	$TMSCH_2$	H	97
4,4'-Bis[(trimethylsilyl)-methyl]-2,2'-bipyridine	$TMSCH_2$	H	H	$TMSCH_2$	97

TABLE II

SYNTHESIS OF HALOMETHYL-2,2'-BIPYRIDINES

(Trimethylsilyl)methyl-
2,2'-bipyridine
$\xrightarrow[\text{DMF or CH}_3\text{CN}]{\text{"C}_2\text{X}_6\text{", CsF}}$

Product	R_1	R_2	R_3	R_4	Yield (%)
4-Chloromethyl-2,2'-bipyridine	$ClCH_2$	H	H	H	94
5-Chloromethyl-2,2'-bipyridine	H	$ClCH_2$	H	H	98
6-Chloromethyl-2,2'-bipyridine	H	H	$ClCH_2$	H	95
4,4'-Bis(chloromethyl)-2,2'-bipyridine	$ClCH_2$	H	H	$ClCH_2$	91
4-Bromomethyl-2,2'-bipyridine	$BrCH_2$	H	H	H	92
5-Bromomethyl-2,2'-bipyridine	H	$BrCH_2$	H	H	98
6-Bromomethyl-2,2'-bipyridine	H	H	$BrCH_2$	H	99
4,4'-Bis(bromomethyl)-2,2'-bipyridine	$BrCH_2$	H	H	$BrCH_2$	97

1. Department of Chemistry, University of Virginia, Charlottesville, VA 22904-4319.

2. Pangborn, A. B.; Giardello, M. A.; Grubbs, R. H.; Rosen, R. K.; Timmers, F. J. *Organometallics* **1996**, *15*, 1518.

3. Burchat, A. F.; Chong, J. M.; Nielsen, N. *J. Organomet. Chem.* **1997**, *542*, 281.

4. (a) Gould, S.; Strouse, G. F.; Meyer, T. J.; Sullivan, B. P. *Inorg. Chem.* **1991**, *30*, 2942 and references therein; (b) Wang, Z.; Reibenspies, J.; Motekaitis, R. J.; Martell, A. E. *J. Chem. Soc., Dalton Trans.* **1995**, 1511; (c) Rodriguez-Ubis, J.-C.; Alpha, B.; Plancherel, D.; Lehn, J.-M. *Helv. Chim. Acta* **1984**, *67*, 2264; (d) Newkome, G. R.; Puckett, W. E.; Kiefer, G. E.; Gupta, V. K.; Xia, Y.; Coreil, M.; Hackney, M. A. *J. Org. Chem.* **1982**, *47*, 4116.

5. (a) Della Ciana, L.; Hamachi, I.; Meyer, T. J. *J. Org. Chem.* **1989**, *54*, 1731; (b) Della Ciana, L.; Dressick, W. J.; Von Zelewsky, A. *J. Heterocycl. Chem.* **1990**, *27*, 163; (c) Newkome, G. R.; Kiefer, G. E.; Kohli, D. K.; Xia, Y.-J.; Fronczek, F. R.; Baker, G. R. *J. Org. Chem.* **1989**, *54*, 5105; (d) Imperiali, B.; Prins, T. J.; Fisher, S. L. *J. Org. Chem.* **1993**, *58*, 1613.

6. Uenishi, J.; Tanaka, T.; Nishiwaki, K.; Wakabayashi, S.; Oae, S.; Tsukube, H. *J. Org. Chem.* **1993**, *58*, 4382.

7. Fraser, C. L.; Anastasi, N. R.; Lamba, J. J. S. *J. Org. Chem.* **1997**, *62*, 9314.

8. Savage, S. A.; Smith, A. P.; Fraser, C. L. *J. Org. Chem.* **1998**, *63*, 10048.

9. (a) Boulas, P. L.; Gómez-Kaifer, M.; Echegoyen, L. *Angew. Chem., Int. Ed. Engl.* **1998**, *37*, 216; (b) Mamula, O.; von Zelewsky, A.; Bernardinelli, G. *Angew. Chem., Int. Ed. Engl.* **1998**, *37*, 290.

10. (a) Gray, H. B.; Winkler, J. R. *Annu. Rev. Biochem.* **1996**, *65*, 537; (b) Dandliker, P. J.; Holmlin, R. E.; Barton, J. K. *Science* **1997**, *275*, 1465.

11. For a recent review see: Matyjaszewski, K., Ed. "Controlled Radical Polymerizations"; American Chemical Society: Washington, DC, 1998.

12. (a) Collins, J. E.; Fraser, C. L. *Macromolecules* **1998**, *31*, 6715; (b) McAlvin, J. E.; Fraser, C. L. *Macromolecules* **1999**, *32*, 1341; (c) Wu, X.; Fraser, C. L. *Macromolecules* **2000**, *33*, 4053.

Appendix
Chemical Abstracts Nomenclature (Collective Index Number);
(Registry Number)

4,4'-Bis(chloromethyl)-2,2'-bipyridines: 2,2'-Bipyridine, 4,4'-bis(chloromethyl)- (13); (138219-98-4)

4,4'-Bis[(trimethylsilyl)methyl]-2,2'-bipyridine: 2,2'-Bipyridine, 4,4'-bis[(trimethylsilyl)methyl]- (14); (199282-52-5)

Diisopropylamine (8); 2-Propanamine, N-(1-methylethyl)- (9); (108-18-9)

Butyllithium: Lithium, butyl- (8,9); (109-72-8)

4,4'-Dimethyl-2,2'-bipyridine: 2,2'-Bipyridine, 4,4'-dimethyl- (9); (1134-35-6)

Chlorotrimethylsilane: Silane, chlorotrimethyl- (8,9); (75-77-4)

Hexachloroethane: Ethane, hexachloro- (8,9); (67-72-1)

Cesium fluoride (8,9); (13400-13-0)

Acetonitrile: TOXIC (8,9); (75-05-8)

N-Benzylbenzamide: Benzamide, N-benzyl- (8); Benzamide, N-(phenylmethyl)- (9); (1485-70-7)

PREPARATION AND USE OF N,N'-DI-BOC-N"-TRIFLYLGUANIDINE

(Carbamic acid, [[(trifluoromethyl)sulfonyl]carbonimidoyl]bis-, bis(1,1-dimethylethyl) ester)

A.

$$\underset{\text{BocHN}\quad\text{NHBoc}}{\overset{\text{NH}}{\|}} \xrightarrow[\text{CH}_2\text{Cl}_2,\ -78°\text{C}]{\text{Tf}_2\text{O, NEt}_3} \underset{\text{BocHN}\quad\text{NHBoc}}{\overset{\text{NTf}}{\|}}$$

B.

$$\text{PhCH}_2\text{NH}_2 + \underset{\text{BocHN}\quad\text{NHBoc}}{\overset{\text{NTf}}{\|}} \xrightarrow[\text{CH}_2\text{Cl}_2,\ 0.5\ \text{hr}]{\text{Et}_3\text{N, rt}} \text{PhCH}_2\text{NH}\underset{\text{H}}{\overset{\text{NBoc}}{\|}}\text{NHBoc}$$

Submitted by Tracy J. Baker, Mika Tomioka, and Murray Goodman.[1]

Checked by Dustin J. Mergott and William R. Roush.

1. Procedure

A. N,N'-di-Boc-N"-triflylguanidine. A 250-mL, two-necked, round-bottomed flask equipped with a 10-mL pressure-equalizing dropping funnel sealed with a rubber septum, gas inlet, and a large football-shaped Teflon-coated magnetic stirring bar is purged with nitrogen (Note 1). The flask is charged with N,N'-di-Boc-guanidine (7.5 g, 29 mmol, Note 2), dichloromethane (100 mL, Note 3), and triethylamine (5.0 mL, 36 mmol, Note 4). The temperature of the mixture is equilibrated to -78°C using a dry ice/isopropyl alcohol bath. Triflic anhydride (5.9 mL, 35 mmol, Note 5) is added dropwise through the dropping funnel over a period of 20 min, and the resulting mixture is allowed to warm to -20°C over 4 hr (Notes 6, 7). A 2 M aqueous sodium bisulfate solution is added to the mixture at -20°C, such that the reaction temperature does not rise above -10°C, and the resulting layers are stirred vigorously for 5 min (longer stir times lead to decreased yields). The layers are immediately separated, and the aqueous phase is extracted with dichloromethane (3 x 50 mL). The combined organic layers are washed with 2 M aqueous sodium bisulfate (80 mL), brine (50 mL), dried (MgSO$_4$), filtered and concentrated under reduced pressure. The crude material is purified by flash column chromatography (Note 8) and dried under reduced pressure to afford N,N'-di-Boc-N"-triflylguanidine (10 g, 90%, mp 124°C, Notes 9, 10).

   B. *N,N'-Bis(tert-butoxycarbonyl)-N"-benzylguanidine.* An oven-dried, 50-mL, round-bottomed flask equipped with a magnetic stirring bar is charged with N,N'-di-Boc-N"-triflylguanidine (1.00 g, 2.55 mmol) and dichloromethane (13 mL, Notes 11, 12). Benzylamine (0.31 mL, 2.8 mmol, Note 13) is added in one portion via syringe at room temperature. After 30 min, the mixture is transferred to a 60-mL separatory funnel and washed with 2 M aqueous sodium bisulfate (10 mL) and saturated sodium bicarbonate (10 mL). Each aqueous layer is extracted with dichloromethane (2 x 10 mL). The combined organic phases are washed with brine (10 mL), dried (MgSO$_4$), filtered and concentrated under reduced pressure to afford N,N'-di-Boc-N"-benzylguanidine in quantitative yield (0.89 g, Notes 14, 15).

2. Notes

   1. All glassware was flame-dried and cooled in a desiccator charged with anhydrous calcium sulfate. Once completely cooled, the glassware was quickly assembled and purged with nitrogen.

   2. N,N'-di-Boc-guanidine [1,3-bis(tert-butoxycarbonyl)guanidine, 98%] was purchased from Aldrich Chemical Company, Inc. (catalog #: 49,687-1) and used as received. Alternatively, N,N'-di-Boc-guanidine can be easily synthesized from guanidine hydrochloride and di-tert-butyl dicarbonate according to the literature procedure.[2]

3. Dichloromethane (Fisher Scientific Company) was predried with calcium chloride and freshly distilled under argon from calcium hydride prior to use.

4. Triethylamine (Aldrich Chemical Company, Inc.) was predried with potassium hydroxide and freshly distilled under argon from calcium hydride prior to use.

5. Triflic anhydride (trifluoromethanesulfonic anhydride, $\geq$ 98%) was purchased from Fluka Chemika and used as received.

6. The reaction mixture turned yellow-orange upon addition of triflic anhydride.

7. The submitters allowed the reaction to warm to -5°C prior to quenching with with bisulfate solution, and obtained the product in 90% yield. They noted that if the reaction mixture was allowed to warm above -5°C or stir longer than 4 hr, N,N'-di-Boc-N"-triflylguanidine degrades to N-mono-Boc-N"-triflylguanidine. The checkers obtained yields of 71-85% when this procedure was followed exactly. However, if the reaction was quenched at -20°C, with care not to allow the internal temperature to rise above -10°C during the quench or to stir longer than 5 min following the quench, the checkers obtained yields of 93-96%.

8. A column (5 cm in diameter) of silica gel (J. T. Baker, 233-400 mesh, 250 g, dry-packed) was equilibrated with 20% hexanes in dichloromethane. The crude material dissolved in a minimal amount of chloroform was loaded on the column and eluted with 20% hexanes in dichloromethane. Fractions were collected in 15 x 160 mm test tubes.

9. The compound has the following characteristics: ^{1}H NMR (0.6 M, DMSO-d_6; 500 MHz) δ: 1.46 (s, 18 H, CH$_3$), 11.06 (br s, 2 H, NH); ^{13}C NMR (0.6 M, DMSO-d_6; 125 MHz) δ: 27.5 (6C), 83.4 (2C), 119.1 (q, J_{CF} = 320 Hz), 150.1 (2C), 152.3; IR (neat film/NaCl plate) cm^{-1}: 1202, 1343, 1557, 1622, 1739, 1788, 3300; FAB-MS m/z (relative intensity) 414 (M+Na)+, 392 (M+H)+, 336, 280, 236. Anal. Calcd for

$C_{12}H_{20}F_3N_3O_6S$: C, 36.83; H, 5.15; N, 10.74; F, 14.56; S, 8.19. Found: C, 36.93; H, 5.21; N, 10.66; F, 14.80; S, 8.33.

10. The submitters report that the production scale of N,N'-di-Boc-N"-triflylguanidine can be increased ten-fold as long as Note 7 is followed.

11. Reagent grade dichloromethane may be used without further purification.

12. Guanidinylations of less reactive amines may require triethylamine (1.1 eq.) freshly distilled from calcium hydride.

13. Benzylamine (Aldrich Chemical Company, Inc., 99%) is stored over potassium hydroxide and used without further purification.

14. The spectral data of N,N'-di-Boc-N"-benzylguanidine matched that reported in the literature.[3] The sample has the following characteristics: ^{1}H NMR (CDCl$_3$, 500 MHz) δ: 1.42 (s, 9 H, CH$_3$), 1.45 (s, 9 H, CH$_3$), 4.57 (d, 2 H, J = 5.2, C$\underline{H}_2$Ph), 7.21-7.20 (m, 5 H, arom.), 8.55 (br s, 1 H, NH), 11.50 (br s, 1 H, NH); ^{13}C NMR (CDCl$_3$, 125 MHz) δ: 28.0 (3C), 28.2 (3C), 45.0, 79.3, 83.1, 127.5, 127.7 (2C), 128.7 (2C), 137.1, 153.1, 156.0, 163.5; IR (KBr) cm[-1]: 1560, 1626, 1654, 1741; FAB-MS m/z (relative intensity) 350 (M+H)+, 238, 194, 91; high resolution mass spectrum, calcd for $C_{18}H_{27}N_3O_4$: m/z 349.2002, found 349.1997. The product was greater than 95% pure as determined by ^{1}H NMR spectral analysis.

15. The production of N,N'-di-Boc-N"-benzylguanidine can be increased ten-fold or more.

Waste Disposal Information

All toxic materials were disposed of in accordance with "Prudent Practices in the Laboratory"; National Academy Press; Washington, DC, 1995.

3. Discussion

Both natural and nonnatural guanidine-containing molecules have important biological activity ranging from antimicrobial, antiviral, and antihypertensive to neurotoxic.[4] Hence, the conversion of amines to guanidines has been a significant synthetic endeavor for many years. The present method, using N,N'-di-Boc-N"-triflylguanidine for the guanidinylation of amines is the most efficient and general approach for most applications in solution and on solid phase.[2,5]

To date, the most commonly used reagents for the guanidinylation of amines are derivatives of protected thioureas in the presence of the Mukaiyama reagent,[3] pyrazole-1-carboxamidines,[6] and S-alkylisothioureas.[7] Protected thioureas in conjunction with the Mukaiyama reagent have displayed the most versatile usage. This combination has been successful in the conversion of sterically demanding and resin-bound amines to protected guanidines.[3] However, this method is limited to the use of highly polar aprotic solvents such as dimethylformamide because of the solubility properties of the Mukaiyama reagent. Guanidinylations with pyrazole-1-carboxamidines,[6] and S-alkylisothioureas[7] are sluggish in comparison to N,N'-di-Boc-N"-triflylguanidine and are incompatible with solid phase application.

The simple preparation of N,N'-di-Boc-N"-triflylguanidine from a commercially available source and its straightforward isolation make this reagent extremely attractive. Guanidinylation using N,N'-di-Boc-N"-triflylguanidine is effective for amines both in solution and on solid phase.[2,5] These reactions may be carried out in a variety of solvents with dichloromethane and chloroform being the most common; however, reaction rates slow with increase in solvent polarity. The use of protected thioureas with the Mukaiyama reagent seems to be superior for guanidinylations of sterically hindered and less reactive amines, but simple product isolation and experimental

95

setup of the N,N'-di-Boc-N"-triflylguanidine method make this the reagent of choice for most applications. Some representative examples are compiled in the Table.[2,5]

1. Department of Chemistry and Biochemistry, University of California, San Diego, La Jolla, CA 92093-0343.

2. Feichtinger, K.; Zapf, C.; Sings, H. L.; Goodman, M. *J. Org. Chem.* **1998**, *63*, 3804.

3. Yong, Y. F.; Kowalski, J. A.; Lipton, M. A. *J. Org. Chem.* **1997**, *62*, 1540.

4. (a) Berlinck, R. G. S. *Prog. Chem. Org. Nat.* **1995**, *66*, 119-295; (b) Berlinck, R. G. S. *Nat. Prod. Rep.* **1996**, *13*, 377-409.

5. (a) Feichtinger, K.; Sings, H. L.; Baker, T. J.; Matthews, K.; Goodman, M. *J. Org. Chem.* **1998**, *63*, 8432; (b) Baker, T. J.; Goodman, M. *Synthesis* **1999**, *(Spec. Iss.)* 1423.

6. (a) Drake, B.; Patek, M.; Lebl, M. *Synthesis* **1994**, 579; (b) Bernatowicz, M. S.; Wu, Y.; Matsueda, G. R. *Tetrahedron Lett.* **1993**, *34*, 3389.

7. Bergeron, R. J.; McManis, J. S. *J. Org. Chem.* **1987**, *52*, 1700.

Appendix

Chemical Abstracts Nomenclature (Collective Index Number); (Registry Number)

N,N'-Di-Boc-N"-triflylguanidine: N,N'-Bis(tert-butoxycarbonyl)-N"-trifluoromethane-sulfonylguanidine: Carbamic acid, [[(trifluoromethyl)sulfonyl]carbonimidoyl]bis-, bis(1,1-dimethylethyl) ester (14); (207857-15-6)

N,N'-Di-Boc-guanidine: 1,3-Bis(tert-butoxycarbonyl)guanidine: Carbamic acid, carbonimidolybis-, bis(1,1-dimethylethyl) ester (13); (154476-57-0)

Triflic anhydride: Methanesulfonic acid, trifluoro-, anhydride (8,9); (358-23-6)

N,N'-Di-Boc-N"-benzylguanidine: N,N'-Bis(tert-butoxycarbonyl-N"-benzylguanidine: Carbamic acid, [[(phenylmethyl)imino]methylene]bis-, bis(1,1-dimethylethyl) ester (13); (145013-06-5)

Benzylamine (8); Benzenemethanamine (9); (100-46-9)

Guanidine hydrochloride: Guanidine, monohydrochloride (9); (50-01-1)

Di-tert-butyl dicarbonate: Formic acid, oxydi-, di-tert-butyl ester (8); Dicarbonic acid, bis(1,1-dimethylethyl) ester (9); (24424-99-5)

TABLE

GUANIDINYLATION OF AMINES IN SOLUTION AND ON SOLID PHASE USING
N,N'-DI-BOC-N"-TRIFLYGUANIDINE

Entry	Amine	Product	Yield (%)
1[a]			100
2[a]			89
3[b]			100
4[b]			100
5[a]			82
6			83[c]
7		 guanoxan · HCl	94[d]

[a]Reference 2. [b]Reference 5a. [c]Total yield of peptide after cleavage from resin. Reference 5a.
[d]Total yield of guanoxan · HCl after guanidinylation and Boc removal. Reference 5b.

REDUCTION OF SULFONYL HALIDES WITH ZINC POWDER:
S-METHYL METHANETHIOSULFONATE
(Methanesulfonothioic acid, S-methyl ester)

$$CH_3-\overset{O}{\underset{O}{\overset{\|}{\underset{\|}{S}}}}-Cl \quad \xrightarrow[\text{Ethyl acetate}]{\text{Zn powder, } CH_3COCl} \quad CH_3-\overset{O}{\underset{O}{\overset{\|}{\underset{\|}{S}}}}-S-CH_3$$

Submitted by Fabrice Chemla[1] and Philippe Karoyan.[2]
Checked by Mitsuru Kitamura and Koichi Narasaka.

1. Procedure

CAUTION: The reaction must be performed in a well-ventilated hood.

A 1-L, three-necked, round-bottomed flask equipped with a large magnetic stirring bar, a thermometer, reflux condenser, and a 100-mL pressure-equalizing addition funnel with a rubber septum is charged with 49 g (0.75 mol) of zinc dust, 1.0 mL (0.012 mol) of 1,2-dibromoethane, 2.0 mL (0.016 mol) of chlorotrimethylsilane, and 500 mL of ethyl acetate (Note 1). After the mixture is stirred for 15 min at room temperature, it is heated to reflux by means of an oil bath and 38.7 mL (0.500 mol) of methanesulfonyl chloride (Note 2) is added dropwise through the addition funnel. The reaction is highly exothermic. At the end of the addition (1 hr), a large part of the zinc is consumed. The resulting grey suspension is stirred for an additional 15 min. Acetyl chloride, 35.7 mL (0.502 mol), (Note 2) is then added dropwise (*CAUTION* : Note 3) through the addition funnel with care taken to maintain vigorous stirring of the reaction mixture. After completion of the addition, the resulting mixture becomes clear and no zinc remains. Heating is maintained for an additional 15 min, and the mixture is allowed to cool to room temperature. The resulting clear solution is poured into a 1 M aqueous hydrochloric acid solution (ca. 200 mL) and the phases are separated. The aqueous phase is extracted with ethyl acetate (ca. 200 mL) and the combined organic phases are washed with brine (ca. 200 mL) and dried over anhydrous magnesium sulfate. The solvent is removed under reduced pressure and the residual yellow oil is distilled under reduced pressure to give 24.1 g (76% yield) of methyl methanethiosulfonate (Note 4) as a colorless liquid, bp 80-83°C/0.5 mm.

2. Notes

1. Zinc dust (<10 micron) was purchased from Aldrich Chemical Company, Inc. 1,2-Dibromoethane and chlorotrimethylsilane were purchased from Acros Chemicals or Tokyo Chemical Industry Co. and used as received. Ethyl acetate was purchased from Merck (HPLC grade) and used without any purification.

2. Methanesulfonyl chloride and acetyl chloride were purchased from Acros Chemicals, Tokyo Chemical Industry Co., and Wako Pure Chemical Industries, Ltd. and used as received.

3. The reaction is highly exothermic. The addition rate is approximately one drop per two seconds, in order to achieve addition within approximately 1 hr. The reaction mixture must be well-stirred and maintained at reflux during this period to avoid any uncontrolled event.

4. The product exhibits the following physical and spectral properties: ^{1}H NMR (CDCl$_3$, 400 MHz) δ: 2.70 (s, 3 H), 3.32 (s, 3 H); ^{13}C NMR (CDCl$_3$, 126 MHz) δ: 18.2. 48.8; IR (film) cm^{-1}: 3030 (weak), 3010 (weak), 2930 (weak), 1430 (medium), 1410 (medium), 1330 (strong), 1300 (strong), 1130 (strong), 960 (strong), 750 (strong). Microanalysis: Calcd for C$_2$H$_6$O$_2$S$_2$: C, 19.04; H, 4.76. Found: C, 18.99; H, 4.91.

Waste Disposal Information

All toxic materials were disposed of in accordance with "Prudent Practices in the Laboratory"; National Academy Press; Washington, DC, 1995.

3. Discussion

Thiosulfonic S-esters[3] are powerful sulfenylating reagents,[4] more reactive than the commonly used disulfides, and more stable than the very reactive sulfenyl halides. In addition, they have found wide industrial applications as biologically active compounds or in polymer production.[3] Their use, however, has been limited by the lack of easy and practical preparations.

S-Methyl methanethiosulfonate is commercially available, but is expensive. Other preparation methods involve oxidation of thiols or disulfides by halogens or peroxides,[3] reduction of sulfinyl halides[5] (which have to be prepared) or sulfonyl halides with potassium iodide[6] or copper/bronze[7] as well as thermolysis of sulfonylhydrazines.[8] Finally, a two-day procedure for the preparation of methyl methanethiosulfonate has been reported from dimethyl sulfoxide.[9]

The reaction described above has been successfully applied to the preparation of various symmetrical thiosulfonic S-esters (Table).[10] The procedure described here gives better and more reproducible results than the one reported before.[10]

1. Laboratoire de Chimie des OrganoEléments, Tour 44-45, Case 183, Université Pierre et Marie Curie, 4 place Jussieu, 75252 Paris Cedex 05, France.

2. Laboratoire de Chimie Organique Biologique, Tour 44-45, Case 182, Université Pierre et Marie Curie, 4 place Jussieu, 75252 Paris Cedex 05, France.

3.   For a review, see: Zefirov, N. S.; Zyk, N. V.; Beloglazkina, E. K.; Kutateladze, A. G. *Sulfur Rep.* **1993**, *14*, 223.

4.   (a) Trost, B. M. *Chem. Rev.* **1978**, *78*, 363; (b) Palumbo, G.; Ferreri, C.; D'Ambrosio, C.; Caputo, R. *Phosphorus Sulfur* **1984**, *19*, 235, and references cited therein.

5.   (a) Freeman, F.; Bartosik, L. G.; van Bui, N.; Keindl, M. C.; Nelson, E. L. *Phosphorus Sulfur* **1988**, *35*, 375; (b) Freeman, F. *Chem. Rev.* **1984**, *84*, 117; (c) Freeman, F.; Keindl, M. C. *Synthesis* **1983**, 913.

6.   Palumbo, G.; Caputo, R. *Synthesis* **1981**, 888.

7.   Karrer, P.; Wehrli, W.; Biedermann, E.; dalla Vedova, M. *Helv. Chim. Acta* **1928**, *11*, 233.

8.   Meier, H.; Menzel, I. *Synthesis* **1972**, 267.

9.   Laszlo, P.; Mathy, A. *J. Org. Chem.* **1984**, *49*, 2281.

10.   Chemla, F. *Synlett* **1998**, 894.

Appendix
Chemical Abstracts Nomenclature (Collective Index Number); (Registry Number)

Zinc (8,9); (7440-66-6)

S-Methyl methanethiosulfonate: Methanesulfonic acid, thio-, S-methyl ester (8); Methanesulfonothioic acid, S-methyl ester (9); (2949-92-0)

1,2-Dibromoethane: Ethane, 1,2-dibromo-, (8,9); (106-93-4)

Chlorotrimethylsilane: Silane, chlorotrimethyl-, (8,9); (75-77-4)

Methanesulfonyl chloride (8,9); (124-63-0)

Acetyl chloride (8,9); (75-36-5)

TABLE[10]

PREPARATION OF SYMMETRICAL THIOSULFONATES FROM SULFONYL

CHLORIDES

	Starting Material	Product	Yield
1	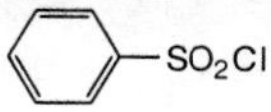 ⟨⟩—SO₂Cl	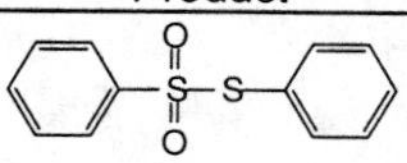 Ph–S(O)₂–S–Ph	90 %
2	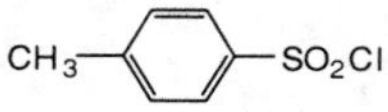 CH₃—⟨⟩—SO₂Cl	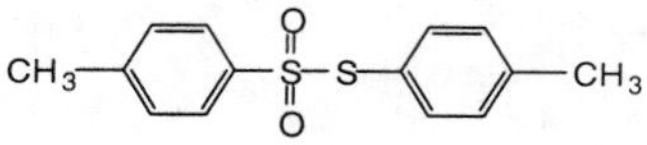 CH₃—⟨⟩—S(O)₂–S—⟨⟩—CH₃	90 %
3	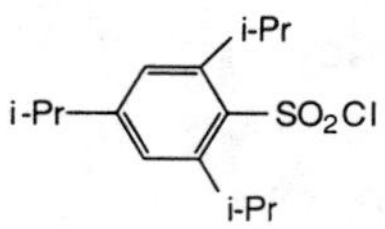 i-Pr, i-Pr, i-Pr—⟨⟩—SO₂Cl	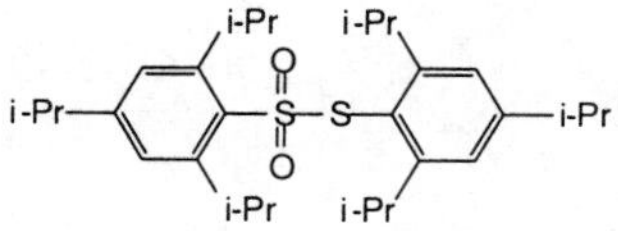 i-Pr...—SO₂—S—...i-Pr	44 %
4	 OH₃C—⟨⟩—SO₂Cl	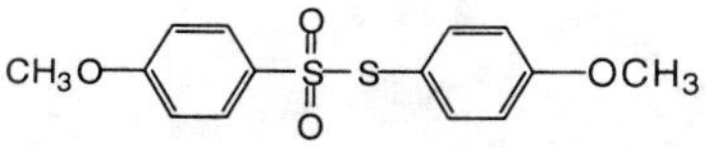 CH₃O—⟨⟩—S(O)₂–S—⟨⟩—OCH₃	65 %
5	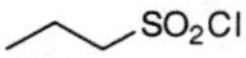 ⁀⁀SO₂Cl	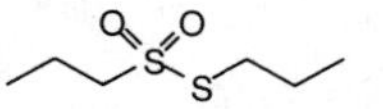 ⁀⁀S(O)₂–S⁀⁀	60 %

(PHENYL)[2-(TRIMETHYLSILYL)PHENYL]IODONIUM TRIFLATE.
AN EFFICIENT AND MILD BENZYNE PRECURSOR
(Iodonium, phenyl-, 2-(trimethylsilyl)phenyl-, salt with trifluoromethane-
sulfonic acid)

Submitted by Tsugio Kitamura, Mitsuru Todaka, and Yuzo Fujiwara.[1]
Checked by Ralf Demuth and Rick L. Danheiser.

1.   Procedure

A.   1,2-Bis(trimethylsilyl)benzene (**1**). A dry, 500-mL, three-necked, round-
bottomed flask is equipped with a large Teflon-covered magnetic stir bar, 100-mL
pressure-equalizing addition funnel, Dimroth condenser (Note 1) fitted with a drying
tube, and a glass stopper. The flask is charged with 9.72 g (0.400 mol) of magnesium

turnings (Note 2), 70 mL of hexamethylphosphoramide (HMPA) (Note 3), 11.25 mL (0.100 mol) of 1,2-dichlorobenzene (Note 4), and 0.254 g (1.00 mmol) of iodine (I_2) (Note 5). The addition funnel is charged with 51.0 mL (0.400 mol) of freshly distilled chlorotrimethylsilane (Note 6). The flask is immersed in an oil bath at 70°C, stirring is initiated, and chlorotrimethylsilane is added slowly, dropwise with vigorous stirring. After completion of the addition, the oil bath is heated to 100°C and the reaction mixture is stirred at this temperature for 2 days. During this time, the reaction mixture becomes viscous and finally separates into two phases. The reaction mixture is cooled to ca. 40°C (Note 7) and poured into a 1-L beaker containing saturated sodium bicarbonate ($NaHCO_3$) solution (200 mL), diethyl ether (100 mL), and ice (ca. 100 g). Solids and unreacted magnesium metal are separated by suction filtration, the filtrate is transferred to a separatory funnel, and the aqueous phase is extracted with three 150-mL portions of ether. The combined ethereal extracts are washed with water (500 mL) and saturated sodium chloride (500 mL), dried over anhydrous sodium sulfate, and filtered. The solvent is evaporated under reduced pressure and the residue is distilled from a 100-mL, round-bottomed flask with a magnetic stir bar through a 20-cm Vigreux column at reduced pressure. The fraction boiling at 128-133°C (20 mm) is collected to afford 16.5-16.7 g (74-75%) of 1,2-bis(trimethylsilyl)benzene (**1**) (Note 8) as a colorless liquid.

   B.   *(Phenyl)[2-(trimethylsilyl)phenyl]iodonium triflate* (**2**).   A 100-mL round-bottomed flask equipped with an argon inlet adapter and a magnetic stir bar is charged with 12.9 g (40.0 mmol) of finely ground (diacetoxyiodo)benzene (Note 9) and 70 mL of dichloromethane (Note 10). The suspension is cooled at 0°C with an ice bath and 6.9 mL (78 mmol) of trifluoromethanesulfonic acid (Note 11) is added in one portion by syringe. The resulting clear yellow solution is stirred at room temperature for 2 hr and a solution of 8.9 mL (40.0 mmol) of 1,2-bis(trimethylsilyl)benzene in 10 mL of dichloromethane is added dropwise by syringe. The resulting mixture is stirred at room temperature for 12 hr, and the solvent is removed by rotary evaporation under reduced pressure to afford the product as colorless crystals. (When an oily residue is obtained, it can be crystallized by triturating with diethyl ether.) The crystals are collected by filtration and washed with 40 mL of diethyl ether to afford 14.7-15.7 g (73-78%) of (phenyl)[2-(trimethylsilyl)phenyl]iodonium triflate (**2**) as colorless needles, mp 142-143°C (Notes 12, 13).

C. Generation of benzyne and trapping with furan. A 50-mL round-bottomed flask fitted with a pressure-equalizing addition funnel equipped with an argon inlet adapter and a magnetic stir bar is charged with 1.51 g (3.00 mmol) of (phenyl)[2-(trimethylsilyl)phenyl]iodonium triflate (**2**), 10 mL of dichloromethane, and 1.10 mL (15.1 mmol) of furan (Note 14). The addition funnel is charged with 3.6 mL (3.6 mmol) of 1.0 M tetrabutylammonium fluoride ($Bu_4N^+F^-$) in tetrahydrofuran (THF) (Note 15). The flask is placed in an ice bath and the tetrabutylammonium fluoride solution is added dropwise over ca. 5 min. The reaction mixture is stirred at room temperature for 30 min, and water (20 mL) is added. The aqueous phase is separated and extracted with three 10-mL portions of dichloromethane. The combined organic extracts are washed with 15 mL of water, dried over anhydrous sodium sulfate, and filtered. The solvent is evaporated by rotary evaporation under reduced pressure, and the residual oil is purified by column chromatography through 60 g of silica gel packed in a 4-cm diameter column (elution with dichloromethane) to give 0.415-0.418 g (96-97%) of 1,4-dihydronaphthalene 1,4-oxide (**3**) as colorless crystals, mp 52-55°C (Notes 16, 17).

2. Notes

1. A highly efficient reflux condenser is required to avoid the loss of chlorotrimethylsilane by evaporation during the reaction.

106

2. Magnesium turnings were purchased from Nacalai Tesque, Inc. or Fisher Scientific Company.

3. HMPA, hexamethylphosphoramide, is toxic and a cancer-suspect agent. It was purchased from Tokyo Kasei Kogyo Co. or Aldrich Chemical Company, Inc. and distilled from calcium hydride under reduced pressure before use. When dimethylpropyleneurea (DMPU) was used in place of HMPA, none of the desired bis(trimethylsilyl)benzene was obtained.

4. 1,2-Dichlorobenzene was purchased by the submitters from Tokyo Kasei Kogyo Co. and distilled under reduced pressure. The checkers used 99% anhydrous 1,2-dichlorobenzene from Aldrich Chemical Company, Inc., without further purification.

5. Iodine was purchased from Tokyo Kasei Kogyo Co. or Aldrich Chemical Company, Inc., and used as received.

6. The submitters purchased chlorotrimethylsilane from Shin-Etsu Chemicals and distilled it prior to use. The checkers used 99+% chlorotrimethylsilane from Aldrich Chemical Company, Inc., without further purification.

7. If cooled to room temperature, the lower layer solidifies.

8. Bis(trimethylsilyl)benzene (1) has the following spectral properties: ^{1}H NMR (300 MHz, CDCl$_3$) δ: 0.36 (s, 18 H), 7.28-7.34 (m, 2 H), 7.64-7.68 (m, 2 H); ^{13}C NMR (75 MHz, CDCl$_3$) δ: 2.0, 127.8, 135.2, 146.0.

9. (Diacetoxyiodo)benzene was purchased from Aldrich Chemical Company, Inc., and was used as received.

10. Dichloromethane was distilled from phosphorus pentoxide (P$_2$O$_5$) or calcium hydride prior to use.

11. Trifluoromethanesulfonic acid from Central Glass Co. or Aldrich Chemical Company, Inc., was employed.

12. Product **2** has the following spectral properties: ^{1}H NMR (400 MHz, CDCl$_3$) δ: 0.42 (s, 9 H), 7.26-8.13 (m, 9 H); ^{13}C NMR (100 MHz, CDCl$_3$) δ: 0.1, 114.0, 121.2, 132.2, 133.2, 133.4, 138.5, 139.1, 147.3.

13. The submitters obtained **2** in 86% yield.

14. Furan was purchased from Tokyo Kasei Kogyo Co. or Aldrich Chemical Company, Inc., and distilled prior to use.

15. 1.0 M Tetrabutylammonium fluoride in THF was obtained from Aldrich Chemical Company, Inc.

16. If the product is obtained as an oil, it is cooled in a -78°C bath to induce crystallization.

17. Product **3** has the following spectral properties: ^{1}H NMR (400 MHz, CDCl$_3$) δ: 5.69 (s, 2 H), 6.94-6.96 (m, 2 H), 7.00 (s, 2 H), 7.22-7.24 (m, 2 H); ^{13}C NMR (100 MHz, CDCl$_3$) δ: 82.2, 120.2, 124.9, 142.9, 148.9.

Waste Disposal Information

All toxic materials were disposed of in accordance with "Prudent Practices in the Laboratory"; National Academy Press; Washington, DC, 1995.

3. Discussion

Benzyne is one of a group of reactive intermediates widely applicable to organic synthesis.[2-5] The title hypervalent iodine-benzyne precursor, (phenyl)[2-(trimethylsilyl)phenyl]iodonium triflate **2**,[6] is prepared by only two steps from commercially available reagents. Products **1** and **2** are stable and easily purified. The hypervalent iodine-benzyne precursor **2** is obtained as a stable solid and handled without any precautions. More importantly, benzyne is generated by using tetrabutylammonium fluoride under mild and neutral conditions. Therefore, compound **2** is useful for reactions of substrates that cannot be conducted at high temperatures or under basic conditions.

The advantages of the use of this hypervalent iodine-benzyne precursor **2** are as follows: (1) The benzyne precursor **2** is a stable crystalline compound up to its melting point, usually to 130°C. (2) The benzyne precursor **2** is not hygroscopic and is stable to air; it can be handled without any special precautions. (3) The generation of benzyne can be conducted under neutral conditions and at room temperature.

The high efficiency of the present precursor **2** is demonstrated by comparison with a similar precursor, 2-(trimethylsilyl)phenyl triflate (**4**), which generates benzyne under mild conditions (room temperature and neutral).[7] Benzyne precursor **2** gives the adduct, 1,4-epoxy-1,4-dihydronaphthalene **3**, quantitatively in the reaction with furan, while the reaction of benzyne precursor **4** under the same conditions leads to a lower yield of adduct **3** and needs longer reaction time.

The reaction of thiobenzophenones with benzyne shows the superiority of the present iodine precursor **2** over benzenediazonium-2-carboxylate (**5**), which is widely used.[8] The reaction of the hypervalent iodine precursor **2** with thiobenzophenones affords [4+2] cycloadducts from benzyne and thiophenzophenones under mild

R = Me, MeO

conditions. However, the reaction with benzenediazonium-2-carboxylate **5** gives no benzyne adducts, but benzoxathianones, which are presumably derived from the reaction of 2-carboxyphenyl cation and cyclization.

1. Department of Chemistry and Biochemistry, Graduate School of Engineering, Kyushu University, Hakozaki, Fukuoka 812-8581, Japan.

2. Hart, H. in "The Chemistry of Triple-Bonded Functional Groups, Supplement C2"; Patai, S., Ed.; John Wiley & Sons: Chichester, 1994; Chapter 18, pp. 1017-1134.

3. Gilchrist, T. L. in "The Chemistry of Triple-Bonded Functional Groups, Supplement C"; Patai, S.; Rappoport, Z., Eds.; John Wiley & Sons: Chichester, 1983; Chapter 11, pp. 383-419.

4. Hoffmann, R. W. in "Chemistry of Acetylenes"; Viehe, H. G., Ed.; Dekker: New York, 1969; pp. 1063-1148.

5.   Hoffmann, R. W. "Dehydrobenzene and Cycloalkynes"; Academic Press: New York, 1967.

6.   Kitamura, T.; Yamane, M. *J. Chem. Soc., Chem. Commun.* **1995**, 983-984.

7.   Himeshima, Y.; Sonoda, T.; Kobayashi, H. *Chem. Lett.* **1983**, 1211-1214.

8.   Okuma, K.; Yamamoto, T.; Shirokawa, T.; Kitamura, T.; Fujiwara, Y. *Tetrahedron Lett.* **1996**, *37*, 8883-8886.

Appendix
Chemical Abstracts Nomenclature (Collective Index Number); (Registry Number)

(Phenyl) [2-(trimethylsilyl)phenyl]iodonium triflate: Iodonium, phenyl-, 2-(trimethylsilyl)-phenyl-, salt with trifluoromethanesulfonic acid (1:1) (13); (164594-13-2)

1,2-Bis(trimethylsilyl)benzene: Silane, o-phenylenebis[trimethyl- (8); Silane, 1,2-phenylenebis[trimethyl- (9); (17151-09-6)

Magnesium (8,9); (7439-95-4)

Hexamethylphosphoramide: HIGHLY TOXIC: CANCER SUSPECT AGENT: Phosphoric triamide, hexamethyl- (8,9) (680-31-9)

l,2-Dichlorobenzene; Benzene, o-dichloro- (8); Benzene, 1,2-dichloro- (9); (95-50-1)

Iodine (8,9); (7553-56-2)

Chlorotrimethylsilane: Silane, chlorotrimethyl- (8,9); (75-77-4)

(Diacetoxyiodo)benzene: Aldrich: Iodobenzene diacetate: Benzene, (diacetoxyiodo)- (8); Iodine, bis(aceto-O)phenyl- (9); (3240-34-4)

Trifluoromethanesulfonic acid: HIGHLY CORROSIVE: Methanesulfonic acid, trifluoro- (8,9); (1493-13-6)

Furan (8,9); (110-00-9)

Tetrabutylammonium fluoride: Ammonium, tetrabutyl-, fluoride (8); 1-Butanaminium, N,N,N-tributyl-, fluoride (9); (429-41-4)

1,4-Dihydronaphthalene 1,4-oxide: 1,4-Epoxy-1,4-dihydronaphthalene: 1,4-Epoxynaphthalene, 1,4-dihydro- (8,9); (573-57-9)

4-HYDROXY[1-¹³C]BENZOIC ACID

(Benzoic-1-¹³C acid, 4-hydroxy-)

A. $H_3{}^{13}C$—C(=O)—OEt

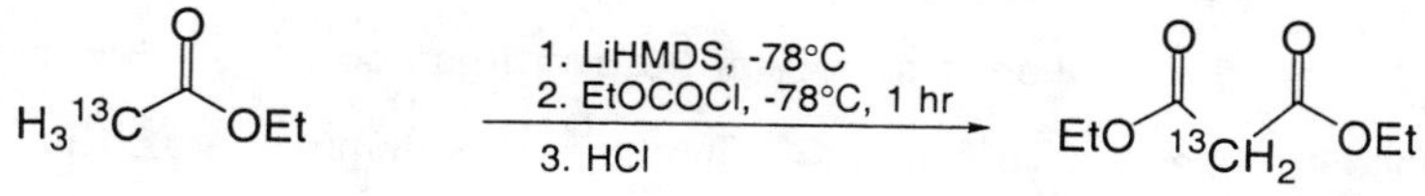

1. LiHMDS, -78°C
2. EtOCOCl, -78°C, 1 hr
3. HCl

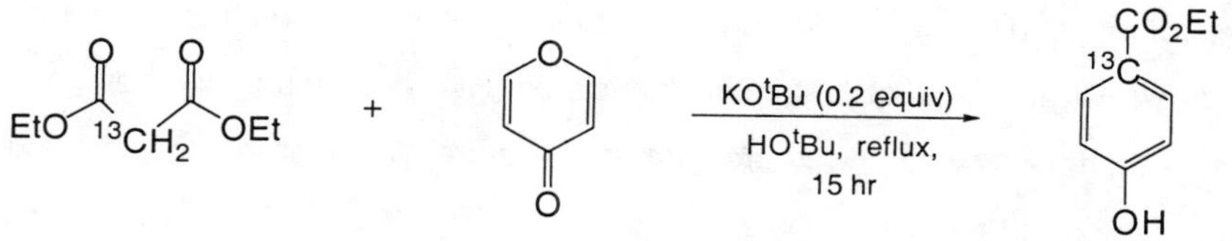

→ EtO—C(=O)—¹³CH₂—C(=O)—OEt

B. EtO—C(=O)—¹³CH₂—C(=O)—OEt + (pyranone)

KO^tBu (0.2 equiv)
HO^tBu, reflux, 15 hr

→ (¹³C—CO₂Et aryl, OH)

C. (¹³C—CO₂Et aryl, OH)

2M NaOH
24 hr

→ (¹³C—CO₂H aryl, OH)

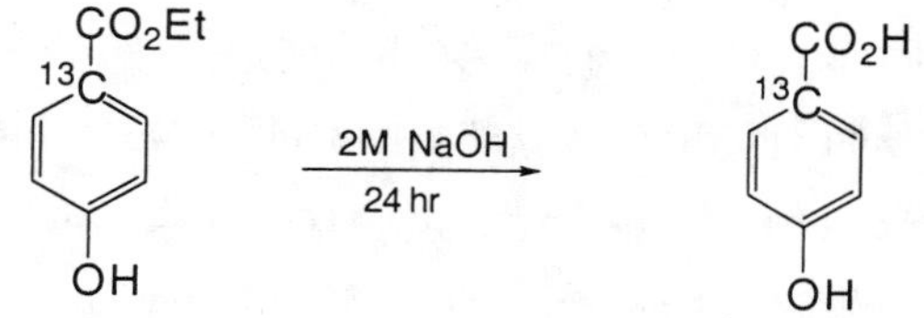

Submitted by Martin Lang, Susanne Lang-Fugmann, and Wolfgang Steglich.[1]
Checked by Tina M. Marks, Nathan X. Yu, and Edward J. J. Grabowski.

1. Procedure

   A. *Diethyl [2-¹³C]malonate* (Note 1). A flame-dried, 100-mL, round-bottomed Schlenk flask equipped with a rubber septum and a magnetic stirring bar is purged with argon. The flask is charged with 15 mL of anhydrous tetrahydrofuran (THF) (Note 2) and 5.8 mL (28 mmol, 2.5 equiv) of hexamethyldisilazane (Note 3). After the solution is cooled to 0°C in an ice-water bath, 9.4 mL (23.5 mmol, 2.1 equiv) of a solution of butyllithium (2.5 M in hexanes) (Note 4) is added slowly via a syringe to the stirred solution. The ice bath is removed and the mixture is allowed to warm to room temperature. After the solution is stirred for 30 min, it is cooled to -78°C using an acetone-dry ice bath and equilibrated for 5 min at the same temperature. Then 1.00 g (11.2 mmol, 1 equiv) of ethyl [2-^{13}C]acetate (Note 5) is added within 5 min via a syringe, and the acetate-containing flask is rinsed with 0.5 mL of anhydrous THF. Stirring is continued at -78°C for 20 min, and 1.07 mL (11.2 mmol, 1 equiv) of ethyl chloroformate (Note 6) is added within 5 min via a syringe. The mixture is stirred for 1 hr at -78°C, and 5 mL of 6 M hydrochloric acid (HCl) is added in one portion (Note 7). The mixture is allowed to warm to room temperature, and after the addition of 20 mL of water, the pH of the solution is adjusted to 1-2 with 2 M HCl. The mixture is extracted with diethyl ether (3 x 50 mL), and the combined organic phases are washed successively with 2 M HCl, water, and brine (30 mL each). The HCl and water phases are combined and reextracted with ether (50 mL). The organic layer is washed with brine (20 mL) and added to the combined organic phases. The combined extracts are dried over anhydrous sodium sulfate (Na_2SO_4), filtered and concentrated under reduced pressure with a rotary evaporator (200 mbar/35°C, 150 mm/35°C). The crude product is distilled in a microdistillation apparatus at 90 mbar (67.5 mm) to give 1.66 g (10.3 mmol, 92%) of diethyl [2-^{13}C]malonate as a colorless liquid (Notes 8 and 9).

B. Ethyl 4-hydroxy[1-^{13}C]benzoate. A 100-mL, single-necked, round-bottomed flask equipped with a magnetic stirring bar, pressure-equalizing addition funnel and a reflux condenser beyond the dropping funnel (Note 10) is charged with 20 mL of tert-butyl alcohol (t-BuOH) (Note 11), 1.01 g (10.5 mmol, 1.05 equiv) of 4H-pyran-4-one (Note 12), and 1.61 g (10.0 mmol, 1 equiv) of diethyl [2-^{13}C]malonate. The condenser is sealed with a silica gel drying tube, and the stirred solution is put in an oil bath at 105°C. A solution of 0.22 g (2.0 mmol, 0.2 equiv) of potassium tert-butoxide (Note 13) in 20 mL of tert-butyl alcohol (t-BuOH) is added dropwise via the funnel during 20 min; the mixture turns red and turbid. After the mixture is heated under reflux for 15 hr, it is allowed to cool to room temperature. Water (30 mL) is added, followed by 5 mL of 2 M HCl. After removal of most of the solvent with a rotary evaporator, the aqueous mixture is extracted with diethyl ether (3 x 50 mL). The combined organic phases are washed with water and brine (each 30 mL). Drying over Na_2SO_4, filtration, and removal of the solvent under reduced pressure with a rotary evaporator affords the crude product. Flash chromatography on silica gel (Note 14) with ethyl acetate/petroleum ether (4:1) as eluent affords 1.34 g (8.01 mmol, 80%) of ethyl 4-hydroxy[1-^{13}C]benzoate, mp 112-113°C (Note 15).

C. 4-Hydroxy[1-^{13}C]benzoic acid. A 25-mL, one-necked, round-bottomed flask equipped with a magnetic stirring bar is charged with 1.25 g (7.50 mmol) of ethyl 4-hydroxy[1-^{13}C]benzoate and 11.3 mL (22.5 mmol, 3 equiv) of 2 M sodium hydroxide (NaOH). After the solution is stirred for 24 hr at room temperature, 17 mL of 2 M HCl is added slowly, whereby the product precipitates. Water is added (10 mL), and the mixture is extracted with diethyl ether (3 x 50 mL). The combined organic phases are washed with 1 M HCl (2 x 30 mL). Removal of the solvent under reduced pressure on a rotary evaporator and drying under reduced pressure affords 1.02 g (7.34 mmol, 98%) of 4-hydroxy[1-^{13}C]benzoic acid (mp 212-213°C) (Note 16), which can be used for feeding experiments without further purification.

2. Notes

1. The procedure follows closely that of Mueller and Leete[2] with slight improvements by the submitters.

2. Tetrahydrofuran was distilled from potassium and benzophenone under an argon atmosphere immediately before use. The checkers used anhydrous THF purchased from Aldrich Chemical Company, Inc.

3. Hexamethyldisilazane (98%), purchased from Lancaster Synthesis Inc. or Aldrich Chemical Company, Inc., was used as received.

4. Butyllithium (2.5 M in hexanes) was purchased from Aldrich Chemical Company, Inc. The actual concentration was determined by titration with diphenylacetic acid or 4-biphenylmethanol.[3]

5. Ethyl [2-^{13}C]acetate is commercially available (Aldrich Chemical Company, Inc.), but expensive. The compound can be prepared by O-ethylation[4,5,6] of the cheaper sodium [2-^{13}C]acetate (Aldrich Chemical Company, Inc.) or via [2-^{13}C]acetyl chloride.[7]

6. Ethyl chloroformate (97%), purchased from Aldrich Chemical Company, Inc., was used as received.

7. The solution should be quenched prior to warming to room temperature. Solutions of lithiated ethyl acetate decompose rapidly at 0°C.[8]

8. The receiver is cooled to -10°C. Cooling to -78°C is not advisable because obstruction may occur. The product (bp 123-125°C/90 mbar, 67.5 mm) can be separated from the hydrolysis products trimethylsilanol and traces of hexamethyldisiloxane.[9] At the end, the apparatus is rinsed with diethyl ether to obtain all the product.

9. The spectral data are as follows: ^{1}H NMR (300 MHz, CDCl$_3$) δ: 1.27 (t, 6 H, J = 7.2), 3.34 (d, 2 H, J = 132), 4.19 (q, 4 H, J = 7.2); ^{13}C NMR (75 MHz, CDCl$_3$) δ: 14.1, 41.7 (^{13}C), 61.5, 166.7 (d, J = 59); MS (EI): 161 [M+] (3), 134 (40), 116 (100), 106 (7), 89 (52), 61 (31); IR (KBr) cm^{-1}: 3465 (br), 2986, 2942, 1754, 1733, 1467, 1448, 1410, 1369, 1319, 1267, 1189, 1151, 1097, 1035, 949, 866, 844, 787, 666, 603.

10. Because of the relatively high melting point of the solvent (23-26°C), it is not advisable to put the condenser directly on the flask. With the addition funnel between the flask and the condenser, most of the tert-butyl alcohol is condensed as a liquid and does not collect as a solid on the cold condenser.

11. tert-Butyl alcohol ($\geq$99.7%), purchased from Fluka Chemical Corp. or Fisher Scientific, was used as received.

12. 4H-Pyran-4-one is commercially available (98+%, Aldrich Chemical Company, Inc.), but the substance is expensive. It can be synthesized by decarboxylation of chelidonic acid monohydrate (Lancaster Synthesis Inc.) following the procedure of De Souza and co-workers.[10]

13. Potassium tert-butoxide (99%), purchased from Fluka Chemical Corp. or Aldrich Chemical Company, Inc., was used as received.

14. Flash chromatography was performed on E. Merck silica gel 230-400 mesh: 150 g of silica gel was loaded on a 7- x 2-in size column using a minimum amount of ethyl acetate as loading solvent. The checkers used a 90-g silica column purchased from Biotage.

15. The spectral data are as follows: ^{1}H NMR (300 MHz, CDCl$_3$) δ: 1.39 (t, 3 H, J = 7.2), 4.36 (q, 2 H, J = 7.2), 6.50 (s, 1 H), 6.85-6.93 (m, 2 H), 7.93-7.99 (m, 2 H); ^{13}C NMR (75 MHz, CDCl$_3$) δ: 14.4, 61.1, 115.3 (d, J = 1.5), 122.7 (^{13}C), 132.0 (d, J = 60), 160.4 (d, J = 9.1), 167.2 (d, J = 77); MS (EI): 167 [M+] (30), 139 (23), 122 (100), 94 (11), 83 (10); IR (KBr) cm^{-1}: 3218 (br), 1672, 1602, 1583, 1441, 1370, 1306, 1286, 1239, 1169, 1104, 1018, 847, 768, 723, 697, 618.

16. The spectral data are as follows: ^{1}H NMR (300 MHz, DMSO-d_6) δ: 6.77-6.85 (m, 2 H), 7.75-7.81 (m, 2 H), 10.2 (s, br, ~0.8 H), 12.4 (s, br, ~0.8 H); (The coupling pattern of the aromatic protons is even at 600 MHz not clearly resolved.) ^{13}C NMR (75 MHz, DMSO-d_6) δ: 115.3, 121.6 (^{13}C), 131.7 (d, J = 59), 161.8 (d, J = 8.5), 167.3 (d, J = 74); MS (EI) 139 [M+] (100), 122 (94), 94 (20); IR (KBr) cm^{-1}: 3394 (br), 2966, 2831, 2660, 2562, 1677, 1602, 1588, 1504, 1440, 1421, 1309, 1282, 1243, 1169, 1100, 933, 852, 766, 617, 548.

Waste Disposal Information

All toxic materials were disposed of in accordance with "Prudent Practices in the Laboratory"; National Academy Press; Washington, DC, 1995.

3. Discussion

4-Hydroxybenzoic acid acts as biosynthetic precursor for several secondary metabolites.[11] Since in many cases decarboxylation takes place on route to the metabolites,[12] ring-^{13}C-labeled 4-hydroxybenzoic acids are in demand for biosynthetic studies.

Previous syntheses of ring-labeled 4-hydroxybenzoic acid use many steps and show low overall yields. 4-Hydroxy[3-^{13}C]benzoic acid was synthesized in six steps from ethyl [1-^{13}C]acetate with 2.8% overall yield.[13] 4-Hydroxy[3,5-^{13}C$_2$]benzoic acid was prepared in five steps from [1,3-^{13}C$_2$]acetone with an overall yield of less than 4.5%.[12b] 4-Hydroxy[2,6-^{13}C$_2$]benzoic acid was generated microbiologically from [1-^{13}C]glucose by using a mutant strain of Klebsiella pneumoniae.[14] For 120 mg of product, 18 g of labeled glucose was necessary. Methyl 4-methoxy[3,5-^{13}C$_2$]benzoate was obtained from [1,3-^{13}C$_2$]acetone in five steps with 32% overall yield.[15] Baldwin and co-workers[16] synthesized methyl 4-methoxy[3,4,5-$^{13}C_3$]benzoate in four steps from [1,2,3-$^{13}C_3$]acetone without indicating the overall yield.

The present procedure affords 4-hydroxy[1-^{13}C]benzoic acid from ethyl [2-^{13}C]acetate in three steps with 72% overall yield. It is based on an observation of Woodward[17] that ethyl 4-hydroxybenzoate is formed by base-catalyzed condensation of 4H-pyran-4-one with diethyl malonate. The submitters studied several solvent-base combinations for this reaction and found that tert-butyl alcohol/potassium tert-butoxide gave the highest yields. When a stoichiometric amount of the base is used, an excess of 4H-pyran-4-one has to be used.[4b] This can be avoided by the use of substoichiometric amounts as given in the procedure.

The use of ethyl [2-^{13}C]acetoacetate instead of diethyl [2-^{13}C]malonate in the condensation reaction with 4H-pyran-4-one afforded ethyl 4-hydroxy[1-^{13}C]benzoate in 87% yield. In this case, 1.1 equiv of 4H-pyran-4-one and 1.1 equiv of potassium tert-butoxide were optimal. The addition of catalytic amounts of the base was not satisfactory. Ethyl [2-^{13}C]acetoacetate was prepared from ethyl [2-^{13}C]acetate as described for diethyl [2-^{13}C]malonate.[18] The maximum yield for this reaction on a 10-mmol scale was only 70% after distillation. 4H-Pyran-4-one reacted with nitromethane and potassium tert-butoxide (each 1.1 equiv) to afford 4-nitrophenol in 75% yield after purification by flash chromatography. This gives easy access to 4-nitro[4-^{13}C]phenol. With 2,4-pentanedione, the condensation with 4H-pyran-4-one under the same reaction conditions gave 4-hydroxyacetophenone in 45-50% yield after purification.

Ethyl 4-hydroxy[1-^{13}C]benzoate can be converted into other ring-^{13}C-labeled compounds like 3,4-dihydroxy[1-^{13}C]benzoic acid, 1,3,4-trihydroxy[1-^{13}C]benzene, and D,L-[1'-^{13}C]tyrosine.[4b]

1. Department Chemie, Universität München, Butenandtstr. 5-13 F, D-81377 München, Germany.

2. Mueller, M. E.; Leete, E. *J. Org. Chem.* **1981**, *46*, 3151-3152.

3. (a) Kofron, W. G.; Baclawski, L. M. *J. Org. Chem.* **1976**, *41*, 1879-1880; (b) Juaristi, E.; Martínez-Richa, A.; García-Rivera, A.; Cruz-Sánchez, J. S. *J. Org. Chem.* **1983**, *48*, 2603-2606.

4. (a) Graebe, C. *Justus Liebiegs Ann. Chem.* **1905**, *340*, 244-249; (b) Beyer, J.; Lang-Fugmann, S.; Mühlbauer, A.; Steglich, W. *Synthesis* **1998**, 1047-1051.

5. (a) Ropp, G. A. *J. Am. Chem. Soc.* **1950**, *72*, 2299; (b) Bodine, R. S.; Hylarides, M.; Daub, G. H.; VanderJagt, D. L. *J. Org. Chem.* **1978**, *43*, 4025-4028.

6. D'Alessandro, G.; Sleiter, G. *J. Labelled Compd. Radiopharm.* **1980**, *17*, 813-824.

7. Sedlmaier, H.; Müller, F.; Keller, P. J.; Bacher, A. Z. *Naturforsch. C: Biosci.* **1987**, *42*, 425-429.

8. Rathke, M. W. *J. Am. Chem. Soc.* **1970**, *92*, 3222-3223.

9. Sauer, R. O. *J. Am. Chem. Soc.* **1944**, *66*, 1707-1710.

10. De Souza, C.; Hajikarimian, Y.; Sheldrake, P. W. *Synth. Commun.* **1992**, *22*, 755-759.

11. Inouye, H.; Leistner, E. In "The Chemistry of Quinonoid Compounds", Patai, S.; Rappoport, Z., Eds.; Wiley: New York, 1988; Vol. 2, pt. 2, p. 1293.

12. E.g. (a) Arbutin: Zenk, M. H. Z. *Naturforsch.* **1964**, *19b*, 856-857; (b) Pentabromopseudilin: Hanefeld, U.; Floss, H. G.; Laatsch, H. *J. Org. Chem.* **1994**, *59*, 3604-3608; (c) Ubiquinones and shikonin: Dewick, P. M. *Nat. Prod. Rep.* **1998**, *15*, 17-58 and literature cited therein; (d) Boviquinones: Mühlbauer, A., Beyer, J.; Steglich, W. *Tetrahedron Lett.* **1998**, *39*, 5167-5170.

13. Vilas Boas, L. F.; Gillard, R. D.; Mitchell, P. R. *Transition Met. Chem.* **1977**, *2*, 80-83.

14. Müller, R.; Wagener, A.; Schmidt, K.; Leistner, E. *Appl. Microbiol. Biotechnol.* **1995**, *43*, 985-988.

15. Viswanatha, V.; Hruby, V. J. *J. Org. Chem.* **1979**, *44*, 2892-2896.

16. Baldwin, J. E.; Bansal, H. S.; Chondrogianni, J. Gallagher, P. T.; Taha, A. A.; Taylor, A.; Thaller, V. *J. Chem. Res., Synop* **1984**, 176-177.

17. Woodward, R. B., private communication, cited in: Bergmann, E. D.; Ginsburg, D.; Pappo, R. *Org. React.* **1959**, *10*, 219.

18. Leete, E.; Bjorklund, J. A.; Couladis, M. M.; Kim, S. H. *J. Am. Chem. Soc.* **1991**, *113*, 9286-9292.

Appendix
Chemical Abstracts Nomenclature (Collective Index Number); (Registry Number)

4-Hydroxybenzoic-1-^{13}C acid: Benzoic-1-^{13}C acid, 4-hydroxy- (14); (211519-30-1)

Diethyl malonate-2-^{13}C: Propanedioic-2-^{13}C acid, diethyl ester (10); (67035-94-3)

1,1,1,3,3,3-Hexamethyldisilazane: Disilazane, 1,1,1,3,3,3-hexamethyl- (8);

Silanamine, 1,1,1-trimethyl-N-(trimethylsilyl)- (9); (999-97-3)

Butyllithium: Lithium, butyl- (8,9); (109-72-8)

Ethyl acetate-2-^{13}C: Acetic-2-^{13}C acid, ethyl ester (9); (58735-82-3)

Ethyl chloroformate: Formic acid, chloro-, ethyl ester (8); Carbonochloridic acid, ethyl ester (9); (541-41-3)

Ethyl 4-hydroxybenzoate-1-^{13}C: Benzoic-1-^{13}C acid, 4-hydroxy-, ethyl ester (14); (211519-29-8)

tert-Butyl alcohol (8); 2-Propanol, 2-methyl- (9); (75-65-0)

4H-Pyran-4-one (8,9); (108-97-4)

Potassium tert-butoxide: tert-Butyl alcohol, potassium salt (8); 2-Propanol, 2-methyl-, potassium salt (9); (865-47-4)

Diphenylacetic acid: Acetic acid, diphenyl- (8); Benzeneacetic acid, α-phenyl- (9); (117-34-0)

4-Biphenylmethanol (9); (3597-91-9)

Sodium acetate-2-^{13}C: Acetic-2-^{13}C acid, sodium salt (9); (13291-89-9)

Acetyl-2-^{13}C chloride: Acetyl-2-^{13}C chloride (8,9); (14770-40-2)

Trimethylsilanol: Silanol, trimethyl- (8,9); (1066-40-6)

Hexamethyldisiloxane: Disiloxane, hexamethyl- (8,9); (107-46-0)

(1'R)-(-)-2,4-O-ETHYLIDENE-D-ERYTHROSE AND ETHYL (E)-(-)-4,6-O-ETHYLIDENE-(4S,5R,1'R)-4,5,6-TRIHYDROXY-2-HEXENOATE

(1,3-Dioxane-(R)-4-carboxaldehyde, 5-hydroxy-2-methyl-, [2R-(2α,4α,5β)]- and D-erythro-Hex-2-enonic acid, 2,3-dideoxy-4,6-O-ethylidene-, ethyl ester [2E,4(S)]-)

Submitted by M. Fengler-Veith, O. Schwardt, U. Kautz, B. Krämer, and V. Jäger.[1]

Checked by Brian Haney, Brian Bucher, and Dennis P. Curran.

1. Procedure

A. (1'R)-(-)-4,6-O-Ethylidene-D-glucose. A 500-mL, round-bottomed flask is charged with D-glucose (81.8 g, 454 mmol, Note 1) and paraldehyde (68.0 g, 514 mmol, Note 2). Concentrated sulfuric acid (0.5 mL) is added dropwise within 30 sec (Note 3) with shaking. The mixture is mechanically shaken for 40 min (Note 4) and then left for 3 days at room temperature. Ethanol (300 mL, Note 5) is added to the adhesive, colorless mass, then the pH is adjusted to 6.5-7 (Note 6) by addition of a 1 N solution of potassium hydroxide in ethanol (Note 7). The residue is dissolved by careful heating. During this procedure the pH is maintained constant by gradual addition of more 1 N ethanolic potassium hydroxide. Charcoal (5 g, Note 8) is added to the yellow solution and the mixture is filtered through a sintered-glass funnel containing a 2-cm pad of Celite. The filter cake is washed with hot ethanol (50 mL). The filtrate, on standing overnight in a freezer at -30°C, deposits a colorless solid material that is recrystallized from ethanol (90 mL) at -30°C (Note 9) to yield 39.4 g (42%) of (-)-4,6-O-ethylidene-D-glucose. The combined mother liquors are concentrated by rotary evaporation (30°C, 30 mm), followed by removal of solvent and excess paraldehyde at room temperature and 0.02 mm (oil vacuum). Recrystallization of the yellowish solid residue from ethanol at -30°C as described above gives another 22.5 g (24%) of product. The combined yield of (1'R)-(-)-4,6-O-ethylidene-D-glucose is 61.9 g (66%), mp 173-174°C (Note 10).

B. (-)-2,4-O-Ethylidene-D-erythrose. A 1-L, three-necked, round-bottomed flask, equipped with a thermometer and two 200-mL pressure-equalizing dropping funnels, is charged with a suspension of sodium metaperiodate (59.2 g, 277 mmol, Note 11) in water (450 mL). The flask is cooled to 0°C using an ice-water bath. A solution of (-)-4,6-O-ethylidene-D-glucose (29.2 g, 142 mmol, Note 12) in water (120 mL) is added dropwise with stirring and under permanent control of the pH (Note 6) and

124

temperature. The temperature in the flask should be kept below 10°C, and the pH is maintained at approximately 4 by dropwise addition of 8 N aqueous sodium hydroxide (Note 13). After stirring for 3 hr at ≤ 10°C (Note 14), the pH is adjusted to 6.5 by addition of more 8 N sodium hydroxide (Note 15) and stirring is continued for another 2 hr at room temperature. The solution is evaporated under reduced pressure (50°C, 30 mm) and the residue dried at room temperature and 0.02 mm (oil vacuum). To the pale-yellow, solid crude product, ethyl acetate (80 mL) is added and the flask is heated for 2 min at 80°C with stirring. The suspension is filtered and the solid residue is treated three times with ethyl acetate (80 mL each) as described above. The combined, filtered extracts are dried over sodium sulfate for 1 hr with stirring and concentrated by rotary evaporation (30°C, 30 mm), followed by slow (Note 16) removal of solvent at 0.02 mm (oil vacuum) at room temperature to yield 20.1 g (97%) of (-)-2,4-O-ethylidene-D-erythrose as a colorless, amorphous solid, mp 120-121°C (Notes 17, 18).

C. *Ethyl (E)-(-)-4,6-O-ethylidene-(4S,5R,1'R)-4,5,6-trihydroxy-2-hexenoate.* A 1-L, two-necked, round-bottomed flask, fitted with a nitrogen inlet (Note 19) and a stopper, is oven-dried (140°C) and flushed with nitrogen. The flask is charged with a suspension of sodium hydride in paraffin (6.20 g), containing 60% of sodium hydride (Note 20). The suspension is washed with pentane (3 x 30 mL, Note 21) and the residue is freed from remaining pentane at 0.01 mm (oil vacuum) to give 3.72 g (ca. 155 mmol) of sodium hydride. Under nitrogen, a magnetic stirring bar and tetrahydrofuran (200 mL, Note 22) are added and the flask is sealed with a septum. The suspension is cooled to 0°C and triethyl phosphonoacetate (39.2 g, 175 mmol, Note 23) is added to the stirred sodium hydride/tetrahydrofuran suspension over a period of 5 - 10 min by means of a 100-mL syringe. The mixture is cooled to -78°C (2-propanol/dry ice) and a solution of (-)-2,4-O-ethylidene-D-erythrose (mainly as dimer, 14.6 g, 100 mmol) in tetrahydrofuran (200 mL, Note 22) is added by means of a syringe over 5-10 min. After the mixture is stirred for 15 min at -78°C, it is allowed to warm to room temperature and stirred for another 45 min. The reaction is quenched with saturated ammonium chloride solution (250 mL) and transferred to a 2.5-L separatory funnel with 1200 mL of ether. The aqueous phase is separated and extracted with ether (4 x 100 mL). The combined organic layers are washed with a mixture of saturated sodium bicarbonate/brine (1:1) (2 x 200 mL), dried for 1 hr over magnesium sulfate with stirring, filtered, and evaporated to dryness at 30°C/30 mm to leave 32.0 g of a yellowish oil (Note 24). The crude product is purified by column chromatography over silica (Note 25) with petroleum ether/ethyl acetate 3/2 as eluent (Note 26) to yield 19.2 g of a pale-yellow solid. Recrystallization of the product from n-hexane (Notes 27, 28) affords 15.4 g (71%) of analytically pure ethyl (E)-(-)-4,6-O-ethylidene-4,5,6-trihydroxy-2-hexenoate; colorless crystals, mp 62-63°C (Notes 29, 30).

2. Notes

1. D-Glucose (BioChemika, ≥ 99.5%) was obtained from Fluka Feinchemikalien GmbH, Neu-Ulm, Germany or Aldrich Chemical Company, Inc.

2. Paraldehyde (≥ 97%) was obtained from Fluka Feinchemikalien GmbH, Neu-Ulm, Germany or Acros Chemical Company, and was used without distillation.

3. If the sulfuric acid is added too fast, the reaction mixture becomes brown, probably from charring of the glucose.

4. The submitters used a shaking machine (IKA Labortechnik KS 250 basic, ca. 500/min). The checkers used a shaking machine from Lab Line Instruments (Model number 4600).

5. The submitters distilled ethanol (technical grade) from sodium and diethyl phthalate. The checkers used absolute ethyl alcohol from Pharmco.

6. The pH was checked with Merck Universal-Indikatorpapier, range 1-14.

7. Approximately 18 mL of 1 N ethanolic potassium hydroxide was needed.

8. Charcoal (powdered) was obtained from E. Merck KGaA, Darmstadt, Germany or J. T. Baker Chemical Company.

9. The solids are dissolved in hot ethanol and then kept overnight in the freezer at -30°C. The checkers found that standing overnight at -5°C gave similar results.

10. In deuterium oxide (D_2O) the product is a 34/66-mixture of α/β-anomers. The spectral properties of (-)-4,6-O-ethylidene-D-glucose are as follows: ^{1}H NMR (250 MHz, D_2O) δ: 1.22 (d, 3 H, J = 5.0, CHC$\underline{H}_3$), 3.03-3.78 (m, 5 H, 2-H, 3-H, 4-H, 5-H, 6-H_a), 3.90-4.12 (m, 1 H, 6-H_b), 4.40-5.13 (m, 5 H, 1-H, 3 OH, C$\underline{H}$CH$_3$);[2a] $[\alpha]_D^{20}$ -2.3° (H_2O, 2 d, c 19.7),[2a] $[\alpha]_D^{20}$ -2.37° (H_2O, equilibrium, c 19.7).[3]

11. Sodium periodate ($NaIO_4$) (98%) was obtained from Fluka Feinchemikalien GmbH, Neu-Ulm, Germany or Aldrich Chemical Company, Inc.

12. In several experiments it was found that the yield of 2,4-O-ethylidene-D-erythrose generally is somewhat lower when the reaction is performed on a larger scale.

13. Approximately 24 mL of 8 N sodium hydroxide was used.

14. The pH was checked every 30 min and, if necessary, more aqueous sodium hydroxide was added to keep the pH at 4.

15. Approximately 16 mL of 8 N sodium hydroxide was needed.

16. *Caution*: The evaporation of the solvent must be done slowly and carefully because the product shows a strong tendency to foam.

17. It is difficult to give exact spectral properties of (-)-2,4-O-ethylidene-D-erythrose because of rapid di- and/or oligomerization. The melting points given in the literature differ from 65-80°C[4] to 150-151°C,[5] depending on the degree of oligomerization of the product. With the present procedure, mainly the dimer is obtained. In order to check the optical purity of the product, it is convenient to compare

127

the equilibrium value of specific rotation, as obtained after 2 days in aqueous solution at room temperature: $[\alpha]_D^{20}$ -39.5° (H$_2$O, 2 d, c 1.00),[6] $[\alpha]_D^{20}$ -36.8° (H$_2$O, equilibrium, c 8.25),[3] $[\alpha]_D^{25}$ -36.2° (H$_2$O, equilibrium, c 8.2).[5] The analytical data of the product were as follows:[2a] Calcd for C$_6$H$_{10}$O$_4$ (146.14): C, 49.31; H, 6.90. Found: C, 49.18; H, 7.07; ^{1}H NMR (200.1 MHz, D$_2$O) δ: 1.38 (d, 3 H, J = 5.0, CHC$\underline{H}_3$), 3.39-3.99 (m,1 H, OCH), 4.06-4.31 (m, 2 H, OCH$_2$), 4.75-5.63 (m, 1 H, OH, C$\underline{H}$CH$_3$); ^{13}C NMR (50.3 MHz, D$_2$O/dioxane) δ: 20.1 (CHC$\underline{H}_3$), 61.2 (CHOH), 67.1, 67.9, 68.4 (CH$_2$O), 70.3, 71.5, 76.0, 78.5, 80.8, 90.6, 91.6 (CHO), 95.6, 97.4, 100.3, 101.2 (O$_2\underline{C}$CHCH$_3$).

18. According to ref. 5, monomeric (-)-2,4-O-ethylidene-D-erythrose may be obtained by heating a solution of the dimer in ethyl acetate with a catalytic amount of glacial acetic acid or 100% phosphoric acid for 20 min at 90°C.

19. Nitrogen was dried by means of a Sicapent® (E. Merck) drying tube. The checkers used dry argon.

20. Sodium hydride (60% sodium hydride in paraffin) was obtained from Fluka Feinchemikalien GmbH, Neu-Ulm, Germany or Aldrich Chemical Company, Inc.

21. Pentane (technical grade) was purified by distillation from sodium.

22. Tetrahydrofuran was purified by distillation under nitrogen from a blue solution of sodium and benzophenone.

23. Triethyl phosphonoacetate (Aldrich Chemical Company, Inc.) was purified by distillation (bp 142°C, 10 mm).

24. Crude product contains triethyl phosphonacetate; the isomeric purity (E/Z) of ethyl 4,6-O-ethylidene-(4S,5R,1'R)-4,5,6-trihydroxy-2-hexenoate was > 95 : 5 according to ^{13}C NMR.

25. A 20 cm x 5 cm column packed with 200 g of Kieselgel 60, (E. Merck, 0.040-0.063 mm, 250-400 mesh) was used. The checkers used a 40 cm x 10 cm column packed with silica gel (Bowman Chemical Co., 60Å) and eluted with 1/1 hexane/ethyl acetate with similar results.

26. Ethyl acetate and petroleum ether (technical grade; boiling range 40-80°C) were purified by distillation.

27. Hexane (technical grade) was distilled before use.

28. Hexane was added to the solid in 5-mL portions (ca. 35 mL were needed) until a single phase was formed. On slowly cooling to room temperature, the liquid again separates into two phases before crystallization starts.

29. The analytical data (after chromatography) were as follows:[6] Calcd for $C_{10}H_{16}O_5$ (216.23): C, 55.55; H, 7.46. Found: C, 55.40; H, 7.43. The E/Z ratio was found to be > 99:1 (determined by HPLC): t_E = 4.50 min; t_Z = 3.40 min, eluent hexane/ethyl acetate 60/40 [LiChrosorb Si 60 column, E. Merck]. (The Z-diastereomer reference sample was prepared as described in Ref. 7). TLC: R_f = 0.38 (petroleum ether/ethyl acetate 60/40). $[\alpha]_D^{20}$ -41.3° (CHCl$_3$; E/Z > 99:1, c 0.500), mp 62-63°C, ref. 7: $[\alpha]_D^{25}$ -35.2° (CHCl$_3$, c 1.21), mp 59-60°C. ^{13}C NMR (75.5 MHz, CDCl$_3$) δ: 14.2 (OCH$_2$$\underline{C}H_3$), 20.4 (O$_2CH\underline{C}H_3$), 60.8 (O$\underline{C}H_2CH_3$), 65.1 (C-5), 70.7 (C-6), 79.9 (C-4), 98.8 (O$_2$$\underline{C}HCH_3$), 122.2 (C-2), 143.7 (C-3), 166.7 (C-1); ^{1}H NMR (300 MHz, CDCl$_3$) δ: 1.30 (t, 3 H, J = 7.1, OCH$_2$C$\underline{H}_3$), 1.36 (d, 3 H, J = 5.1, O$_2$CHC$\underline{H}_3$), 2.85 (bs, 1 H, OH), 3.44 (t, 1 H, J = 9.5, 6-H$_a$), 3.52 (dt, 1 H, J = 4.4, J = 9.5, 5-H), 4.01 (ddd, 1 H, ^{4}J = 1.7, J = 4.5, J = 9.5, 4-H), 4.14 (dd, 1 H, J = 4.4, J = 9.9, 6-H$_b$), 4.20 (q, 2 H, J = 7.1, OC$\underline{H}_2$CH$_3$), 4.74 (q, 1 H, J = 5.1, O$_2$C$\underline{H}$CH$_3$), 6.16 (dd, 1 H, ^{4}J = 1.7, J = 15.8, 2-H), 7.09 (dd, 1 H, J = 4.5, J = 15.8, 3-H).

30. In various runs, 20 to 100 mmol of D-erythrose acetal were used, with yields ranging from 65 to 73%.

Waste Disposal Information

All toxic materials were disposed of in accordance with "Prudent Practices in the Laboratory"; National Academy Press; Washington, DC, 1995 and "Neue Datenblätter für gefährliche Arbeitsstoffe nach der Gefahrstoffverordnung", Welzbacher, U. (Ed.); WEKA Fachverlage, Kissing, 1991.

3. Discussion

Optically active C_4-building blocks of type **1** are versatile starting compounds in organic synthesis. Important members of this class, among numerous others, are derivatives of threose[8,9] and erythrose, respectively, such as 2,4-O-ethylidene-D-erythrose **2**, the corresponding "Horner enoate" **3**, or the erythritol **4**.

As described here, **2** can be prepared in two steps from commercially available D-glucose in up to 65% overall yield. Procedures to obtain the ethylidene glucose[10-12] and the ensuing oxidative degradation[3,5,12-16] are based on earlier literature reports. The D-erythrose acetal **2** has also been prepared from D-mannitol in three steps, with an overall yield of 7%.[17] The erythritol **4** can be synthesized from **2** by reduction with sodium borohydride.[3] A mixture of Z/E-**3** (ca. 2 : 1) is known to result from a Witting reaction of **2** with ethoxycarbonylmethylenetriphenylphosphorane.[2a,6,7]

Erythrose derivatives such as **2-4**, with a free hydroxy function, offer many possibilities for regioselective conversions; in addition, the free hydroxy group in **2** or **3** does or may influence the regio- and stereoselectivity of additions to the carbonyl group. 2,4-O-Ethylideneerythrose, because of its di- or oligomeric form, is configurationally stable at room temperature and can be stored for several months at room temperature.[9-18]

Both enantiomers of 2,4-O-ethylideneerythose have been used as intermediates in the preparation of free D- and L-erythrose.[12,13,15-17] In some reactions it proved advantageous to promote monomer formation of **2** from the dimer/oligomers by addition of 2-pyridone.[2,19] The N-benzylimine[20] and some hydrazones[20,21] of **2** have been described earlier in the literature. Imines, nitrones, oximes, and nitrile oxides derived from **2** were recently employed in a variety of additions and cycloadditions.[2,22,23] Aldehyde **2** has been transformed in various other Wittig reactions[24-26] and in an Abramov reaction with dimethyl phosphite.[27] Formation of the diethyldithioacetal,[28] the dimethyl phosphonate,[29] or the condensation with nitromethane[4,12,30] represent other uses of **2**. 2,3-Epoxyamides were prepared by treating **2** with stabilized sulfur ylides generated in situ.[31] (-)-2,4-O-Ethylidene-D-erythrose **2** has been used for the preparation of 2-deoxy-D-ribose via addition of stabilized ylides and subsequent hydrolysis in the presence of mercuric ion.[5] Further, diastereoselective propargyl addition to the aldehyde **2** was recently performed with propargyl bromide and zinc.[32]

1.   Institut für Organische Chemie der Universität Stuttgart, Pfaffenwaldring 55, D-70569 Stuttgart.

2.   (a) Fengler, M. Diplomarbeit, Universität Würzburg, 1992; Fengler-Veith, M. Dissertation, Universität Stuttgart, 1996; (b) Fišera, L.; Jäger, V.; Jarošková, L.; Kuban, J.; Ondruš, V. *Khim. Geterotsikl. Soedin,* **1995**, *10*, 1350; *Chem. Abstr.*

1996, *125*: 10640x; (c) Kubán, J.; Blanáriková, I.; Fengler-Veith, M.; Jäger, V.; Fišera, L. *Chem. Pap.* **1998** *52*, 780; (d) Kubán, J.; Blanáriková, I.; Fišera, L.; Fengler-Veith, M.; Jäger, V.; Kozišek, J.; Humpa, O.; Prónayova, N. *Tetrahedron* **1999**, *55*, 9501.

3. Barker, R.; MacDonald, D. L. *J. Am. Chem. Soc.* **1960**, *82*, 2301.

4. Carlson, K. D.; Smith, C. R., Jr.; Wolff, I. A. *Carbohydr. Res.* **1970**, *13*, 391.

5. Hauske, J. R.; Rapoport, H. *J. Org. Chem.* **1979**, *44*, 2472.

6. Kautz, U. Dissertation, Universität Stuttgart, 2000.

7. Tadano, K.-i.; Minami, M.; Ogawa, S. *J. Org. Chem.* **1990**, *55*, 2108.

8. Synthesis of 2-O-benzyl-L-threitol: Steuer, B.; Wehner, V.; Lieberknecht, A.; Jäger, V. *Org. Synth.* **1997**, *74*, 1, and lit. given therein.

9. Synthesis and use of 2-O-benzyl-3,4-O-isopropylidenethreose: (a) Müller, R.; Leibold, T.; Pätzel, M.; Jäger, V. *Angew. Chem., Int. Ed. Engl.* **1994**, *33*, 1295; (b) Jäger, V.; Müller, R.; Leibold, T.; Hein, M.; Schwarz, M.; Fengler, M.; Jaroskova, L.; Pätzel, M.; LeRoy, P.-Y. *Bull. Soc. Chim. Belg.* **1994**, *103*, 491; (c) Veith, U.; Schwardt, O.; Jäger, V. *Synlett* **1996**, 1181; (d) Schwardt, O.; Veith, U.; Gaspard, C.; Jäger, V. *Synthesis* **1999**, 1473.

10. Helferich, B.-H.; Appel, H. *Ber. Dtsch. B ChemGesl.***1931**, *64*, 1841.

11. Hockett, R. C.; Collins, D. V.; Scattergood. A. *J. Am. Chem. Soc.* **1951**, *73*, 599.

12. Rappoport, D. A.; Hassid, W. Z. *J. Am. Chem. Soc.* **1951**, *73*, 5524.

13. Neish, A. C. *Can. J. Chem.* **1954**, *32*, 334.

14. Schaffer, R. *J. Am. Chem. Soc.* **1959**, *81*, 2838.

15. Perlin, A. S. *Methods Carbohydr. Chem.* **1962**, *1*, 64.

16. Andersson, R.; Theander, O.; Westerlund, E. *Carbohydr. Res.* **1978**, *61*, 501.

17. Bourne, E. J.; Bruce, G. T.; Wiggins, L. F. *J. Chem. Soc.* **1951**, 2708.

18. cf. Analogous behavior of 2-O-benzylglyceraldehyde: Jäger, V.; Wehner, V. *Angew. Chem., Int. Ed. Engl.* **1989**, *28*, 469; see also Ref. 8.

132

19. cf. Swain, C. G.; Brown, J. F., Jr. *J. Am. Chem. Soc.* **1952**, *74*, 2534.

20. Ziderman, I.; Dimant, E. *J. Org. Chem.* **1966**, *31*, 223.

21. El Khadem, H. S.; El Shafei, Z.; El Ashry, E. S.; El Sadek, M. *Carbohydr. Res.* **1976**, *49*, 185.

22. Krämer, B. Dissertation, Universität Stuttgart, 1998.

23. (a) Schwardt, O. Dissertation, Universität Stuttgart, 1999; (b) Baur, M. Dissertation, Universität Stuttgart in preparation.

24. Cohen, N.; Banner, B. L.; Lopresti, R. J.; Wong, F.; Rosenberger, M.; Liu, Y.-Y.; Thom, E.; Liebman, A. A. *J. Am. Chem. Soc.* **1983**, *105*, 3661.

25. Izquierdo, I.; Rodriguez, M.; Plaza, M. T.; Vallejo, I. *Tetrahedron: Asymmetry* **1993**, *4*, 2535.

26. Li, Y.-L.; Sun, X.-L.; Wu, Y.-L. *Tetrahedron* **1994**, *50*, 10727.

27. Paulsen, H.; Bartsch, W.; Thiem, J. *Chem. Ber.* **1971**, *104*, 2545.

28. Lopez Aparicio, F. J.; Zorrilla Benitez, F.; Santoyo Gonzalez, F.; Asensio Rosell, J. L. *Carbohydr. Res.* **1987**, *163*, 29.

29. (a) Paulsen, H.; Kuhne, H. *Chem. Ber.* **1975**, *108*, 1239; (b) Wróblewski, A. E. *Liebigs Ann. Chem.* **1986**, 1854.

30. cf. Jäger, V.; Öhrlein, R.; Wehner, V.; Poggendorf, P.; Steuer, B.; Raczko, J.; Griesser, H.; Kiess, F.-M.; Menzel, A. *Enantiomer* **1999**, *4*, 205.

31. López-Herrera, F. J.; Pino-González, M. S.; Sarabia-Garcia, F.; Heras-López, A.; Ortega-Alcántara, J. J.; Pedraza-Cerbrián, M. G. *Tetrahedron: Asymmetry* **1996**, *7*, 2065.

32. (a) Li, Y.-L.; Mao, X.-H.; Wu, Y.-L. *J. Chem. Soc., Perkin Trans. I* **1995**, 1559; (b) Wu, W.-L.; Yao, Z.-J.; Li, Y.-L.; Li, J.-C.; Xia, Y.; Wu, Y.-L. *J. Org. Chem.* **1995**, *60*, 3257.

Appendix

Chemical Abstracts Nomenclature (Collective Index Number); (Registry Number)

(1'R)-(-)-2,4-O-Ethylidene-D-erythrose: 1,3-Dioxane-4-carboxaldehyde, 5-hydroxy-2-methyl-, [2R-(2α,4α,5β)]- (10); (70377-89-8)

Ethyl (E)-(-)-4,6-O-ethylidene-(4S,5R,1'R)-4,5,6-trihydroxy-2-hexenoate: D-erythro-Hex-2-enonic acid, 2,3-dideoxy-4,6-O-ethylidene-, ethyl ester,

Ref. 7: [2E,4(R)]- (12); (125567-87-5)

Ref. 7: [2Z,4(S)]- (12); (125567-86-4)

This prep: [2E,4(S)]-

(1'R)-(-)-4,6-O-Ethylidene-D-glucose: Glucopyranose, 4,6-O-ethylidene- (8);

D-Glucopyranose, 4,6-O-ethylidene- (9); (18465-50-4)

D-Glucose: α-D-Glucopyranose (8,9); (492-62-6)

Paraldehyde: s-Trioxane, 2,4,6-trimethyl- (8); 1,3,5-Trioxane, 2,4,6-trimethyl- (9); (123-53-7)

Sodium periodate: Periodic acid, sodium salt (8,9); (7790-28-5)

Sodium hydride (8,9); (7646-69-7)

Triethyl phosphonoacetate: Acetic acid, phosphono-, triethyl ester (8); Acetic acid, (diethoxyphosphinyl)-, ethyl ester (9); (867-13-0)

SYNTHESIS OF PENTA-1,2-DIEN-4-ONE (ACETYLALLENE)

(3,4-Pentadien-2-one)

$$\text{(reaction scheme)} \quad \xrightarrow[\text{— Ph}_3\text{PO} \cdot \text{HBr}]{\text{PPh}_3\text{Br}_2} \quad \begin{array}{c}\text{Me} \qquad \text{H} \\ \diagdown \diagup \\ \text{Br} \qquad \text{COMe}\end{array} \quad \xrightarrow[\text{— Et}_3\text{NH}^+ \ \text{Br}^-]{\text{Et}_3\text{N}} \quad \text{H}_2\text{C}=\text{C}=\diagup^{\text{CH}_3}_{\text{O}}$$

(Z + E)   **1**

Submitted by Thierry Constantieux and Gérard Buono.[1]
Checked by Dawn M. Bennett and Rick L. Danheiser.

1.   Procedure

A 1-L, three-necked flask, equipped with a mechanical stirrer, reflux condenser fitted with an argon inlet adapter, and pressure-equalizing dropping funnel is charged with 151 g (0.58 mol) of triphenylphosphine (Note 1) and 350 mL of dichloromethane (Note 2). The mixture is cooled with an ice-salt bath to -5°C, and maintained under an argon atmosphere.

A solution of 92 g (0.58 mol) of bromine (Note 3) in 60 mL of dichloromethane is added dropwise over 1 hr while the reaction mixture is vigorous stirred. Instantaneous decoloration of bromine and formation of a precipitate of dibromotriphenyl-phosphorane (Ph_3PBr_2) is observed. After the addition, the reaction mixture is stirred for an additional 30 min while being cooled in the ice bath.

A solution of 57.2 g (0.570 mol) of acetylacetone (Note 4) in 60 mL of dichloromethane is then added dropwise over 1 hr. An exothermic reaction occurs. At the end of the addition, the solution is allowed to warm very slowly to room temperature (Note 5) and stirred at that temperature for 17 hr (Note 6).

The resulting clear yellow-orange solution is transferred to a 1-L, one-necked, round-bottomed flask and concentrated at ca. 20 mm with a rotary evaporator. Anhydrous diethyl ether (230 mL) is added to precipitate triphenylphosphine oxide hydrobromide; the solid is separated by suction filtration and washed with two 100-mL portions of anhydrous ether. The filtrate is concentrated under reduced pressure and the resulting orange liquid is taken up in 350 mL of anhydrous diethyl ether. This solution is filtered to separate any remaining salt and transferred to a 500-mL, three-necked flask equipped with a magnetic stir bar, reflux condenser fitted with an argon inlet adapter, a rubber septum, and a pressure-equalizing dropping funnel.

A solution of 56.7 g (0.56 mol) of triethylamine (Note 7) in 60 mL of anhydrous diethyl ether is added dropwise to the reaction mixture over 1 hr, and the resulting mixture is stirred at room temperature for 12 hr.

The triethylamine hydrobromide precipitate is filtered and washed with two 60-mL portions of diethyl ether. The filtrate is washed with three 30-mL portions of 5% hydrochloric acid to remove unreacted triethylamine, and then washed with 25 mL of cold water (Note 8), dried over anhydrous magnesium sulfate, and filtered.

Diethyl ether is removed by distillation (Note 9) and the residual product is distilled under reduced pressure (Note 10) to afford 30.5 g (65%) of acetylallene (Note 11) as a colorless liquid.

2. Notes

1. Triphenylphosphine was purchased by the submitters from Fluka Chemical Corp. or Aldrich Chemical Company, Inc., and used without further purification.

2. Dichloromethane was purchased from SDS Co. or Mallinckrodt Inc. and distilled from calcium hydride. The distilled solvent was passed through a plug of silica gel immediately before use.

3. Bromine obtained from Janssen Chimica or Aldrich Chemical Company, Inc., was used as received.

4. Acetylacetone (obtained from Labosi Co. or Aldrich Chemical Company, Inc.) was distilled prior to use.

5. If the internal temperature is allowed to rise too quickly, rapid decomposition of 2-bromo-4-oxo-pent-2-ene occurs.

6. The reaction was monitored by ^{31}P NMR spectroscopic analysis. The two organophosphorus compounds in the reaction mixture show the following spectral properties (40.54 MHz, CDCl$_3$, external reference: H$_3$PO$_4$, 85% aqueous solution, δ ppm): PPh$_3$Br$_2$, δ = 50.2 and (PPH$_3$OH$^+$, Br $^-$), δ = 47.4. Complete formation of (PPh$_3$OH$^+$, Br $^-$) was observed after about 12 hr.

7. Triethylamine from Fluka Chemical Corp. or Aldrich Chemical Company, Inc., was distilled from potassium hydroxide prior to use.

8. It is essential to use a minimum of water for these washes because of the high solubility of acetylallene.

9. To minimize polymerization of acetylallene, ca. 100 mg of hydroquinone is added to the ether solution prior to concentration.

10. *Caution*: The highly volatile product must be trapped in a flask cooled in liquid nitrogen. Acetylallene distills at 48-50°C (60 mm) and is obtained with a purity of 96.5% as determined by gas chromatographic analysis using a 25-m SE30 capillary column at 80°C (Vector gas : He, 1 bar; retention time: 4.09 min).

11. The checkers obtained acetylallene in 59-65% yield, while the submitters report isolating the product in 75% yield. *Caution*: Acetylallene is an allergen and is highly lachrymatory. The product exhibits the following spectral properties: IR (film) cm^{-1}: 3025, 1940, 1660, 850; ^{1}H NMR (500 MHz, CDCl$_3$) δ: 2.26 (s, 3 H), 5.25 (d, 2 H, J = 6.4), 5.77 (t, 1 H, J = 6.4); ^{13}C NMR (125 MHz, CDCl$_3$) δ: 27.0, 79.9, 97.7, 198.6, 217.7.

Waste Disposal Information

All toxic materials were disposed of in accordance with "Prudent Practices in the Laboratory"; National Academy Press; Washington, DC, 1995.

3. Discussion

Acetylallene (1) behaves as an excellent dienophile in Diels-Alder reactions and induces peculiar orienting effects, allowing reactions with high regio- and stereo-selectivities.[2] Retro Diels-Alder reactions of modified adducts of furan and 1 afford a general method of synthesis of α-functionalized allenes.[3]

Hydrochlorination of acetylallene in the presence of N,N'-dimethylhydrazine dihydrochloride leads stereoselectively to the corresponding β-chloroenone that is a valuable intermediate in organic synthesis.[4]

Transition metal-catalyzed dimerization of acetylallene leads to the expected dimeric product in mixture with 2-methylfuran.[5] The latter compound may be also obtained by thermal intramolecular cyclization of 1.[6]

1,3-Dipolar cycloaddition of diazoalkanes to acetylallene leads to five-membered heterocycles containing two nitrogen atoms,[7] e.g., pyrazole and pyrazoline derivatives.   Addition of trivalent phosphorus reagents to 1 allows entry to the exomethylene 1,2-oxaphospholene ring system.[8] Triphenylphosphonio groups may also be used as umpolung agents to change the regioselectivity of the addition of nucleophilic compounds on acetylallene.   In this case, α,β-unsaturated ketones are obtained with heteroatomic substituents in the γ-position.[9]

Acetylallene is a valuable starting material in α,β-unsaturated γ-lactones synthesis: the tandem nucleophilic addition-aldol reaction of 1, iodide ion and aldehydes gives 3-iodohomoallylic alcohols in good yields, which can be further transformed to α,β-unsaturated γ-lactones by palladium-catalyzed cyclocarboxylation.[10]

Various complex and low yield procedures for the preparation of acetylallene have been described: oxidation of homopropargylic alcohol with chromium trioxide in sulfuric acid,[11] mild acid hydrolysis of conjugated ethoxyenyne,[12] reaction of propargyltrimethylsilane with acyl halide,[13] flash vacuum thermolysis of β-keto trimethylsilyl enol ether[14] and cycloelimination of β-silylethyl sulfoxide.[15]

The method reported here is a modification of a previously published procedure by Buono.[16] The yields have been increased by control of the reaction temperature during the first step, i.e., zero to room temperature instead of heating, and by direct dehydrobromination of the non-purified bromo intermediate. By monitoring the reaction by ^{31}P NMR spectroscopy, the submitters have determined precisely the end time of the first step, and that the triphenylphosphine oxide generated in the medium is immediately protonated by the hydrogen bromide. One of the major advantages of this procedure lies in the commercial availability of the starting materials. Moreover, only two steps, without purification of the intermediates, are required. Thus, the submitters have developed a new, efficient and cheap procedure for the preparation of acetylallene in large scale (0.5 mol up to 1 mol).

1. Ecole Nationale Supérieure de Synthèse, de Procédés et d'Ingénierie Chimiques d'Aix-Marseille (ENSSPICAM), U.M.R. 6516 Synthèse, Catalyse, Chiralité, Faculté des Sciences et Techniques de St. Jérôme, Université Aix-Marseille, Avenue Escadrille Normandie-Niemen, 13397 Marseille Cedex 20, France.

2. Gras, J. L. *J. Chem. Res., Synop.* **1982**, 300.

3. Bertrand, M.; Gras, J.-L.; Galledou, B. S. *Tetrahedron Lett.* **1978**, 2873; Gras, J. L.; Galledou, B. S.; Bertrand, M. *Bull. Soc. Chim. Fr.* **1988**, *4*, 757.

4. Gras, J. L.; Galledou, B. S. *Bull. Soc. Chim. Fr.* **1983**, (3-4, Pt. 2), 89.

5. Hashmi, A. S. K. *Angew. Chem., Int. Ed. Engl.* **1995**, *34*, 1581.

6. Hunstman, W. D.; Yin, T.-K. *J. Org. Chem.* **1983**, *48*, 1813.

7. Battioni, P.; Vo Quang, L.; Vo Quang, Y. *Bull. Soc. Chim. Fr.* **1978**, (7-8, Pt. 2), 401.

8. Buono, G.; Llinas, J. R. *J. Am. Chem. Soc.* **1981**, *103*, 4532.

9. Cristau, H. J.; Viala, J.; Christol, H. *Bull. Soc. Chim. Fr.* **1985**, 980.

10. Zhang, C.; Lu, X. *Tetrahedron Lett.* **1997**, *38*, 4831.

11. Bertrand, M. *C.R. Acad. Sci., Ser. C* **1957**, *244*, 1790; Bertrand, M.; Le Gras, J. *Bull. Soc. Chim. Fr.* **1962**, 2136; Bardone-Gaudemar, F. *Ann. Chim. (Paris)* **1958**, *3*, 52.

12. Bertrand, M.; Rouvier, C. *Bull. Soc. Chim. Fr.* **1968**, 2533.

13. Pillot, J.-P.; Bennetau, B.; Dunoguès, J.; Calas, R. *Tetrahedron Lett.* **1981**, *22*, 3401.

14. Jullien, J.; Pechine, J. M.; Perez, F.; Piade, J. J. *Tetrahedron* **1982**, *38*, 1413.

15. Fleming, I.; Goldhill, J.; Perry, D. A. *J. Chem. Soc., Perkin Trans. I* **1982**, *7*, 1563.

16. Buono, G. *Synthesis* **1981**, 872.

Appendix

Chemical Abstracts Nomenclature (Collective Index Number); (Registry Number)

Penta-1,2-dien-4-one: Acetylallene: 3,4-Pentadien-2-one (8,9); (2200-53-5)

Triphenylphosphine: Phosphine, triphenyl- (8,9); (603-35-0)

Bromine (8,9); (7726-95-6)

Dibromotriphenylphosphorane: Phosphorane, dibromotriphenyl- (8,9); (1034-39-5)

Acetylacetone: Aldrich: 2,4-Pentanedione (8,9); (123-54-6)

Triphenylphosphine oxide hydrobromide: Phosphine oxide, triphenyl-, compd. with hydrobromic acid (1:1) (9); (13273-31-9)

Triethylamine (8); Ethanamine, N,N-diethyl- (9); (121-44-8)

Triethylamine hydrobromide (8); Ethanamine, N,N-diethyl-, hydrobromide (9); (636-70-4)

BICYCLOPROPYLIDENE

(Cyclopropane, cyclopropylidene)

A. ▷—CO₂Me $\xrightarrow[20°C]{\text{EtMgBr, Ti(i-PrO)}_4,\ \text{Et}_2\text{O}}$ ▷⧄OH

B. ▷⧄OH $\xrightarrow[-15°C \rightarrow 20°C]{\text{Ph}_3\text{P} \cdot \text{Br}_2,\ \text{Py},\ \text{CH}_2\text{Cl}_2}$ ▷⧄Br

C. ▷⧄Br $\xrightarrow[20°C]{\text{t-BuOK, DMSO}}$ ▷=◁

Submitted by Armin de Meijere, Sergei I. Kozhushkov, and Thomas Späth.[1]

Checked by Florence Geneste and Andrew B. Holmes.

1. Procedure

A. 1-Cyclopropylcyclopropanol. A 4-L, four-necked, round-bottomed flask equipped with a mechanical stirrer, thermometer and reflux condenser is charged under nitrogen with 120.2 g (122 mL, 1.2 mol) of methyl cyclopropanecarboxylate (Note 1), 85.3 g (89.3 mL; 0.30 mol) of titanium tetraisopropoxide, and 1.45 L of anhydrous ether (Note 2). To the well-stirred solution, 840 mL (2.52 mol) of ethylmagnesium bromide as a 3 M solution in ether is added over a period of 4 hr (Notes 3, 4). The temperature is maintained between 20°C and 25°C with a water bath. After the addition is complete the black or dark-brown mixture is stirred for an additional 0.5 hr at the same temperature, then cooled to -5°C, and the reaction is quenched by careful addition of 1.56 L of ice-cold aqueous 10% sulfuric acid while the temperature is maintained between -5°C and 0°C with an acetone-dry ice bath. The mixture is stirred at 0°C for an additional 1 hr, in which time the precipitate should have completely dissolved (Note 5), and then is transferred to a 4-L separatory funnel. The inorganic phase is extracted with 360 mL of ether (Note 6). The combined ethereal phases are washed with two 600-mL portions of saturated sodium hydrogen carbonate solution, 600 mL of saturated brine, dried over anhydrous magnesium sulfate, and filtered. The solvent is removed from the filtrate under water-aspirator vacuum at 20°C (Note 7) to give 116 g (99%) of 1-cyclopropylcyclopropanol that is used without further purification (Note 8).

B. *1-Bromo-1-cyclopropylcyclopropane.* In a well-ventilated hood, a 2-L, three-necked, round-bottomed flask, equipped with a magnetic stirrer, thermometer, dropping funnel, and a reflux condenser, is charged under nitrogen with 327 g (1.24 mol) of triphenylphosphine, and 1.24 L of anhydrous dichloromethane. The solution is vigorously stirred under acetone-dry ice cooling, as 199 g (64.1 mL, 1.25 mol) of bromine (Note 9) is added over a period of 0.5 hr at -30°C to -15°C under nitrogen (Note 10). After an additional 15 min of stirring, a mixture of 116 g (1.18 mol) of 1-cyclopropylcyclopropanol and 93.5 g (95.6 mL, 1.18 mol) of anhydrous pyridine is added dropwise at -15°C over a period of 2 hr. The mixture is stirred at 20°C for an additional 24 hr under nitrogen. The reflux condenser and the dropping funnel are removed. The flask is immersed in an oil bath and connected to a 2-L, two-necked, round-bottomed flask via a 90° angle glass tube (Note 11). The second flask is cooled with acetone-dry ice. All the volatile material is bulb-to-bulb distilled, at first under water-aspirator vacuum and 30°C oil bath temperature, and then under further reduced pressure (0.1 mm) with a 100°C oil bath. The distillation is continued until the temperature in the first flask reaches 80°C (Note 12). The receiver flask is allowed to warm to 20°C, and the solvent is removed by distillation at atmospheric pressure using

143

a 30-cm Vigreux column. The residue is distilled under reduced pressure to give 117.1 g (62%) of 1-bromo-1-cyclopropylcyclopropane (Note 13).

   C. *Bicyclopropylidene*. A 2-L, three-necked, round-bottomed flask equipped with a magnetic stirrer, thermometer, dropping funnel, and a reflux condenser is charged under nitrogen with 109 g (0.971 mol) of potassium tert-butoxide (Note 14), and 1 L of dimethyl sulfoxide (Note 15). The solution is vigorously stirred as 105 g (0.65 mol) of neat 1-bromo-1-cyclopropylcyclopropane is added over a period of 2 hr under nitrogen. The temperature is maintained between 20°C and 25°C with a water bath. The mixture is stirred under a blanket of nitrogen (Note 16) at 20°C for an additional 24 hr. The reflux condenser and the dropping funnel are removed. The flask is immersed in a water bath and connected via a 90° angle glass tube to a 200-mL trap cooled with acetone-dry ice. All the volatile material is bulb-to-bulb distilled into the cold trap under reduced pressure (0.1 mm) at a maximurn temperature of 35°C to 40°C inside the flask (Note 17). The contents of the cold trap are allowed to warm to 20°C, transferred into a 150-mL separatory funnel, washed with four 50-mL portions of ice-cold water, and then transferred into a preweighed bottle containing 2-3 g of 4 Å molecular sieves. The yield is: 44 g (84%) of bicyclopropylidene (Note 18).

2. Notes

   1. Methyl cyclopropanecarboxylate is obtained from Aldrich Chemical Company, Inc. The corresponding ethyl ester can also be used, but esters of higher alcohols are undesirable because the isolation of 1-cyclopropylcyclopropanol becomes difficult.

   2. Ether is obtained from E. Merck or Aldrich Chemical Company, Inc., and dried using a common procedure. Titanium tetraisopropoxide is obtained from ABCR or Aldrich Chemical Company, Inc. (97% purity).

144

3. Ethylmagnesium bromide (EtMgBr) in ether (3 M) was purchased from Aldrich Chemical Company, Inc. The submitters also prepared it from bromoethane and magnesium. The concentration of EtMgBr was determined as follows: 1 mL of the Grignard solution was poured into 30 mL of ice-cold water; 4.00 mL of 1 N HCl solution was added, and the mixture was stirred until a clear solution was formed. Excess HCl was measured by titration with 0.1 N NaOH solution.

4. In the course of addition, the extensive evolution of ethane is observed, and therefore it is not necessary to maintain a stream of nitrogen.

5. While the precipitate is dissolving, an ice bath should be used to keep the mixture cool. The dissolution proceeds more rapidly when the precipitate is scraped off the walls of the flask from time to time with a spatula (the stirrer should be stopped for this operation).

6. Only one extraction with ether is necessary.

7. It is desirable to monitor this evaporation by ^{1}H NMR spectroscopy to remove as much 2-propanol (originating from titanium tetraisopropoxide) as possible.

8. 1-Cyclopropylcyclopropanol is a rather unstable compound that rapidly rearranges into cyclopropyl ethyl ketone (IR 1707 cm^{-1}) in the presence of acids and bases or upon heating. Therefore, it should be used immediately. The checkers stored the sample overnight at -78°C. If necessary, it should be stored no longer than 2-3 days at -78°C. Quantities of 20-30 g can be distilled, bp 65-68°C (32 mm); distillation of larger quantities is accompanied by the formation of a significant amount of cyclopropyl ethyl ketone. The physical properties are as follows: IR (film) cm^{-1}: 3332, 3084, 3007, 1217, 1009 and 935; ^{1}H NMR (250 MHz, CDCl$_3$) δ: 0-0.9 (m, 2 H), 0.21 (dd, 2 H, J = 6.5, 5.1), 0.25-0.33 (m, 2 H), 0.50 (dd, 2 H, J = 6.5, 5.1), 1.1-1.2 (m, 1 H), 3.83 (s, 1 H), [impurities at 0.6-0.9 (m), 1.1 (d, J = 6.2), 3.83 (m)]; ^{13}C NMR (62.5 MHz, CDCl$_3$) δ: 2.51, 11.20, 16.20, 56.37, [impurities at 8.07, 21.59, 24.89, 63.84]; MS

(ES+) m/z (rel intensity) 121 [100, (M + Na)+]. Anal. Calcd for $C_6H_{10}O$: C, 73.4; H, 10.3. Found: C, 71.7; H, 10.4.

9. Technical grade triphenylphosphine as obtained from BASF or Aldrich Chemical Company, Inc., is satisfactory. Dichloromethane is obtained from E. Merck or Fluka Chemical Corp. and dried according to commonly used procedures. Pyridine is obtained from Aldrich Chemical Company, Inc., or J. T. Baker ("Baker" grade) and dried by distillation over calcium hydride. Bromine was obtained from Aldrich Chemical Company, Inc., and used as supplied. **CAUTION**: Bromine is a severe irritant, causes burns, and is very poisonous.

10. The checkers added pellets of dry ice continuously to the acetone, cooling the bath to maintain the required temperature.

11. The checkers bent Pyrex glass tubing (o.d. 2 cm) to a 90° angle. The arm lengths were 17 and 16 cm and terminated in a male 24/29 ground glass joint. The longer arm was inserted in the distillation flask to accommodate bumping during the distillation.

12. The conditions of this bulb-to-bulb distillation are of prime importance: the oil bath must not be overheated, and the distillation performed with strict temperature control inside the first flask, otherwise lower yields are obtained. This operation takes 2-3 hr.

13. The submitters obtained 99.4-120.9 g (66-79%) from 1 mol of cyclopropylcyclopropanol; they found it essential to carry out the last 5 min of the bulb-to-bulb distillation under full pumping power with the valve between the pump and the receiving flask completely open; bp 69-73°C (88 mm). The physical properties are as follows: IR (film) cm^{-1}: 3084, 3039, 1724, 1446, 1415, 1381, 1188, 1112, 1099, 1020 and 955; ^{1}H NMR (250 MHz, CDCl$_3$) δ: 0.25-0.30 (m, 2 H), 0.56-0.60 (m, 2 H), 0.70 (dd, 2 H, J = 6.5, 6.5), 1.03 (dd, 2 H, J = 6.5, 6.5), 1.51-1.66 (m, 1 H); ^{13}C NMR (62.5 MHz, CDCl$_3$) δ: 6.07, 14.24, 20.58, 36.80. High resolution mass spectrum is as

follows: (EI) m/z 160.9965 [(M+H)+; calcd for $C_6H_{10}{}^{79}Br$: 160.9966]. Anal. Calcd for C_6H_9Br: C, 44.75 ; H, 5.6. Found: C, 45.3, H. 5.7. The mixture has a tendency to foam upon distillation; if desired, a vacuum distillation capillary can be used to facilitate the distillation.

14. Potassium tert-butoxide is obtained from ABCR GmbH & Co., Karlsruhe, Germany or Aldrich Chemical Company, Inc. (95% purity).

15. The checkers purchased dimethyl sulfoxide from Aldrich Chemical Company, Inc., and dried it by distillation (water aspirator) from calcium hydride. The submitters obtained dimethyl sulfoxide from J. T. Baker ("Baker" grade) and used it without purification.

16. During this stirring, the nitrogen inlet and outlet are at the top of the reflux condenser to avoid losses of the volatile product by evaporation. The nitrogen flow rate was kept to a minimum.

17. In several cases, the bulb-to-bulb distillation of bicyclopropylidene from the reaction mixture was incomplete when the system was evacuated only once. Therefore, it is recommended that distillation be interrupted as soon as no more condensation of product in the receiving flask is observed, that the apparatus be flushed with nitrogen, and then evacuation of the apparatus be repeated to continue bulb-to-bulb distillation. This procedure may have to be repeated one or two more times. Because the melting point of bicyclopropylidene is only -10°C, care must be taken to prevent the connecting tube from clogging with solid bicyclopropylidene. The checkers used a cold trap similar in design to the Aldrich cold trap, catalog No Z10, 310-1.

18. The submitters obtained a yield of 39.1-42.2 g (75-81%) from 1 mol of 1-bromo-1-cyclopropylcyclopropane. Bicyclopropylidene can be distilled (bp 100-102°C); however, polymerization may occur in the distillation flask reducing the yield. The physical properties are as follows: IR (film) cm^{-1}: 3051, 2983, 1410, 1247, 1071

1046, 1016 and 994; ^{1}H NMR (250 MHz, CDCl$_3$) δ: 1.17 (s, 8 H); ^{13}C NMR (62.5 MHz, CDCl$_3$) δ: 2.8, 110.2. The ^{1}H and ^{13}C NMR spectra indicated a sample purity ≥ 95% (the submitters reported 95-97%). Anal. Calcd for C$_6$H$_8$: C, 89.94 ; H, 10.06. Found: C, 89.03, H. 10.04. The sample was analyzed by gas chromatography (retention time 2.8 min) on an SGE capillary column BP5 (5% methylphenylsiloxane), internal diameter 0.32 mm, film thickness 0.25 μm, length 25 m, flow velocity 30 cm/s, injection temp 150°C, flame ionization detector 300°C. The column temperature was raised from 50°C at 10°C/min to 250°C. The checkers found that within a retention time of 3 min the sample purity ranged from 91-97%, but over 25 min the purity ranged from 88-91%. The checkers were unable to record a mass spectrum.

Waste Disposal Information

All toxic materials were disposed of in accordance with "Prudent Practices in the Laboratory"; National Academy Press; Washington, DC, 1995.

3. Discussion

This procedure is also applicable to the production of a number of spirocyclopropanated bicyclopropylidenes[2] in good yields. Bicyclopropylidene with its high-lying HOMO[3] was demonstrated to possess unique reactivity toward a wide range of electrophiles and cyclophiles including nucleophilic carbenes to give complex skeletons and heterocycles in high yields.[4] Functional derivatives can also be prepared directly from bicyclopropylidene.[5] Transition metal-catalyzed reactions of bicyclopropylidene[6] including Heck coupling[7] allow the preparation of carbocycles, polyenes and heterocyclic compounds[8] that are not readily available by other synthetic methods. Bicyclopropylidene has also been prepared by Simmons-Smith monocyclopropanation of the terminal double bond in ethenylidenecyclopropane (15% yield),[9] dimerization of 1-lithiocyclopropene in the presence of lithium amide in liquid ammonia over a period of 1 month (30% yield),[10] and several multistep syntheses starting from acetylcyclopropane (15-30% overall yields).[11-14]

1. Institut für Organische Chemie der Georg-August-Universität, Tammannstrasse 2, D-37077 Göttingen, Germany.

2. de Meijere, A.; Kozhushkov, S. I.; Spaeth, T.; Zefirov, N. S. *J. Org. Chem.* **1993**, *58*, 502.

3. Gleiter, R.; Haider, R.; Conia, J.-M.; Barnier, J.-P.; de Meijere, A.; Weber, W. *J. Chem. Soc., Chem. Commun.* **1979**, 130.

4. Reviews: de Meijere, A.; Kozhushkov, S. I.; Khlebnikov, A. F. *Zh. Org Khim.* **1996**, *32*, 1607; *Russ. J. Org Chem. (Engl. Transl.)* **1996**, *32,* 1555; *Top Curr. Chem.* **1999**, *207*, 89.

5. de Meijere, A.; Kozhushkov, S. I.; Zefirov, N. S. *Synthesis* **1993**, 681.

6. Binger, P.; Wedemann, P.; Kozhushkov, S. I.; de Meijere, A. *Eur. J. Org. Chem.* **1998**, 113; Binger, P.; Schmidt, Th. In "Houben-Weyl"; de Meijere, A., Ed.;
. Thieme: Stuttgart, 1997; Vol. E 17c, pp. 2217.

7. Bräse, S.; de Meijere, A. *Angew. Chem.* **1995**, *107*, 2741; *Angew. Chem., Int. Ed. Engl.* **1995**, *34*, 2545; de Meijere, A.; Nüske, H.; Es-Sayed, M.; Labahn, T.; Schroen, M.; Bräse, S. *Angew. Chem.* **1999**, *111*, 3881; *Angew. Chem., Int. Ed. Engl.* **1999**, *38*, 3669.

8. Kozhushkov, S. I.; Brandl, M.; Yufit, D. S.; Machinek, R.; de Meijere, A. *Liebigs Ann./Recl.* **1997**, 2197.

9. Le Perchec, P.; Conia, J. M. *Tetrahedron Lett.* **1970**, 1587; Denis, J. M.; Le Perchec, P.; Conia, J. M. *Tetrahedron* **1977**, *33*, 399.

10. Schipperijn, A. J. *Recl. Trav. Chim. Pays-Bas* **1971**, *90,* 1110; Schipperijn, A. J.; Smael, P. *Recl. Trav. Chim. Pays-Bas* **1973**, *92,* 1121.

11. Fitjer, L.; Conia, J. M. *Angew. Chem.* **1973**, *85,* 347; *Angew. Chem., Int. Ed. Engl.* **1973**, *12,* 332; Schmidt, A. H.; Schirmer, U.; Conia, J. M. *Chem. Ber.* **1976**, *109,* 2588.

12. Weber, W.; de Meijere, A. *Synth. Commun.* **1986**, *16,* 837.

13. Hofland, A.; Steinberg, H.; de Boer, T. J. *Recl. Trav. Chim. Pays-Bas* **1985**, *104,* 350.

14. Lukin, K. A.; Kuznetsova, T. S.; Kozhushkov, S. I.; Piven, V. A.; Zefirov, N. S. *Zh. Org. Khim.* **1988**, *24,* 1644; *J. Org. Chem. USSR (Engl. Transl.)* **1988**, *24,* 1483.

Appendix

Chemical Abstracts Nomenclature (Collective Index Number); (Registry Number)

Bicyclopropylidene (8); Cyclopropane, cyclopropylidene- (9); (27567-82-4)

1-Cyclopropylcyclopropanol: [1,1'-Bicyclopropyl]-1-ol (9); (54251-80-8)

Methyl cyclopropanecarboxylate: Cyclopropanecarboxylic acid, methyl ester (8,9); (2868-37-3)

Titanium tetraisopropoxide: Isopropyl alcohol, titanium(4+) salt (8); 2-Propanol, titanium(4+) salt (9); (546-68-9)

Ethylmagnesium bromide: Magnesium, bromoethyl- (9); (925-90-6)

1-Bromo-1-cyclopropylcyclopropane: 1,1'-Bicyclopropyl, 1-bromo- (9); (60629-95-0)

Triphenylphosphine: Phosphine, triphenyl- (8,9); (603-35-0)

Bromine (8,9); (7726-95-6)

Pyridine (8,9); (110-86-1)

Potassium tert-butoxide: tert-Butyl alcohol, potassium salt (8); 2-Propanol, 2-methyl-, potassium salt (9); (865-47-4)

Dimethyl sulfoxide: Methyl sulfoxide (8): Methane, sulfinyl bis- (9); (67-68-5)

PREPARATION OF (E)-1-DIMETHYLAMINO-3-tert-BUTYLDIMETHYLSILOXY-1,3-BUTADIENE

(1,3-Butadien-1-amine, 3-[[(1,1-dimethylethyl)dimethylsilyl]oxy]-N,N-dimethyl-,)

Submitted by Sergey A. Kozmin, Shuwen He, and Viresh H. Rawal.[1]
Checked by Ruth Figueroa and David J. Hart.

1. Procedure

A.   (E)-4-Dimethylamino-3-buten-2-one, **1**.[2]   A 250-mL, round-bottomed flask equipped with a magnetic stirring bar is charged with acetylacetaldehyde dimethyl acetal (Note 1) (19.8 g, 0.15 mol), freshly distilled prior to use (67°C, 12 mm).   A 2.0 M solution of dimethylamine in methanol (Note 1) (85 mL, 0.17 mol) is then added in one portion.   The resulting yellow solution is stirred at room temperature for 4 hr, and concentrated on a rotary evaporator.   The resulting oil is purified by bulb-to-bulb distillation (0.25 mm, oven temp 100-120°C) (Note 2) to afford 15.3 g (90%) of the desired vinylogous amide as a pale-orange oil (Note 3).

B.   (E)-1-Dimethylamino-3-tert-butyldimethylsiloxy-1,3-butadiene, **2**.   A dry, 500-mL, three-necked, round-bottomed flask is equipped with a pressure equalizing addition funnel, a large egg-shaped magnetic stirring bar, and a nitrogen/vacuum adapter. The apparatus is evacuated and flushed with nitrogen. The flask is charged with a 1.0 M solution of sodium bis(trimethylsilyl)amide (NaHMDS) in tetrahydrofuran (THF) (Note 1) (100 mL, 0.100 mol) and the flask is cooled in a dry ice-acetone bath (-70°C bath temp), causing a viscous, yellowish-white suspension to form. To this suspension is added, over a period of 30 min via an addition funnel, a solution of (E)-4-dimethylamino-3-buten-2-one (11.3 g, 0.100 mol) in THF (50 mL). The funnel is rinsed with a small amount of THF, and the resulting clear-yellow solution is stirred for 1.0 hr at -78°C. A solution of tert-butylchlorodimethylsilane (Note 4) (15.8 g, 0.105 mol) in THF (50 mL) is added over a 5-min period, via an addition funnel. The funnel is again rinsed with a small amount of THF. The cooling bath is removed and the reaction mixture is allowed to reach room temperature, which requires about 1.5 hr. The reaction mixture is poured into a 1-L Erlenmeyer flask containing 600 mL of anhydrous ether (Note 5). The resulting suspension is allowed to stand for 30 min and then suction filtered through a pad of dry Celite (60 g) (Note 6) packed in a 600-mL sintered glass filter funnel (Note 7). The filter cake is washed with three 50-mL portions of ether (Note 8), and the filtrate is concentrated on a rotary evaporator (heating bath temp <45°C). The resulting dark orange oil, containing the diene and hexamethyldisilazane, is subjected to bulb-to-bulb distillation (110-120°C, 0.3 mm) (Note 9) to yield 20.4 g (90%) of the desired 1-amino-3-siloxy-1,3-butadiene (Note 10) as a light-yellow oil.

2. Notes

1. This reagent was purchased from the Aldrich Chemical Company, Inc.

2. The receiver bulb was cooled with ice as soon as the product started to distill.

3. On occasion the vinylogous amide was obtained as a dark oil, but exhibited good spectroscopic properties. A cleaner-looking sample of the product was obtained by resubjecting the dark oil to bulb-to-bulb distillation. The checkers used a coffee-maker bulb-to-bulb distillation apparatus and recorded a bp of 60-80°C at 0.1 mm. Characterization data follow: IR (neat) cm^{-1}: 1660, 1575, 1436, 1356, 1258, 1112, 962; ^{1}H NMR (300 MHz, CDCl$_3$) δ: 2.10 (s, 3 H), 2.88 (br s, 3 H), 2.99 (br s, 3 H), 5.05 (d, 1 H, J = 12.8), 7.47 (d, 1 H, J = 12.8); ^{13}C NMR (75 MHz, CDCl$_3$) δ: 28.0, 36.9, 44.5, 96.6, 152.6, 195.2; mass spectrum (EI) 113 (C$_6$H$_{11}$NO), 98 (base).

4. This reagent was purchased from Lithco, a division of the FMC Corporation.

5. Anhydrous ether was purchased from Fisher Scientific Company and used without further purification.

6. Celite 545 was purchased from Fisher Scientific Company and was flame-dried under vacuum just prior to filtration. The Celite was packed tightly into the funnel using the bottom of a beaker.

7. A sintered-glass Büchner funnel having a C-porous frit was employed.

8. When the filtration became very slow, the filter cake was stirred with a spatula to break up the pasty layer on top. The filtration must be done carefully to minimize transfer of sodium chloride (NaCl) to the filtrate. Lower yields of less pure product are obtained if considerable amounts of NaCl are present during the subsequent distillation. If a gel is obtained after removal of solvent on the rotary evaporator, too much NaCl is present.

9. A 250-mL flask is used as the pot, connected to a 100-mL collection bulb, connected to a cold trap (dry ice-acetone) to protect the vacuum pump. Hexamethyldisilazane is collected first in the cold finger (-78°C). Then, as soon as the diene starts to distill, the collection bulb is cooled with ice. The checkers recorded a bp of 60-80°C at 0.07 mm.

10. The diene displays the following spectral data: IR (neat) cm^{-1}: 1648; ^{1}H NMR (500 MHz, CDCl$_3$) δ: 0.19 (s, 6 H), 0.98 (s, 9 H), 2.70 (s, 6 H), 3.84 (s, 1 H), 3.92 (s, 1 H), 4.78 (d, 1 H, J = 13.2), 6.57 (d, 1 H, J = 13.2); ^{13}C NMR (75 MHz, CDCl$_3$) δ: -4.6, 18.3, 25.9, 40.5, 85.8, 95.9, 140.9, 156.4; mass spectrum (EI) 227 (C$_{12}$H$_{25}$NOSi), 156 (base). This material contains trace impurities by ^{1}H and ^{13}C NMR.

Waste Disposal Information

All toxic materials were disposed of in accordance with "Prudent Practices in the Laboratory"; National Academy Press; Washington, DC, 1995.

3. Discussion

The usefulness of the Diels-Alder reaction continues to grow as new dienes and dienophiles are developed.[3] For the normal demand Diels-Alder reaction, it is well recognized that electron-donating groups render a diene more reactive toward electron-poor dienophiles. Moreover, the cycloadditions take place with excellent regioselectivity and give products possessing useful functional groups.[4] The submitters recently reported the development of 1-amino-3-siloxy-1,3-butadienes, a new class of highly-reactive heteroatom-containing dienes.[5] Clearly related to dialkoxybutadienes such as the widely used 1-methoxy-3-trimethylsiloxy-1,3-butadiene, known also as Danishefsky's diene,[6] the 1-amino-3-siloxy-1,3-butadienes possess several properties that make them synthetically attractive. They are conveniently prepared, as illustrated above, by deprotonation of readily available vinylogous amides with potassium bis(trimethylsilyl)amide (KHMDS) or NaHMDS, followed by silylation of the corresponding enolates. The dienes also exhibit very high reactivity toward a wide range of dienophiles.[5,7]

The submitters have investigated several procedures for the preparation of vinylogous amide 1, the precursor to the aminosiloxydiene. They first prepared this compound by an Eschenmoser sulfide contraction between dimethylthioformamide and bromoacetone.[5a] While effective, the procedure was not convenient for the preparation of multigram quantities of the vinylogous amide, because of the difficulty associated with removal of the triphenylphosphine sulfide by-product. A better alternative is to react a secondary amine with 4-methoxy-2-butenone.[7] This addition-elimination proceeds well with a wide range of secondary amines. The cost associated with 4-methoxy-2-butenone prompted the investigation of acetylacetaldehyde dimethyl acetal as a starting material. Not only can this compound be converted to 4-methoxy-2-butenone, but it can also be treated directly with secondary amines to yield the desired vinylogous amide. Clean 4-methoxy-2-butenone can be distilled in 89% yield by heating acetylacetaldehyde dimethyl acetal containing a catalytic amount of NaOAc to 160-170°C. The procedure described above was adapted from that reported by Maggiulli and Tang.[2]

The second step in the above sequence, deprotonation followed by silylation of the resulting enolate, was not successful under standard lithium diisopropylamide (LDA) conditions, presumably because silylation of the lithium enolate was slow. The deprotonation/silylation can be carried out effectively using KHMDS, which is available from Aldrich Chemical Company, Inc., as a 0.5 M solution in toluene. This protocol is quite general for the preparation of various dienes containing different silyl and amino groups as illustrated in Table I.[5,7] For preparative scale reactions, such as that described above, the use of NaHMDS was preferred as it is available from Aldrich Chemical Company, Inc., as 1.0 M solution in THF. The procedure described here also provides a convenient and high-yielding preparation of Danishefsky's diene (1-methoxy-3-trimethylsiloxy-1,3-butadiene).[8]

TABLE I

PREPARATION OF VARIOUS 1-AMINO-3-SILOXY-1,3-DIENES

| (90%) | (95%) | (89%) | (90%) | (96%) |

Aminosiloxy dienes are highly reactive in Diels-Alder reactions, considerably more so than the analogous dialkoxy dienes.[5a,9] They undergo [4+2] cycloadditions with a broad range of electron-deficient dienophiles.[5,7] The reactions generally occur under very mild conditions and afford the corresponding cycloadducts in good yields and with complete regioselectivity. A full study on the preparation and cycloadditions of amino siloxy dienes has been carried out.[7] In the procedure that follows, a preparative scale procedure is described for the Diels-Alder reaction of an aminosiloxy diene, reduction of the electron-withdrawing group in the adduct, and hydrolysis of the β-aminoenolsilyl ether moiety to the 4-substituted cyclohexenone.

1. Department of Chemistry, The University of Chicago, 5735 S. Ellis Avenue, Chicago, IL 60637.

2. Maggiulli, C. A.; Tang, P.-W. *Org. Prep. Proc. Int.* **1984**, *16*, 31.

3. Oppolzer, W. In "Comprehensive Organic Synthesis"; Pergamon Press: New York, 1991; Vol. 5, pp. 315-400.

4. (a) Petrzilka, M.; Grayson, J. I. *Synthesis* **1981**, 753; (b) Fringuelli, F.; Taticchi, A. "Dienes in the Diels-Alder Reaction"; Wiley: New York, 1990.

5. (a) Kozmin, S. A.; Rawal, V. H. *J. Org. Chem.* **1997**, *62*, 5252; (b) Kozmin, S. A.; Rawal, V. H. *J. Am. Chem. Soc.* **1997**, *119*, 7165; (c) Kozmin, S. A.; Rawal, V. H. *J. Am. Chem. Soc.* **1998**, *120*, 13523; (d) Kozmin, S. A.; Rawal, V. H. *J. Am. Chem. Soc.* **1999**, *121*, 9562.

6. Reviews: (a) Danishefsky, S. *Acc. Chem. Res.* **1981**, *14*, 400; (b) Danishefsky, S. J.; DeNinno, M. P. *Angew. Chem., Int. Ed. Engl.* **1987**, *26*, 15; (c) Danishefsky, S. *Chemtracs: Org. Chem.* **1989**, *2*, 273.

7. Kozmin, S. A.; Janey, J. M.; Rawal, V. H., *J. Org. Chem.* **1999**, *64*, 3039.

8. Danishefsky, S.; Kitahara, T.; Schuda, P. F. *Org. Synth., Coll. Vol. VII* **1990**, 312.

9. Kozmin, S. A.; Green, M. T.; Rawal, V. H. *J. Org. Chem.* **1999**, *64*, 8045.

Appendix
Chemical Abstracts Nomenclature (Collective Index Number);
(Registry Number)

(E)-1-Dimethylamino-3-tert-butyldimethylsiloxy-1,3-butadiene: 1,3-Butadien-1-amine, 3-[[(1,1-dimethylethyl)dimethylsilyl]oxy]-N,N-dimethyl-, (E)- (14); (194233-66-4)

(E)-4-Dimethylamino-3-buten-2-one: 3-Buten-2-one, 4-(dimethylamino)- (E)-; (2802-08-6)

Acetylacetaldehyde dimethyl acetal: 2-Butanone, 4,4-dimethoxy- (9); (5436-21-5)

Dimethylamine (8); Methanamine, N-methyl- (9); (124-40-3)

Sodium bis(trimethylsilyl)amide: NaHMDS: Disilazane, 1,1,1,3,3,3-hexamethyl-, sodium salt (8); Silanamine, 1,1,1-trimethyl-N-(trimethylsilyl)-, sodium salt (9); (1070-89-9)

tert-Butyldimethylsilyl chloride: CORROSIVE: Silane, chloro(1,1-dimethylethyl)-dimethyl- (9); (18162-48-6)

[4+2] CYCLOADDITION OF 1-DIMETHYLAMINO-3-tert-BUTYLDIMETHYLSILOXY-1,3-BUTADIENE WITH METHYL ACRYLATE: 4-HYDROXYMETHYL-2-CYCLOHEXEN-1-ONE

(2-Cyclohexen-1-one, 4-(hydroxymethyl)-)

Submitted by Sergey A. Kozmin, Shuwen He, and Viresh H. Rawal.[1]

Checked by Ruth Figueroa and David J. Hart.

1. Procedure

A. 4-Carbomethoxy-3-dimethylamino-1-tert-butyldimethylsiloxy-1-cyclohexene,
2. A dry, 250-mL, round-bottomed flask is charged with 1-dimethylamino-3-tert-butyldimethylsiloxy-1,3-butadiene (16.6 g, 0.073 mol) and anhydrous ether (30 mL) (Note 1). Methyl acrylate (13.5 mL, 0.15 mol) (Note 2) is added in one portion and the resulting light yellow solution is allowed to stand for 20 hr at room temperature. The solution is concentrated on a rotary evaporator, and the residual methyl acrylate is removed under high vacuum to afford 22.4 g (98%) of the desired cycloadduct **2**, in good purity, as a pale-yellow oil (Note 3).

B. 4-Hydroxymethyl-3-dimethylamino-1-tert-butyldimethylsiloxy-1-cyclohexene,
3. A dry, 500-mL, three-necked, round-bottomed flask is equipped with a 150-mL pressure-equalizing addition funnel, a reflux condenser topped with an inert atmosphere line, a glass stopper, and an egg-shaped magnetic stirring bar. The flask is charged with lithium aluminum hydride (2.66 g, 0.070 mol) and anhydrous ether (50 mL) (Note 1). The flask is cooled to 5°C in an ice-water bath. A solution of cycloadduct **2** (22.4 g, 0.072 mol) in ether (80 mL) is transferred to the addition funnel and added over a 30-min period. The reaction mixture is stirred for another 15 min, diluted with ether (100 mL), and quenched by dropwise addition of water (9 mL). The ice-water bath is removed, the resulting gray suspension is allowed to reach room temperature, and the mixture is stirred vigorously for an additional 60 min. The mixture is transferred to a 1.0-L Erlenmeyer flask and diluted with 350 mL of ether. Anhydrous sodium sulfate (Na_2SO_4) (60 g) is added, the suspension is stirred for 30 min, and filtered. The filter cake is washed twice with ether (50 mL each time). The solvent is removed on a rotary evaporator and the remaining volatile material is removed under high vacuum to afford 18.6 g (91%) of the desired amino alcohol **3** as a viscous clear oil (Note 4).

*C. 4-Hydroxymethyl-2-cyclohexen-1-one, 4. (Caution: Hydrofluoric acid (HF)
is extremely toxic!)* A 500-mL, round-bottomed flask equipped with a magnetic stirring
bar is charged with amino alcohol **3** (16.0 g, 0.056 mol) and tetrahydrofuran (THF)
(300 mL). The flask is cooled to 5°C in an ice-water bath. A 4.0 M solution of HF in
THF (32 mL, 0.128 mol) (Note 5) is added over 15 min to the cold solution of the amino
alcohol. The reaction mixture is stirred at room temperature for 20 hr. THF is removed
on a rotary evaporator, maintaining the bath temperature below 40°C. Ethyl acetate
(250 mL) is added, and the organic layer is washed with 1.2 M aqueous hydrochloric
acid (HCl) solution (40 mL). The aqueous layer is reextracted with ethyl acetate (5 x
100 mL each). Solid sodium bicarbonate ($NaHCO_3$) (15 g) is added to the combined
organic layers, followed by stirring for 1 hr and subsequent addition of anhydrous
magnesium sulfate ($MgSO_4$). The solution is filtered and solvent is removed on a
rotary evaporator to afford the crude enone, which is further purified by bulb-to-bulb
distillation (0.2 mm, oven temperature 140-180°C, Note 6) to give 5.90 g (84%) of 4-
hydroxymethyl-2-cyclohexen-1-one (Note 7) as a viscous pale-yellow oil.

2. Notes

1. Anhydrous ether was purchased from Fisher Scientific Company and was
not further purified.

2. Methyl acrylate was purchased from Aldrich Chemical Company, Inc., and
used without further purification.

3. The cycloadduct consists of a 1.5:1.0 mixture of trans- and cis-
diastereomers: ^{1}H NMR of cis-diastereomer (300 MHz, $CDCl_3$) δ: 0.16 (s, 6 H), 0.93
(s, 9 H), 1.83 (m, 1 H), 2.0-2.2 (m, 3 H), 2.27 (s, 6 H), 2.62 (m, 1 H), 3.56 (dd, 1 H, J =
5.0, 5.0), 3.70 (s, 3 H), 4.95 (d, 1 H, J = 5.0); ^{13}C NMR (75 MHz, $CDCl_3$) δ: -4.4, 18.0,
20.9, 25.6, 28.9, 43.5, 44.8, 51.3, 59.2 100.7, 153.7, 174.6; ^{1}H NMR of trans-

diastereomer (300 MHz, CDCl$_3$) δ: 0.14 (s, 3 H), 0.15 (s, 3 H), 0.92 (s, 9 H), 1.7-2.0 (m, 2 H), 2.0-2.2 (m, 2 H), 2.24 (s, 6 H), 2.52 (ddd, 1 H, J = 12.7, 9.2, 3.8), 3.66 (m, 1 H), 3.70 (s, 3 H), 4.84 (dd, 1 H, J = 2.0, 1.8); ^{13}C NMR (75 MHz, CDCl$_3$) δ: -4.5, -4.3, 18.0, 25.5, 25.6, 29.0, 40.6, 42.3, 51.7, 61.9, 102.6, 152.4, 176.0.

4. This material is a mixture of diastereomers. Selected peaks from each diastereomer follow: ^{1}H NMR of major diastereomer (400 MHz, C$_6$D$_6$) δ: 0.98 (s, 9 H, CMe$_3$), 1.99 (s, 6 H, NMe$_2$), 4.81 (broad s, 1 H, =CH); ^{13}C NMR (75 MHz, CDCl$_3$) δ: -4.4, 18.0, 20.9, 25.6, 28.9, 43.5, 44.8, 51.3, 59.2, 100.7, 153.7, 174.6; ^{1}H NMR of minor diastereomer (400 MHz, CDCl$_3$) δ: 1.01 (s, 9 H, CMe$_3$), 2.05 (s, 6 H, NMe$_2$), 4.99 (d, 1 H, J = 4, =CH); ^{13}C NMR (75 MHz, CDCl$_3$) δ: -4.0, -3.8, 18.5, 25.2, 30.1, 39.5, 44.4, 62.0, 44.2, 100.4, 155.0 (one peak not observed).

5. A 4.0 M solution of HF in THF was prepared in a 50-mL plastic bottle by the addition of 4.1 mL of the 49% aqueous HF solution (using a disposable plastic syringe) to THF (28 mL).

6. In order to minimize bumping, a 250-mL, round-bottomed flask was used for the distillation.

7. The checkers obtained 5.72 g (81%) of the enone. This compound displays the following spectral data: IR (neat) cm^{-1}: 3420, 1675; ^{1}H NMR (300 MHz, CDCl$_3$) δ: 1.60 (m, 1 H, OH), 1.82 (dddd, 1 H, J = 12.6, 12.6, 9.8, 5.0), 2.14 (dddd, 1 H, J = 12.6, 9.7, 5.0, 1.3), 2.42 (ddd, 1 H, J = 16.8, 12.6, 5.0), 2.56 (ddd, 1 H, J = 16.8, 5.0, 5.0), 2.65 (m, 1 H), 3.71 (m, 2 H), 6.08 (dd, 1 H, J = 10.0, 2.2), 6.96 (ddd, 1 H, J = 10.0, 2.7, 1.3); ^{13}C NMR (75 MHz, CDCl$_3$) δ: 25.3, 36.6, 38.9, 65.1, 130.2, 151.5, 199.9; mass spectrum (EI) 126 (C$_7$H$_{10}$O$_2$); 96 (base).

Waste Disposal Information

All toxic materials were disposed of in accordance with "Prudent Practices in the Laboratory"; National Academy Press; Washington, DC, 1995.

3. Discussion

Aminosiloxy dienes have been found to exhibit unusually high reactivity in Diels-Alder reactions with a wide range of electron-deficient dienophiles.[2-5] Determination of the second order rate constants with methacrolein showed the parent aminosiloxy diene to be about 10^3 more reactive than the analogous 1-alkoxy-3-siloxy diene.[4] The reactions of aminosiloxy dienes with various dienophiles occur under mild conditions and afford the corresponding cycloadducts in high yields and with complete regioselectivity. An extensive study on the preparation and cycloadditions of amino siloxy dienes has been carried out.[3]

The three-step procedure above illustrates the usefulness of a 1-amino-3-siloxy diene for the synthesis of functionalized cyclohexenones and is representative of the general usefulness of the dienes. The first step in the sequence, the Diels-Alder reaction between the parent aminosiloxy diene and methyl acrylate, cleanly produces the cycloadduct as a mixture of endo and exo diastereomers. This diene also reacts with many other dienophiles and the results are summarized in Table I.[3] The cycloadducts are generally obtained as mixtures of endo-exo diastereomers, except for the reactions with N-phenylmaleimide and methacrolein, which proceed with high endo selectivity, and with dimethyl maleate, which gives predominantly the exo adduct.

The products of the Diels-Alder reactions of aminosiloxy dienes are versatile synthetic intermediates, as they can be subjected to reduction, Wittig olefination, or

164

deprotonation without any hydrolysis or elimination.[2,3] On the other hand, it is possible under acidic conditions both to hydrolyze the silyl enol ether and β-eliminate the amino group to afford the corresponding enones.

The usefulness of aminosiloxy diene Diels-Alder chemistry to the preparation of different substituted cyclohexenones is demonstrated in Table II.[3] The functionality at the 4 and 5-positions of the cyclohexenones can be easily controlled by the substitution pattern in the dienophile. The differing endo-exo selectivity found in the initial cycloadducts does not impact the usefulness of this route to cyclohexenones, since the amino group is eliminated in the last step. Chiral versions of aminosiloxy dienes provide the opportunity for asymmetric synthesis. Indeed, the diphenylpyrrolidine-substituted diene allows the synthesis of a variety of cyclohexenones, with good to excellent ee's.[2b] The usefulness of aminosiloxy diene Diels-Alder reactions to natural product synthesis is exemplified through the stereocontrolled synthesis of the pentacyclic indole alkaloid tabersonine.[2c]

TABLE I

CYCLOADDITIONS OF 1-AMINO-3-SILOXY-1,3-DIENES

Dienophile	Temp (°C)	Yield (%)[a]	endo:exo[b]
N-phenylmaleimide (NPh)	-70	96	>98:2
EtO₂C–CH=CH–CO₂Et	20	100	1.4:1
cis-CH(CO₂Me)=CH(CO₂Me)	20	94	>2:98
CH₂=C(CH₃)–CHO	20	87	>98:2
CH₂=CH–CO₂Me	20	92	1:1.5
CH₂=CH–C(O)CH₃	20	97	1.1:1
CH₂=CH–CN	20	85	1:4
CH₃CH=CH–CO₂Me	90	87	1:3
Ph–CH=CH–CO₂Me	90	90	1:3

[a]Refers to an isolated yield after silica gel chromatography.
[b]Determined by NMR analysis of the crude reaction mixtures.

TABLE II

PREPARATION OF SUBSTITUTED CYCLOHEXENONES

Entry	Cycloadduct	Product	Overall Yield (%)
1	TBSO, CO₂t-Bu, NMe₂	O, OH	90
2	TBSO, CH₃, CHO, NMe₂	O, OH	85
3	TBSO, Ph, CO₂Me, NMe₂	O, Ph, OH	86
4	TBSO, CO₂Et, CO₂Et, R-N-R	O, OH, OH	82
5	TBSO, CO₂Me, CO₂Me, NMe₂	O, OH, OH	85

1. Department of Chemistry, The University of Chicago, 5735 S. Ellis Avenue, Chicago, IL 60637.

2. (a) Kozmin, S. A.; Rawal, V. H. *J. Org. Chem.* **1997**, *62,* 5252; (b) Kozmin, S. A.; Rawal, V. H. *J. Am. Chem. Soc.* **1997**, *119,* 7165; (c) Kozmin, S. A.; Rawal, V. H. *J. Am. Chem. Soc.* **1998**, *120,* 13523.

3. Kozmin, S. A.; Janey, J. M.; Rawal, V. H. *J. Org. Chem.* **1999**, *64,* 3039.

4. Kozmin, S. A.; Green, M. T.; Rawal, V. H. *J. Org. Chem.* **1999**, *64,* 8045.

5. Reviews: (a) Danishefsky, S. *Acc. Chem. Res.* **1981**, *14,* 400; (b) Danishefsky, S. J.; DeNinno, M. P. *Angew. Chem., Int. Ed. Engl.* **1987**, *26,* 15; (c) Danishefsky, S. *Chemtracs: Org. Chem.* **1989**, *2,* 273.

Appendix

Chemical Abstracts Nomenclature (Collective Index Number); (Registry Number)

(E)-1-Dimethylamino-3-tert-butyldimethylsiloxy-1,3-butadiene: 1,3-Butadien-1-amine, 3-[[(1,1-dimethylethyl)dimethylsilyl]oxy]-N,N-dimethyl-, (E)- (14); (194233-66-4)

Methyl acrylate: Acrylic acid, methyl ester (8); 2-Propenoic acid, methyl ester (9); (96-33-3)

4-Hydroxymethyl-2-cyclohexen-1-one: 2-Cyclohexen-1-one, 4-(hydroxymethyl)-, (14); (224578-91-0)

4-Carbomethoxy-3-dimethylamino-1-tert-butyldimethylsiloxy-1-cyclohexene: 3-Cyclohexene-1-carboxylic acid, 2-(dimethylamino)-4-[[(1,1-dimethylethyl)dimethyl-silyl]oxy]-, methyl ester, (14); cis- (194233-86-8); trans- (194233-84-6)

Lithium aluminum hydride: Aluminate(1-), tetrahydro-, lithium (8); Aluminate(1-), tetrahydro-, lithium (T-4)- (9); (16853-85-3)

Hydrofluoric acid: CORROSIVE: (8,9); (7664-39-3)

DIETHYL [(PHENYLSULFONYL)METHYL]PHOSPHONATE
(Phosphonic acid, [(phenylsulfonyl)methyl]-, diethyl ester)

Submitted by D. Enders,[1] S. von Berg,[1] and B. Jandeleit.[2]
Checked by Jianming Cheng and William R. Roush.

1. Procedure

A. Chloromethyl phenyl sulfide. As described in ref. 3, a 1-L, three-necked, round-bottomed flask fitted with a reflux condenser, 250-mL pressure-equalizing dropping funnel, and equipped with a large magnetic stirring bar is charged with 120 mL of toluene and 18.9 g (0.63 mol) of paraformaldehyde (Note 1). While the solution is stirred, 500 mL of concd hydrochloric acid is added to the suspension. The reaction mixture is heated to 50°C and the dropping funnel is charged with a solution of 55.1 g (51.3 mL, 0.5 mol) of thiophenol in 120 mL of toluene. The thiophenol solution is added dropwise over 1 hr. The reaction mixture is stirred at 50°C for 1 hr and subsequently at room temperature for an additional 3 hr. The reaction mixture is transferred to a separatory funnel and the organic phase is separated. The aqueous phase is extracted three times with 50-mL portions of toluene and the combined organic phase is washed with 100 mL of aqueous saturated sodium chloride solution. Most of the solvent is removed by distillation at atmospheric pressure and the remaining crude product is purified by fractional distillation under reduced pressure using a Vigreux column (15-20 cm) to yield 57.9 g (73%) of chloromethyl phenyl sulfide as a colorless liquid, bp 106-107°C (11 mm) (Note 2).

B. Diethyl [(phenylthio)methyl]phosphonate. As described in ref. 4, a 1-L, three-necked, round-bottomed flask fitted with a thermometer, a warm water (ca. 40-60°C) reflux condenser connected to an empty safety bottle and a safety bottle charged with ethanol (Note 3), a 200-mL pressure-equalizing dropping funnel, and equipped with a large magnetic stirring bar, is charged with 116.0 g (0.7 mol) of triethyl phosphite (Note 4). The dropping funnel is charged with 55.5 g (0.35 mol) of chloromethyl phenyl sulfide and the flask is heated under stirring to 130°C (internal temperature). The chloromethyl phenyl sulfide is added slowly at a constant rate over 40 min and the internal temperature is allowed to rise to 150-160°C. After complete addition, the resulting reaction mixture is stirred for an additional 12 hr under reflux and then cooled to room temperature. Excess triethyl phosphite is removed under reduced pressure (11 mm) through a short path distillation apparatus. The crude product is purified by fractional distillation under reduced pressure using a Vigreux column (20 cm) to afford 82.2 g (91%) of diethyl [(phenylthio)methyl]phosphonate as a colorless liquid, bp 130-135°C (0.08 mm) (Notes 5 and 6).

C. Diethyl [(phenylsulfonyl)methyl]phosphonate. A 1-L, three-necked, round-bottomed flask, fitted with a reflux condenser, thermometer, pressure-equalizing dropping funnel and equipped with a large magnetic stirring bar is charged with 300 mL of acetic acid and 75.0 g (0.29 mol) of diethyl [(phenylthio)methyl]phosphonate. The dropping funnel is charged with 90 mL of an aqueous hydrogen peroxide solution (30%) and the reaction mixture is heated under stirring to 50°C (internal temperature). The hydrogen peroxide solution is added slowly so that the internal temperature does not rise above 80°C (*Caution: This very exothermic reaction has an induction period!*) (Note 7). The reaction mixture is heated to 85°C for an additional 3 hr (Note 8), cooled to room temperature, and transferred to a 4-L beaker. Ice (1000 g) and then a concd sodium hydroxide solution (10 M) is added until the solution is basic, pH 8-9 (*Caution: Exothermic reaction!*) The reaction mixture is transferred to a separatory funnel, and the water phase is extracted five times with 200-mL portions of dichloromethane. The combined organic phase is washed with 50-mL portions of aqueous sodium hydrogen sulfite solution (10%) until no oxidizing agent remains (Note 9), and then dried over anhydrous magnesium sulfate. After removal of the magnesium sulfate by filtration, the solvents are evaporated under reduced pressure using a rotary evaporator. Purification of the crude product by flash chromatography (5 x 30-cm column) using 1:1 diethyl ether-hexanes to elute the non-polar impurities followed by 100% ethyl acetate to elute the product provides diethyl [(phenylsulfonyl)methyl]phosphonate (76.3 g, 90%) as colorless crystals, mp 51°C (Note 10).

2.   Notes

1. Paraformaldehyde (96%) was purchased from Fisher Scientific Company and thiophenol (99%) was purchased from Sigma Chemical Company. Both reagents were used as received.

2. Chloromethyl phenyl sulfide shows the following physical data: ^{1}H NMR (400 MHz, CDCl$_3$) δ: 4.96 (s, 2 H), 7.35 (m, 3 H), 7.51 (m, 2 H); ^{13}C NMR (100 MHz, CDCl$_3$) δ: 51.0, 127.9, 129.2, 130.9, 133.2; IR cm^{-1}: 1584, 1482, 1440, 1228; HRMS

calcd for $C_7H_7{}^{35}ClS$ (M+) 157.9957, found 157.9953. Anal. Calcd for C_7H_7ClS: C, 53.00; H, 4.45. Found: C, 52.97; H, 4.49. Chloromethyl phenyl sulfide is also commercially available (Aldrich Chemical Company, Inc.).

3. Volatile ethyl chloride is formed during the reaction, which evaporates through the reflux condenser and is trapped in an ethanol-filled safety bottle.

4. Triethyl phosphite was purchased from Aldrich Chemical Company, Inc., and used without further purification.

5. Diethyl [(phenylthio)methyl]phosphonate shows the following physical data: ^{1}H NMR (400 MHz, CDCl$_3$) δ: 1.27 (t, 6 H, J = 7.1), 3.19 (d, 2 H, J = 14.1), 4.14 (m, 4 H), 7.26 (m, 1 H), 7.30 (m, 2 H), 7.45 (m, 2 H); ^{13}C NMR (100 MHz, CDCl$_3$) δ: 16.2, 28.4 (J = 48), 62.5, 126.6, 128.8, 129.4, 135.4; IR cm^{-1}: 2982, 1582, 1482, 1440, 1392, 1255; HRMS calcd for $C_{11}H_{17}O_3PS$ (M+) 260.0636, found 260.0630. Anal. Calcd for $C_{11}H_{17}O_3PS$: C, 50.76; H, 6.58. Found: C, 50.49; H, 6.51.

6. If the product does not show the required purity, a second distillation may be necessary.

7. The checkers used a water bath as heat source and did not observe a significant increase in reaction temperature during the addition of hydrogen peroxide.

8. The internal temperature may rise for a short time up to 100°C on heating.

9. Commercially available peroxide test strips (Aldrich Chemical Company, Inc.) are used to verify the presence of oxidizing agents.

10. The compound shows the following physical data: R_f = 0.31 (0.25-mm silica gel on glass, diethyl ether/methanol, 20:1, (v/v); ^{1}H NMR (400 MHz, CDCl$_3$) δ: 1.29 (td, 6 H, J = 7.0, 0.5), 3.77 (d, 2 H, J = 16.7), 4.16 (m, 4 H), 7.58 (m, 2 H), 7.68 (m, 1 H), 8.00 (m, 2 H); ^{13}C NMR (100 MHz, CDCl$_3$) δ: 16.1, 53.7 (J = 38), 63.3, 128.2, 129.0, 134.0, 139.9; IR cm^{-1}: 2984, 1586, 1480, 1448, 1394, 1369, 1324, 1263, 1157; HRMS calcd for $C_{11}H_{18}O_5PS$ (M+ +H) 292.0613, found 293.0602. Anal. Calcd for $C_{11}H_{17}O_5PS$: C, 45.20; H, 5.86. Found: C, 45.35; H, 6.13.

Waste Disposal Information

All toxic materials were disposed of in accordance with "Prudent Practices in the Laboratory"; National Academy Press; Washington, DC, 1995.

3. Discussion

The sulfone group is well known as an activating group and enables the preparation of a vast array of functionalized products. In addition, this functional group is of enormous value in total synthesis. These compounds are easily prepared by a range of mild and high-yield routes. The sulfone is a robust group and frequently confers useful properties such as crystallinity. Of importance is the ease of formation of carbanions α to the sulfone group. This enables efficient C-C bond formation via alkylation and aldol-like processes.[5]

Diethyl [(phenylsulfonyl)methyl]phosphonate (1) has served in many procedures as a versatile intermediate in synthesis. Thus 1 has been used in alkylation reactions by reaction of appropriate electrophiles with the metalated sulfone.[6] In addition 1 may be used as the phosphonate component in Horner-Wadsworth-Emmons olefination reactions.[7,8,9] The olefination step can be carried out with high (E)-selectivity, as illustrated in the accompanying procedure.[10] α,β-Unsaturated phenyl sulfones are in general useful as Michael acceptors in reactions with a host of nucleophilic partners. In contrast to unsaturated carbonyl compounds, competing addition to the sulfone functionality cannot occur and the vinyl phenyl sulfone itself acts as a "two-carbon acceptor," which is not available using carbonyl-type Michael acceptors.[5] Furthermore, unsaturated sulfones in which the sulfone is directly attached to an alkenyl or alkynyl group undergo a range of cycloadditions including [2+2], [3+2][8] and [4+2][9] processes. However, the sulfone is rarely required

173

in the final target molecule and so methods for removal of the sulfone group have been developed.[5] Finally, α,β-unsaturated sulfones are also an effective class of cysteine protease inhibitors.[11]

1. Institut für Organische Chemie, Technical University of Aachen, Professor Pirlet-Straße 1, D-52074 Aachen, Germany.

2. Symyx Technologies, 3100 Central Expressway, Santa Clara, CA 95051.

3. Fancher, L. W., German Patent, 1112735, 1958; *Chem. Abstr.* **1962**, *56*, 11499b.

4. (a) Shahak, I.; Almog, J. *Synthesis* **1969**, 170; (b) Shahak, I.; Almog, J. *Synthesis* **1970**, 145.

5. Reviews: (a) Simpkins, N. S. In "Sulphones in Organic Chemistry, Tetrahedron Organic Chemistry Series", 1st ed.; Pergamon: Oxford, 1993; Vol. 10; (b) Simpkins, N. S. *Tetrahedron* **1990**, *46*, 6951; (c) "The Chemistry of Sulphones and Sulphoxides"; Patai, S.; Rappoport, Z.; Stirling, C., Eds.; Wiley: Chichester, 1988; (d) Fuchs, P. L.; Braish, T. F. *Chem. Rev.* **1986**, *86*, 903; (e) Solladie, G. In "Comprehensive Organic Synthesis"; Trost, B. M., Ed.; Pergamon Press: Oxford, 1991; Vol 6, 133; (f) Schank, K. In "Methoden der Organischen Chemie, Houben-Weyl, 4th ed., E11"; Thieme: Stuttgart, 1985, 1129.

6. (a) Kim, D. Y.; Suh, K. H. *Synth. Commun.* **1998**, *28*, 83; (b) Nasser, J.; About-Jaudet, E.; Collignon, N. *Phosphorus, Sulfur Silicon Relat. Elem.* **1990**, *54*, 171; (c) Ellingsen, P. O.; Undheim, K. *Acta. Chem. Scand., Ser. B.* **1979**, *B33*, 528; (d) Blumenkopf, T. A. *Synth. Commun.* **1986**, *16*, 139.

7. (a) Toyooka, N.; Yotsui, Y.; Yoshida, Y.; Momose, T. *J. Org. Chem.* **1996**, *61*, 4882; (b) Arce, E.; Carmen Carreño, M.; Belen Cid, M. B.; Garcia-Ruano, J. L. *Tetrahedron: Asymmetry* **1995**, *6*, 1757; (c) Mbongo, A.; Frechou, C.; Beaupere, D.; Uzan, R.; Demailly, G. *Carbohydr. Res.* **1993**, *246*, 361; (d) Trost, B. M.;

Seoane, P; Mignani, S.; Acemoglu, M. *J. Am. Chem. Soc.* **1989**, *111*, 7487; (e) Heathcock, C. H.; Blumenkopf, T. A.; Smith, K. M. *J. Org. Chem.* **1989**, *54*, 1548; (f) Fillion, H.; Hseine, A.; Pera, M.-H.; Dufaud, V.; Refouvelet, B. *Synthesis* **1987**, 708; (g) Inomata, K.; Sasaoka, K.; Kobayashi, T.; Tanaka, Y.; Igarashi, S.; Otani, T.; Kinoshita, H.; Kotake, H. *Bull. Chem. Soc. Jpn.* **1987**, *60*, 1767; (h) Haebich, D.; Metzger, K. *Heterocycles* **1986**, *24*, 289; (i) De Jong, B. E.; De Koning, H.; Huisman, H. O. *Recl. J. R. Neth. Chem. Soc.* **1981**, *100*, 410; (j) Padwa, A.; Murphree, S. S.; Ni, Z.; Watterson, S. H. *J. Org. Chem.* **1996**, *61*, 3829; (k) Flitsch, W.; Lubisch, W. *Chem. Ber.* **1984**, *117*, 1424.

8. Reactions involving **1** include: (a) Trost, B. M.; Higuchi, R. I. *J. Am. Chem. Soc.* **1996**, *118*, 10094; (b) Trost, B. M.; Grese, T. A. *J. Org. Chem.* **1992**, *57*, 686; (c) Trost, B. M.; Grese, T. A.; Chan, D. M. T. *J. Am. Chem. Soc.* **1991**, *113*, 7350; (d) Trost, B. M.; Grese, T. A. *J. Am. Chem. Soc.* **1991**, *113*, 7363.

9. Reactions involving **1** include: Blades, K.; Lequeux, T. P.; Percy, J. M.; *J. Chem. Soc., Chem. Commun.* **1996**, 1457.

10. Enders, D.; von Berg, S.; Jandeleit, B. *Org. Synth.* **2000**, *78*, 177.

11. Palmer, J. T.; Rasnick, D.; Klaus, J. L.; Brömme, D. *J. Med. Chem.* **1995**, *38*, 3193.

Appendix

Chemical Abstracts Nomenclature (Collective Index Number); (Registry Number)

Diethyl [(phenylsulfonyl)methyl]phosphonate: Phosphonic acid, [(phenylsulfonyl)methyl]-, diethyl ester (9); (56069-39-7)

Chloromethyl phenyl sulfide: Sulfide, chloromethyl phenyl (8); Benzene, [(chloromethyl)thio]- (9); (7205-91-6)

Paraformaldehyde (9); (30525-89-4)

Thiophenol: Aldrich: Benzethiol (8,9); (108-98-5)

Diethyl [(phenylthio)methyl]phosphonate: Phosphonic acid, [(phenylthio)methyl]-, diethyl ester (9); (38066-16-9)

Triethyl phosphite: Phosphorous acid, triethyl ester (8,9); (122-52-1)

Hydrogen peroxide (8,9); (7722-84-1)

Sodium hydrogen sulfite: Sulfurous acid, monosodium salt (8,9); (7631-90-5)

SYNTHESIS OF (-)-(E,S)-3-(BENZYLOXY)-1-BUTENYL PHENYL SULFONE VIA A HORNER-WADSWORTH-EMMONS REACTION OF (-)-(S)-2-(BENZYLOXY)PROPANAL

(Benzene, [[[1-methyl-3-(phenylsulfonyl)-2-propenyl]oxy]methyl]-, [S-(E)]-) from (Propanal, 2-(phenylmethoxy)-, (S)-)

Submitted by D. Enders,[1] S. von Berg,[1] and B. Jandeleit.[2]

Checked by Brad M. Savall and William R. Roush.

1. Procedure

A. *(-)-(S)-Ethyl 2-(benzyloxy)propanoate.* As described in ref. 3, a flame-dried, 500-mL Schlenk flask equipped with a magnetic stirring bar, rubber septum, and an argon balloon is charged with 11.8 g (100 mmol) of (S)-ethyl 2-hydroxypropanoate [(S)-ethyl lactate] (Note 1) and 50.9 g (200 mmol) of O-benzyl-2,2,2-trichloroacetimidate (Note 2). The reagents are dissolved in 250 mL of a mixture of anhydrous cyclohexane (Note 3) and anhydrous dichloromethane (Note 4) (7:1 v/v) under an atmosphere of argon. Neat trifluoromethanesulfonic acid (Note 5) (0.4 mL, 4.53 mmol) is added dropwise by means of a syringe while the mixture is stirred rapidly (Note 6). The reaction mixture is stirred for 48 to 60 hr at room temperature (Note 7) and subsequently diluted with water (100 mL) and hexane (300 mL). Stirring is continued for an additional 3 hr at room temperature. The precipitated colorless trichloroacetamide is filtered off by means of a Büchner funnel. The aqueous phase is separated and extracted three times with 50-mL portions of hexane. The combined organic extracts are washed with 50 mL of aqueous saturated sodium bicarbonate (NaHCO$_3$) solution and finally with 50 mL of aqueous saturated sodium chloride (NaCl) solution. After drying over magnesium sulfate (MgSO$_4$), filtration and removal of the solvents under reduced pressure by means of a rotary evaporator, the residue is purified by fractional distillation using a Vigreux column (15-20 cm) to yield 18.7 g (90%) of a colorless slightly turbid liquid (Notes 8, 9).

B. *(-)-(S)-2-(Benzyloxy)propanal.* A flame-dried, 500-mL Schlenk flask equipped with a magnetic stirring bar, dropping funnel sealed with a rubber septum, and an argon balloon is loaded under an atmosphere of argon with 18.7 g (90 mmol) of (-)-(S)-ethyl 2-(benzyloxy)propanoate and the compound is dissolved in anhydrous diethyl ether (180 mL) (Note 10). The reaction mixture is cooled to -78°C by means of a cooling bath (dry ice/ethanol). A 1 M solution of diisobutylaluminum hydride (DIBAH)

178

in hexane (126 mL, 126 mmol) (Note 11) is added very slowly dropwise to the solution of the ester and stirring is continued for at least 1 hr after the complete addition of the DIBAH solution (Note 12). Upon complete consumption of the ester, the crude reaction mixture is poured directly with vigorous stirring into 360 mL of ice cold 4 N hydrochloric acid (Note 13). The aqueous phase is extracted with diethyl ether (4 x 180 mL) and the combined organic extracts are washed with 50 mL of aqueous saturated NaCl solution. After drying over MgSO$_4$, filtration and removal of the solvents under reduced pressure by means of a rotary evaporator, 14.4 g (98%) of the crude aldehyde is obtained (Note 14).

C. (-)-(E,S)-3-(Benzyloxy)-1-butenyl phenyl sulfone. As described in ref. 4, a flame-dried, 500-mL Schlenk flask (or three-necked flask with a thermometer) equipped with a large magnetic stirring bar (Note 15), dropping funnel sealed with a rubber septum, and an argon balloon is charged under an atmosphere of argon with 13.1 g (151 mmol) of lithium bromide (Note 16) and 36.6 g (125 mmol) of diethyl [(phenylsulfonyl)methyl]phosphonate (Note 17). The reagents are suspended in 250 mL of anhydrous acetonitrile (Note 18) and 19.1 mL (13.9 g, 138 mmol) of triethylamine (Note 19) is then added. The reaction mixture is stirred at room temperature until it becomes homogeneous (Note 20) and is then cooled to 0°C (an ice-salt bath is used to maintain the internal temperature at 0°C). The dropping funnel is charged with a solution of 20.6 g (125 mmol) of (-)-(S)-2-(benzyloxy)propanal in 50 mL of anhydrous acetonitrile. The aldehyde solution is added dropwise at 0°C with vigorous stirring. After complete addition the reaction mixture is stirred for ca. 12 hr and allowed to warm to room temperature during this period. The reaction is monitored by TLC and is halted by the addition of 0.1 N hydrochloric acid (150 mL) and water (150 mL). The reaction mixture is diluted with diethyl ether (200 mL). After phase separation, the aqueous phase is reextracted with diethyl ether (4 x 200 mL) and the combined organic extracts are washed with aqueous saturated NaCl solution

179

(200 mL). After drying over MgSO$_4$, filtration, and removal of the solvents under reduced pressure using a rotary evaporator, the crude product is purified by column chromatography using a 15 : 1 (w/w) ratio of silica gel to crude product and 1:2 → 1:1 (v/v) diethyl ether/light petroleum as eluent to yield 34.2 g (90%) of the pure (E)-isomer as a colorless viscous oil (Notes 21 and 22).

Waste Disposal Information

All toxic materials were disposed of in accordance with "Prudent Practices in the Laboratory"; National Academy Press; Washington, DC, 1995.

2. Notes

1. (S)-Ethyl lactate was purchased from Merck, Darmstadt, Germany, in enantiomerically pure form (ee >> 99%) and was used without further purification.

2. O-Benzyl 2,2,2-trichloroacetimidate was synthesized in 98% yield according to ref. 5 by addition of 1.0 equiv of benzyl alcohol (Aldrich Chemical Company, Inc.; previously distilled over calcium hydride (CaH$_2$) under an atmosphere of argon), to 1.0 equiv of trichloroacetonitrile (Aldrich Chemical Company, Inc.; no previous purification) in the presence of 0.1 equiv of sodium hydride in anhydrous diethyl ether. The resulting viscous, dark brown crude compound showed sufficient purity (> 97%) by ^{1}H NMR and GLC analysis and was used without further purification. The benzylating agent can be stored at 4°C under an atmosphere of argon for several weeks without decomposition or loss of quality. However, the compound can be further purified by distillation to yield a colorless viscous liquid, or, alternatively, can be purchased from Aldrich Chemical Company, Inc.

3. Cyclohexane was purified by distillation from CaH$_2$ under argon.

4. Dichloromethane was purified by distillation from CaH$_2$ under argon.

5. Trifluoromethanesulfonic acid was purchased from Aldrich Chemical Company, Inc., stored and handled under an atmosphere of argon, and used without further purification.

6. During the addition of the catalyst an almost colorless precipitate of trichloroacetamide forms, which might dissolve again after a few minutes. If this happens or if the reaction mixture does not become turbid at all, more (0.4 mL) trifluoromethanesulfonic acid can be added.

7. The submitters indicated that extended reaction times are essential to obtain complete consumption of the starting material and good yields. However, the checkers observed that the reaction appeared to be complete within several hours according to TLC analysis.

8. The checkers observed that a crystalline solid (trichloroacetamide) formed in the condenser at the beginning of the distillation, and that the distillate contained small amounts of a precipitated crystalline solid (trichloroacetamide) that could be removed by filtration through a sintered glass funnel. The checkers obtained 78-83% yields of product, with material collected from 125 - 135° to maximize the yield of product.

9. The compound shows the following physical data: R_f = 0.40 (0.25 mm silica gel on glass, diethyl ether/light petroleum = 1:5); bp: 124-127°C/6 mm, α_D^{21}: -73.6° (neat), {$[\alpha]_D^{20}$: -74.5 (CHCl$_3$, c 2.94)};[3] ^{1}H NMR (300 MHz, CDCl$_3$) δ: 1.30 (t, 3 H, J = 7.2), 1.44 (d, 3 H, J = 6.9), 4.05 (q, 1 H, J = 6.8), 4.22 (m, 2 H), 4.45 (d, 1 H, J = 11.8), 4.69 (d, 1 H, J = 11.8), 7.20-7.40 (m, 5 H); ^{13}C NMR (75 MHz, CDCl$_3$) δ: 14.2, 18.7, 60.8, 71.9, 74.0, 127.8, 127.9, 128.4, 136.6, 173.2; IR (film) cm^{-1}: 2985, 1746, 1455, 1200, 1143; HRMS calcd for $C_{12}H_{17}O_3$ [M+H] 209.1178, found 209.1183. Anal. Calcd for $C_{12}H_{16}O_3$: C, 69.21; H, 7.73, Found: C, 69.17; H, 7.74. The O-Bn protected (S)-ethyl lactate can be stored without special precautions.

10. Diethyl ether was purified by distillation from sodium benzophenone ketyl under argon.

11. Diisobutylaluminum hydride (DIBAH) (1 M in hexane) was purchased from Aldrich Chemical Company, Inc., and used without further purification. DIBAH should be handled with caution and all operations should be performed by employing the usual inert gas techniques (cannula, syringe, etc.).

12. To avoid reduction of the generated aldehyde that is more prone to reduction than the corresponding ester, the reducing agent should be added very slowly, avoiding any local temperature increase. The reaction mixture should be kept below -70°C. The checkers used a three-necked flask equipped with a low temperature thermometer to monitor the internal reaction temperature.

13. It is strongly recommended that the reaction mixture be added to the beaker containing the ice-cold acid solution with vigorous stirring because of the strong evolution of hydrogen gas. Reversing the order of addition causes freezing of the acid and effects a vigorous evolution of hydrogen gas on warming.

14. The resulting viscous colorless crude compound obtained by the submitters showed sufficient purity (> 97%) by ^{1}H NMR and GLC analysis to be used without further purification. The submitters report that the aldehyde can be stored at -20°C under an atmosphere of argon for several days without detectable racemization or loss of quality. However, the checkers observed that some crystalline material resembling aluminum salts formed when the crude aldehyde was stored in the refrigerator. The salts could be removed by dissolving the product in diethyl ether and washing with 1 N HCl (2 x 50 mL). The purity of the aldehyde obtained by the checkers was 80% by GC analysis. Nevertheless, this material gave acceptable results in the following Horner-Wadsworth-Emmons reaction. If desired, the aldehyde can be purified by distillation to yield a colorless viscous liquid.[3] The compound shows the following physical data: R_f = 0.27 (0.25 mm silica gel on glass, diethyl ether/light petroleum = 1:3); α_D^{21}: -65.3° (neat) {α_D^{20}: -65.9° (neat)};[3] ^{1}H NMR (300 MHz, CDCl$_3$) δ: 1.31 (d, 3 H, J = 7.2), 3.88 (qd, 1 H, J = 6.9, 1.9), 4.57 (d, 1 H, J = 11.8), 4.67 (d, 1 H, J = 11.8), 7.35 (m, 5 H), 9.64

(d, 1 H, $J = 1.6$); ^{13}C NMR (75 MHz, $CDCl_3$) δ: 15.5, 72.0, 79.4, 127.9, 128.6, 137.3, 203.4; IR (film) cm[-1]: 3448, 2870, 1733, 1455, 1375, 1094; HRMS calcd for $C_{10}H_{16}NO_2$ [M+NH$_4^+$] 182.1181, found 182.1176. Anal. Calcd for $C_{10}H_{12}O_2$: C, 73.14; H, 7.37. Found: C, 68.98; H, 7.52.

15. Because of precipitating lithium phosphates during the course of the reaction the mixture forms a sticky slurry and so the use of a large stirring bar is recommended.

16. Lithium bromide was purchased from Fluka Chemical Corp. and dried for ca. 12 hr at 100-110°C under reduced pressure using a high vacuum pump. The salt was handled and stored under an atmosphere of argon with the exclusion of moisture.

17. Diethyl [(phenylsulfonyl)methyl]phosphonate was synthesized according to the accompanying procedure: *Org. Synth.* **2000**, *78*, 169.

18. Acetonitrile was purified by distillation from CaH_2 under argon.

19. Triethylamine was purified by distillation from CaH_2 under argon.

20. To obtain good chemical yields and a very high (E)-selectivity in the olefination step it seems to be crucial that all of the phosphonate must be transformed into its chelated lithium derivative before the aldehyde is added. Sometimes the reaction mixture becomes turbid during cooling but this does not affect reactivity.

21. During the olefination step only very small amounts of the corresponding (Z)-isomer are formed, which are easily removed by column chromatography on silica gel. The (Z)-isomer has a slightly higher R_f-value than the (E)-isomer: $R_f(Z) = 0.49$ (0.25 mm silica gel on glass, diethyl ether/light petroleum (v/v) 1:1).

22. The compound shows the following physical data: $R_f = 0.44$ (0.25 mm silica gel on glass, dimethyl ether/light petroleum (v/v) 1:1); $[\alpha]_D^{23}$: -31.9° ($CHCl_3$, *c* 1.10); ^{1}H NMR (300 MHz, $CDCl_3$) δ: 1.31 (d, 3 H, $J = 6.6$), 4.18 (qdd, 1 H, $J = 6.6$, 4.8, 1.6), 4.44 (d, 1 H, $J = 12.0$), 4.48 (d, 1 H, $J = 12.0$), 6.57 (dd, 1 H, $J = 15$, 1.5), 6.94 (dd, 1 H, $J = 15.3$, 4.9), 7.21-7.34 (m, 5 H), 7.48-7.55 (m, 2 H), 7.57-7.64 (m, 1 H), 7.85-7.90 (m, 2

H); ^{13}C NMR (75 MHz, CDCl$_3$) δ: 20.0, 71.0, 72.9, 127.5, 127.6, 127.8, 128.4, 129.3, 130.3, 133.4, 137.6, 140.3, 147.0; IR (film) cm^{-1}: 3063, 1447, 1307, 1147, 1086, 834; HRMS calcd for C$_{17}$H$_{22}$O$_3$NS [M+NH$_4$+] 320.1320, found 320.1310. Anal. Calcd for C$_{17}$H$_{18}$O$_3$S: C, 67.43; H, 6.00. Found: C, 66.45; H, 6.09.

The checkers found that the product solidified when stored at -20°C, cracking the glass bottle.

3. Discussion

(-)-(S)-2-(Benzyloxy)propanal (**2**) has attracted considerable interest since it is readily available in optically pure form from inexpensive starting materials. This useful aldehyde is widely used in organic synthesis[6] and has therefore been synthesized many times employing several approaches.[3,7-12]

The most frequently used procedures to obtain aldehyde **2** employ commercially available ethyl (S)-lactate. The α-hydroxy ester is converted in the first step to ethyl (S)-2-(benzyloxy)propionate (**1**) by O-benzylation. Alkylation of the ester can be accomplished by using freshly prepared silver(I) oxide and benzyl bromide according to a procedure described by Mislow.[13] This method is high yielding, but it is not amenable to large scale preparation of **1**. Benzylation using benzyl bromide and sodium hydride is also described in the literature. It is reported that this approach affords the product in low yield[14] and that the alkylation results in considerable racemization.[7] However, Varelis and Johnson[15] reported that the latter method is also suited to large-scale preparation of virtually enantiopure **1**.

Using the acid-catalyzed benzylation of the ethyl lactate as described here avoids racemization during the reaction course and affords the product in high yield. In addition all reagents are commercially available and it is possible to carry out the reaction on a large scale.

Besides the one-step procedure described here, conversion of ester **1** to aldehyde **2** can be accomplished by reduction of the ester with lithium aluminum hydride, and subsequent oxidation of the alcohol produced using a Swern-oxidation protocol.[8,14,16]

The highly enantioenriched vinyl sulfone **3** has been used in the synthesis of the highly diastereomerically and enantiomerically enriched tetracarbonyl π–allyl iron(1+) complex **4**. Nucleophilic attack on this electrophilic organometal complex occurs regioselectively at the γ-position with respect to the sulfone functionality. In addition, the reaction proceeds with conservation of the double bond configuration allowing syntheses of highly enantioenriched (E)-alkenyl sulfones **5** with a wide range of substitution patterns at the allylic position[17,18,19] (Scheme 1). The value of **3** and the related methodology of the iron-mediated chirality transfer has also been demonstrated in the syntheses of methyl-branched natural products in high enantiomeric purity and in their naturally occurring absolute configuration.[20,21,22]

Scheme 1

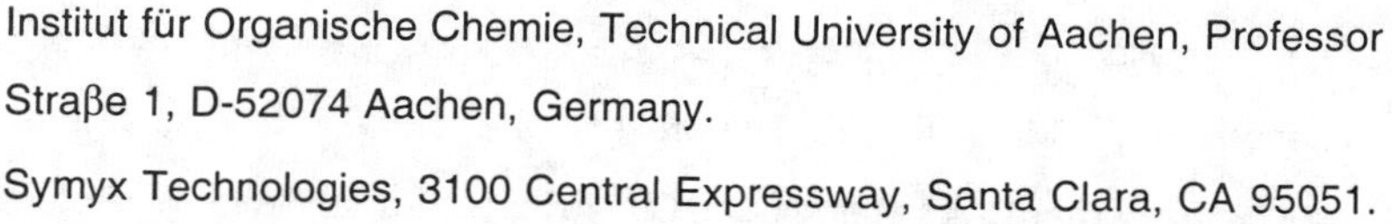

| **3** | (1R,2S,3R)-**4** | **5**: ee > 96- > 97% |

1. Institut für Organische Chemie, Technical University of Aachen, Professor Pirlet-Straße 1, D-52074 Aachen, Germany.

2. Symyx Technologies, 3100 Central Expressway, Santa Clara, CA 95051.

3. Ito, Y.; Kobayashi, Y.; Kawabata, T.; Takase, M.; Terashima, S. *Tetrahedron* **1989**, *45*, 5767.

4. Rathke, M. W.; Nowak, M. *J. Org. Chem.* **1985**, *50*, 2624.

5. Wessel, H.-P.; Iversen, T.; Bundle, D. R. *J. Chem. Soc., Perkin Trans. I* **1985**, 2247.

6. Some recent publications include: (a) Ayerbe, M.; Arrieta, A.; Cossio, F. P.; Linden, A. *J. Org. Chem.* **1998**, *63*, 1795; (b) Yang, H. W.; Romo, D. *J. Org. Chem.* **1998**, *63*, 1344; (c) Reetz, M.; Haning, H. *J. Organomet. Chem.* **1997**, *541*, 117; (d) Almendros, P.; Thomas, E. J. *J. Chem. Soc., Perkin Trans. I* **1997**, 2561; (e) Jacobi, P. A.; Herradura, P. *Tetrahedron Lett.* **1997**, *38*, 6621; (f) Szymoniak, J.; Thery, N.; Moise, C. *Bull. Soc. Chim. Fr.* **1997**, *134*, 85; (g) Mikami, K.; Matsukawa, S.; Sawa, E.; Harada, A.; Koga, N. *Tetrahedron Lett.* **1997**, *38*, 1951; (h) Hoppe, D.; Tebben, P.; Reggelin, M.; Bolte, M. *Synthesis* **1997**, 183; (i) Cormick, R.; Lofstedt, J.; Perlmutter, P.; Westman, G. *Tetrahedron Lett.* **1997**, *38*, 2737; (j) Gennari, C.; Moresca, D.; Vulpetti, A.; Pain, G. *Tetrahedron* **1997**, *53*, 5593; (k) Enders, D.; Jandeleit, B.; von Berg, S. *J. Organomet. Chem.* **1997**, *533*, 219; (l) Enders, D.; Jandeleit, B.; von Berg, S. *Synlett* **1997**, 421; (m) Enders, D.; Frank, U.; Fey, P.; Jandeleit, B.; Lohray, B. B. *J. Organomet. Chem.* **1996**, *519*, 147; (n) Klute, W.; Kruger, M.; Hoffmann, R. W. *Chem. Ber.* **1996**, *129*, 633; (o) Evans, D. A.; Murry, J. A.; Kozlowski, M. C. *J. Am. Chem. Soc.* **1996**, *118*, 5814; (p) Peng, Z.-H.; Li, Y.-L.; Wu, W.-L.; Liu, C.-X.; Wu, Y.-L. *J. Chem. Soc., Perkin Trans. I* **1996**, 1057; (q) Marshall, J. A.; Jablonowski, J. A.; Welmaker, G. S. *J. Org. Chem.* **1996**, *61*, 2904; (r) Solladie-Cavallo, A.; Roche, D.; Fischer, J.; De Cian, A. *J. Org. Chem.* **1996**, *61*, 2690; (s) Jacobi, P. A.; Murphree, S.; Rupprecht, F.; Zheng, W. *J. Org. Chem.* **1996**, *61*, 2413; (t) Bernardi, A.; Marchionni, C.; Novo, B.; Karamfilova, K.; Potenza, D.; Scolastico, C.; Roversi, P. *Tetrahedron* **1996**, *52*, 3497; (u) Solladie-Cavallo, A.; Bonne, F. *Tetrahedron: Asymmetry* **1996**, *7*, 171; (v) Jain, N. F.; Cirillo, P. F.; Pelletier, R.; Panek, J. S. *Tetrahedron Lett.* **1995**, *36*, 8727; (w) Marshall, J. A.; Jablonowski, J. A.; Luke, G. P. *J. Org. Chem.* **1994**, *59*, 7825; (x) Roush, W. R.;

VanNieuwenhze, M. S. *J. Am. Chem. Soc.* **1994**, *116*, 8536; (y) Shanmuganathan, K.; French, L. G.; Jensen, B. L. *Tetrahedron: Asymmetry* **1994**, *5*, 797; (z) Gennari, C.; Moresca, D.; Vulpetti, A.; Pain, G. *Tetrahedron Lett.* **1994**, *35*, 4623.

7.   Kobayashi, Y.; Takase, M.; Ito, Y.; Terashima, S. *Bull. Chem. Soc. Jpn.* **1989**, *62*, 3038.

8.   Takai, K.; Heathcock, C. H. *J. Org. Chem.* **1985**, *50*, 3247.

9.   Baker, D. C.; Hawkins, L. D. *J. Org. Chem.* **1982**, *47*, 2179.

10.   Mulzer, J.; Angermann, A. *Tetrahedron Lett.* **1983**, *24*, 2843.

11.   Guanti, G.; Banfi, L.; Guaragna, A.; Narisano, E. *J. Chem. Soc., Chem. Commun.* **1986**, 138.

12.   Bianchi, D.; Cesti, P.; Golini, P. *Tetrahedron* **1989**, *45*, 869.

13.   Mislow, K.; O'Brien, R. E.; Schaefer, H. *J. Am. Chem. Soc.* **1962**, *84*, 1940.

14.   Massad, S. K.; Hawkins, L. D.; Baker, D. C. *J. Org. Chem.* **1983**, *48*, 5180.

15.   Varelis, P.; Johnson, B. L. *Aust. J. Chem.* **1995**, *48*, 1775.

16.   Wuts, P. G. M.; Bigelow, S. S. *J. Org. Chem.* **1983**, *48*, 3489.

17.   For accounts and reviews see: (a) Ref. 6 (l); (b) Enders, D.; Jandeleit, B.; von Berg, S. In "Organic Synthesis via Organometallics, OSM 5"; Helmchen, G., Ed.; Vieweg: Braunschweig, 1997, 279.

18.   Enders, D.; Jandeleit, B.; Raabe, G. *Angew. Chem.* **1994**, *106*, 2033; *Angew. Chem., Int. Ed. Engl.* **1994**, *33*, 1949.

19.   (a) Enders, D.; von Berg, S.; Jandeleit, B. *Synlett* **1996**, 18; (b) Jackson, R. F. W.; Turner, D.; Block, M. H. *Synlett* **1997**, 789.

20.   Enders, D.; Jandeleit, B. *Synthesis* **1994**, 1327.

21.   Enders, D.; Jandeleit, B. *Liebigs Ann.* **1995**, 1173.

22.   Enders, D.; Jandeleit, B.; Prokopenko, O. F. *Tetrahedron* **1995**, *51*, 6273.

Appendix

Chemical Abstracts Nomenclature (Collective Index Number); (Registry Number)

(-)-(E,S)-3-(Benzyloxy)-1-butenyl phenyl sulfone: Benzene, [[[1-methyl-3-(phenylsulfonyl)-2-propenyl]oxy]methyl]-, [S-(E)]- (13); (168431-27-4)

(-)-(S)-Ethyl 2-(benzyloxy)propanoate: Propanoic acid, 2-(phenylmethoxy)-, ethyl ester, (S)- (9); (54783-72-1)

(-)-(S)-2-(Benzyloxy)propanal: Propanal, 2-(phenylmethoxy)-, (S)- (11); (81445-44-5)

(S)-Ethyl hydroxypropanoate: (S)-Ethyl lactate: Lactic acic, ethyl ester, L- (8); Propanoic acid, 2-hydroxy-, ethyl ester, (S)- (9); (687-47-8)

O-Benzyl-2,2,2-trichloroacetimidate: Ethanimidic acid, 2,2,2-trichloro-, phenylmethyl ester (11); (81927-55-1)

Trifluoromethanesulfonic acid: HIGHLY CORROSIVE: Methanesulfonic acid, trifluoro- (8,9); (1493-13-6)

Diisobutylaluminum hydride: Aluminum, hydrodiisobutyl- (8); Aluminum, hydrobis(2-methylpropyl)- (9); (1191-15-7)

Lithium bromide (8,9); (7550-35-8)

Diethyl[(phenylsulfonyl)methyl]phosphonate: Phosphonic acid, [(phenylsulfonyl)methyl]-, diethyl ester (9); (56069-39-7)

Acetonitrile: TOXIC (8,9); (75-05-8)

Triethylamine (8); Ethanamine, N,N-diethyl- (9); (121-44-8)

Benzyl alcohol (8); Benzenemethanol (9); (100-51-6)

Trichloroacetonitrile: Acetonitrile, trichloro- (8,9); (545-06-2)

Sodium hydride (8,9); (7646-69-7)

(+)-(1R,2S,3R)-TETRACARBONYL[(1-3η)-1-(PHENYLSULFONYL)-BUT-2-EN-1-YL]IRON(1+) TETRAFLUOROBORATE

(Iron(1+), tetracarbonyl[(1,2,3-η)-1-(phenylsulfonyl)-2-butenyl]-, stereoisomer, tetrafluoroborate(1)-)

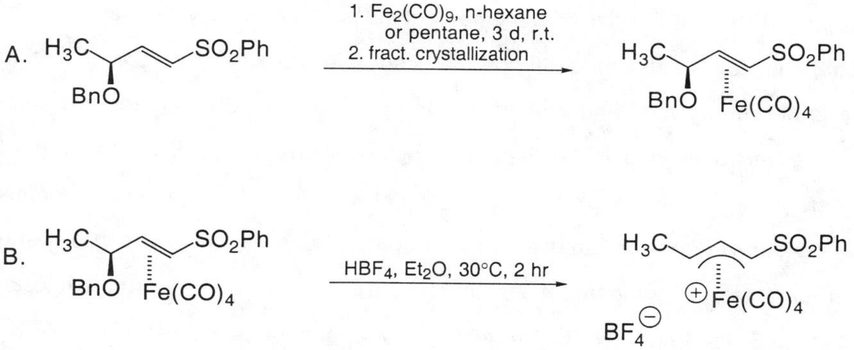

A.

1. Fe$_2$(CO)$_9$, n-hexane
   or pentane, 3 d, r.t.
2. fract. crystallization

B.

HBF$_4$, Et$_2$O, 30°C, 2 hr

(1R,2S,3R)-1

Submitted by D. Enders,[1] B. Jandeleit,[2] and S. von Berg.[1]
Checked by Matthew J. Schnaderbeck and William R. Roush.

1. Procedure

A. (+)-(E,1R,3S)-Tetracarbonyl[(3-(benzyloxy)-1-(phenylsulfonyl)-η^2-but-1-ene]-iron(0). A flame-dried, 250-mL Schlenk-flask equipped with a septum, magnetic stirring bar and a balloon filled with carbon monoxide (Note 1) is charged with 3.03 g (10.0 mmol) of (-)-(E,S)-3-(benzyloxy)-1-butenyl phenyl sulfone (Note 2) and 4.73 g (13.0 mmol) of nonacarbonyldiiron [Fe$_2$(CO)$_9$] (Note 3). The reagents are suspended in 150 mL of anhydrous hexane (Note 4) and the orange suspension is stirred for 3 days at room temperature under an atmosphere of carbon monoxide and with

189

exclusion of light (Note 5). The reaction mixture is diluted with anhydrous diethyl ether (ca. 100 mL) (Note 6) and transferred via canula to an air-free filter containing a short column of Celite (Note 7) by means of a flame-dried inert gas frit (Note 8) and filtered under an atmosphere of argon. The bright yellow filtrate is collected in a 1000-mL, flame-dried Schlenk-flask (Note 5). The filter column is washed with anhydrous diethyl ether (Note 6) until the filtrate becomes colorless. The clear yellow solution of the neutral complex is concentrated under reduced pressure in a room temperature water bath to between a third and a quarter of the original volume by means of a medium pressure vacuum pump (Note 9) and is then fractionally crystallized at -25°C in a freezer. The pale yellow precipitate is washed with anhydrous hexane pre-cooled to -25°C (Note 4) under an atmosphere of argon and dried under reduced pressure at room temperature by means of a high vacuum pump (Note 9) with the exclusion of light (Note 5) to afford 1.88-3.05 g (40-65%) of a moderately air sensitive pale yellow crystalline solid (Notes 10, 11).

   B. *(+)-(1R,2S,3R)-Tetracarbonyl[(1-3η)-1-(phenylsulfonyl)but-2-en-1-yl]iron(1+) tetrafluoroborate* **1**. A flame-dried, 50-mL Schlenk-flask equipped with a septum and an argon balloon is charged with 1.88 g (4.0 mmol) of (+)-(E,1R,3S)-tetracarbonyl[(1-2η)-3-(benzyloxy)-1-(phenylsulfonyl)-but-1-ene]iron(0) and the complex is dissolved in 15 mL of anhydrous diethyl ether (Note 6) under an atmosphere of argon. The bright yellow solution is taken up into a syringe under an atmosphere of argon and filtered through a PTFE-syringe filter into a flame-dried, 100-mL Schlenk flask equipped with a magnetic stirring bar, a septum and an argon balloon (Note 12). The solution is diluted with anhydrous diethyl ether (Note 6) to give a total volume of 50-70 mL and then warmed to 30°C by means of a water bath. Fluoroboric acid, (HBF$_4$), 0.7 mL (4.8 mmol) of a 54% solution of HBF$_4$ in diethyl ether (Note 13) is added dropwise by means of a syringe to the rapidly stirring solution. After ca. 2 hr, the precipitate is filtered off under an atmosphere of argon using an inert gas frit. The precipitate is washed with anhydrous diethyl ether (Note 6) until the filtrate becomes colorless (3 x 10-20 mL). The residue is dried for ca. 12 hr under reduced pressure at room temperature using a high vacuum pump (Notes 8, 9) to yield 1.73 g (96%) of a pale yellow solid (Note 14). Normally the complex is obtained in spectroscopically and analytically pure form (by ^{1}H and ^{13}C NMR, elemental analysis) and can be used without further purification for nucleophilic addition reactions. If necessary, the complex can be further purified by re-precipitation from nitromethane solution with the addition of excess cold diethyl ether. The dry complex should be stored at -25°C in a freezer under an atmosphere of argon and may be handled for short periods in air.

2. Notes

1. Carbon monoxide was purchased from Linde, Germany (99%) and was used without further purification.

2. Prepared according to the accompanying procedure: Enders, D.; von Berg, S.; Jandeleit, B. *Org. Synth.* **2000**, *78*, 177.

3. According to ref. 3, nonacarbonyldiiron [Fe$_2$(CO)$_9$] was synthesized by photochemical dimerization of pentacarbonyliron [Fe(CO)$_5$] in a mixture of glacial acetic acid and acetic acid anhydride (10:1) at room temperature employing a Dema irradiation apparatus with a Philips HPK 125 W or TQ 150 W medium pressure mercury lamp. The material should be stored at -25°C in a freezer and handled under an atmosphere of argon. Pentacarbonyliron [Fe(CO)$_5$] was a gift from BASF AG, Germany and was used without further purification.

4. Hexane was purified by distillation from calcium hydride under argon.

5. Aluminum foil should be wrapped around the Schlenk flask to exclude sunlight.

6. Diethyl ether was purified by distillation from sodium and benzophenone.

7. Celite was purchased from Fluka, Germany and dried in an oven for 4 hr at ca. 110°C. The dried material was degassed three times in an inert gas frit prior to use by an evacuation/argon purge cycle. After this procedure the Celite was compressed to a ca. 4-cm deep layer and then covered by an 1-cm layer of previously dried sea sand (Riedel de Haën, Netherlands) to avoid a disturbance to the Celite layer during manipulations.

8. Aluminum foil should be wrapped around the inert gas frit to exclude sunlight.

9. An additional effective cooling trap should be installed to condense any of the highly toxic pentacarbonyliron [$Fe(CO)_5$].

10. Initially, a diastereomeric mixture of olefinic iron complexes (de ≈ 70%) is obtained from which the desired major diastereomer can be separated in a highly diastereo- and enantiomerically enriched form following the crystallization procedure described. A second crop can be obtained from the mother liquor to increase the chemical yield. However, additional fractions may not be as diastereomerically pure as the first fraction, giving rise to cationic complexes of lower enantiomeric purity in the next step.

11. The compound shows the following analytical and spectroscopic data: R_f = 0.43 (0.25-mm silica gel on glass, diethyl ether/light petroleum, 1:2); mp: 103°C (dec.); de = ee > 99% (by ^{1}H NMR, 500 MHz, signals: CHC$\underline{H}_3$, ortho-C-H); $[\alpha]_D^{26}$ +171.8 (benzene, c 1.05); ^{1}H NMR (500 MHz, C_6D_6) δ: 0.91 (d, 3 H, J = 6.1), 3.04 (qdd, 1 H, J = 6.1, 5.8, 0.3), 3.29 (dd, 1 H, J = 10.4, 5.8), 3.79 (d, 1 H, J = 12.1), 3.86 (dd, 1 H, J = 10.2, 0.3), 3.93 (d, 1 H, J = 12.1), 6.87-7.13 (m, 8 H), 7.85-7.91 (m, 2 H); ^{13}C NMR (125 MHz, C_6D_6) δ: 21.73, 57.96, 66.70, 70.15, 76.31, 127.84, 128.04, 128.23, 128.47, 129.13, 132.54, 137.97, 142.66, 207.25 ppm; IR (KBr) cm^{-1}: 3085, 3056, 3032 (w, Ar-C-H), 2969, 2873 (w, CH, CH_2, CH_3), 2103 (vs, apical-Fe-CO), 2045, 2022, 1988 (vs, Fe-CO), 1585, 1496, 1479 (vw, Ar-C=C), 1446 (m), 1385, 1377 (w, CH_3), 1326 (m),

1300 (s, S=O), 1262 (w-m), 1191 (w), 1144 (s, S=O), 1084 (s, C-O-C), 1041, 1026 (m), 807, 752, 734, 718, 689 (m), 624 (vs), 591, 562 (s); IR (hexane) cm^{-1}: 2104, 2035, 2026, 2002 (vs, Fe-CO). MS (70 eV): m/z (%): 414 (3) [M$^{+\cdot}$-2CO], 386 (4) [M$^{+\cdot}$-3CO], 359 (20), 358 (99) [M$^{+\cdot}$-4CO], 303 (3) [M$^{+\cdot}$+1-Fe(CO)$_4$], 268 (14), 267 (100) [358-C$_7$H$_7$], 250 (14), 239 (17), 217 (4) [358-SO$_2$C$_6$H$_5$], 198 (10), 186 (10), 184 (12), 161 (27), 143 (3) [H$_2$SO$_2$C$_6$H$_5^+$], 141 (2) [SO$_2$C$_6$H$_5^+$], 134 (12), 133 (53), 107 (6) [C$_7$H$_8$O$^+$], 91 (79) [C$_7$H$_7^+$], 77 (20) [C$_6$H$_5^+$], 65 (16) [C$_5$H$_5^+$], 56 (63) [Fe$^+$], 55 (10), 53 (17), 51 (10); calculated for C$_{21}$H$_{18}$FeO$_7$S (M$_r$: 470.28): C 53.63, H 3.86, found C 53.60, H 3.89.

12. A stainless steel filtration device and PTFE-filters (Satorius, Germany, diameter: 25 mm, pore size: 0.45 µm) were used to purify the solution by removing paramagnetic impurities from the neutral complex. If the solution is not filtered, paramagnetic impurities may be found in the solidified complex but these should not affect reactivity of the resulting cationic complex.

13. The 54% solution of HBF$_4$ in diethyl ether was purchased from Merck, Germany and was used without further purification. The acid solution should be stored in a refrigerator under an atmosphere of argon to avoid colorization and any loss of quality.

14. The compound shows the following analytical and spectroscopic data: mp: 163°C (crystals yellowed), 173°C (dec.); de > 99% (3-syn/3-anti: >> 99: << 1 by ^{1}H NMR, 500 MHz, signals: CHC$\underline{H}_3$, C$\underline{H}$-C$\underline{H}$SO$_2$, C$\underline{H}$CH$_3$); ee > 99%; $[\alpha]_D^{21}$ +169.1 (acetone, c 1.14); ^{1}H NMR (500 MHz, CD$_3$NO$_2$) δ: 2.13 (d, 3 H, J = 6.4), 4.64 (dd, 1 H, J = 10.1, 0.6), 4.93 (dqd, 1 H, J = 12.4, 6.4, 0.6), 6.23 (dd, 1 H, J = 12.4, 10.1), 7.72-7.80 (m, 2 H), 7.82-7.91 (m, 1 H), 8.07-8.14 (m, 2 H); better ^{1}H NMR data were obtained in d$_6$-acetone: ^{1}H NMR (500 MHz, d$_6$-acetone) δ: 2.20 (d, 3 H, J = 6.1), 5.10 (dd, 1 H, J = 10.0, 0.6), 5.17 (dq, 1 H, J = 12.4, 6.3), 6.59 (ddd, 1 H, J = 12.4, 10.0, 0.6), 7.72-7.82 (m, 2 H), 7.82-7.9 (m, 1 H), 8.1-8.2 (m, 2 H); ^{13}C NMR (125 MHz, CD$_3$NO$_2$) δ:

20.77, 73.73, 90.58, 97.65, 129.46, 131.54, 136.61, 139.57, 195.35, 196.10, 197.43, 197.67; IR (KBr) cm^{-1}: 3067, 3007 (w, Ar-CH), 2929, 2857 (w), 2162, 2142, 2125, 2100, 2030, 2006 (s-vs, Fe-CO), 1642, 1585, 1521 (w, Ar-C=C), 1448 (m), 1386 (w, CH$_3$), 1303, 1148 (s, S=O), 1084, 1057 (vs, br.), 810, 756, 727, 719, 689 (w), 628, 612, 594 (s), 555 (w); IR (CH$_2$Cl$_2$) cm^{-1}: 2102 (vs, Fe-CO), 2032, 2000 (vs, Fe-CO). MS (70 eV): m/z (%): 446 (2) [M$^{+\cdot}$+1], 321 (2), 306 (3) [321-CH$_3$], 278 (13), 250 (13) [278– CO], 196 (25), 195 (14) [M$^{+\cdot}$-BF$_4^-$, -Fe(CO)$_4$], 186 (7), 184 (21), 161 (7), 143 (4) [H$_2$SO$_2$C$_6$H$_5^+$], 141 (3) [SO$_2$C$_6$H$_5^+$], 133 (9), 129 (13), 126 (15), 125 (28), 115 (6), 108 (6), 107 (24), 105 (6), 97 (10), 95 (9), 94 (9), 93 (11), 91 (19), 79 (18), 78 (20), 77 (41) [C$_6$H$_5^+$], 56 (49) [Fe$^+$], 55 (100), 53 (18), 50 (8), 48 (8), 43 (10), 41 (15), 39 (24); calculated for C$_{14}$H$_{11}$BF$_4$FeO$_6$S (M$_r$: 449.95): C 37.37, H 3.14, found C 36.93, H 2.76.

Waste Disposal Information

All toxic materials were disposed of in accordance with "Prudent Practices in the Laboratory"; National Academy Press; Washington, DC, 1995.

3. Discussion

Among the various carbon-carbon and carbon-hetero atom bond forming reactions promoted or catalyzed by transition metals, allylic substitution via electrophilic π-allyl-complexes is of utmost importance. Studies focused on the synthetic potential of alkyl or aryl substituted (η^3-allyl)Fe(CO)$_4$(1+) complexes have shown that nucleophilic attack by soft carbon and hetero atom nucleophiles preferentially proceeds regioselectively at the less or syn-substituted allyl terminus.[4] Additionally, polar effects on the regioselectivity of this reaction caused by electron-withdrawing functionalities (e.g., CO$_2$R, CONR$_2$) have been examined by the

194

submitters' group,[4] Green, et al.,[6] and Speckamp, et al.[7] It has been demonstrated that the reaction affords allyl-coupled addition products with complete γ-regioselectivity with respect to the electron-withdrawing functionality. During the submitters' efforts[3] devoted to developing a useful methodology for the synthesis of highly enantiomerically enriched compounds via iron-mediated allylic substitutions, the "chirality transfer" approach turned out to be a practical solution. Based on the chiral pool precursor lactic acid, an efficient approach to the phenylsulfonyl-functionalized planar chiral (η^3-allyl)Fe(CO)$_4$(1+) complex (1R,2S,3R)-1 in virtually diastereo- and enantiomerically pure form (de, ee > 99%) has been developed.[8] Complex (1R,2S,3R)-1 represents a synthetic equivalent of an a^4-synthon A with planar chirality (Scheme 2) allowing homologous (1,5)-Michael additions[9] or an "Umpolung" of classical d^4-chemistry.[10] Regioselective addition of carbon and heteroatom nucleophiles to (1R,2S,3R)-1 (Scheme 2) provides an efficient access to highly enantiomerically enriched alkenyl sulfones with a wide range of substitution patterns at the allylic position; a class of compounds which is of increasing importance.[11] As depicted in Scheme 2, (1R,2S,3R)-1 can be combined with various nucleophiles.[8,12,13] These reactions proceed with virtually no loss of chirality information from central (C-O) to planar (C-Fe) and back to central chirality (C-C or C-X) affording products 2-8 with high enantiomeric purity (ee > 95 - > 99%) and overall retention (double inversion) with respect to the starting material. In addition, the reaction proceeds with complete γ-regioselectivity and conservation of the (E)-double bond geometry leading to highly functionalized molecules of well-defined stereochemistry.

Scheme 2

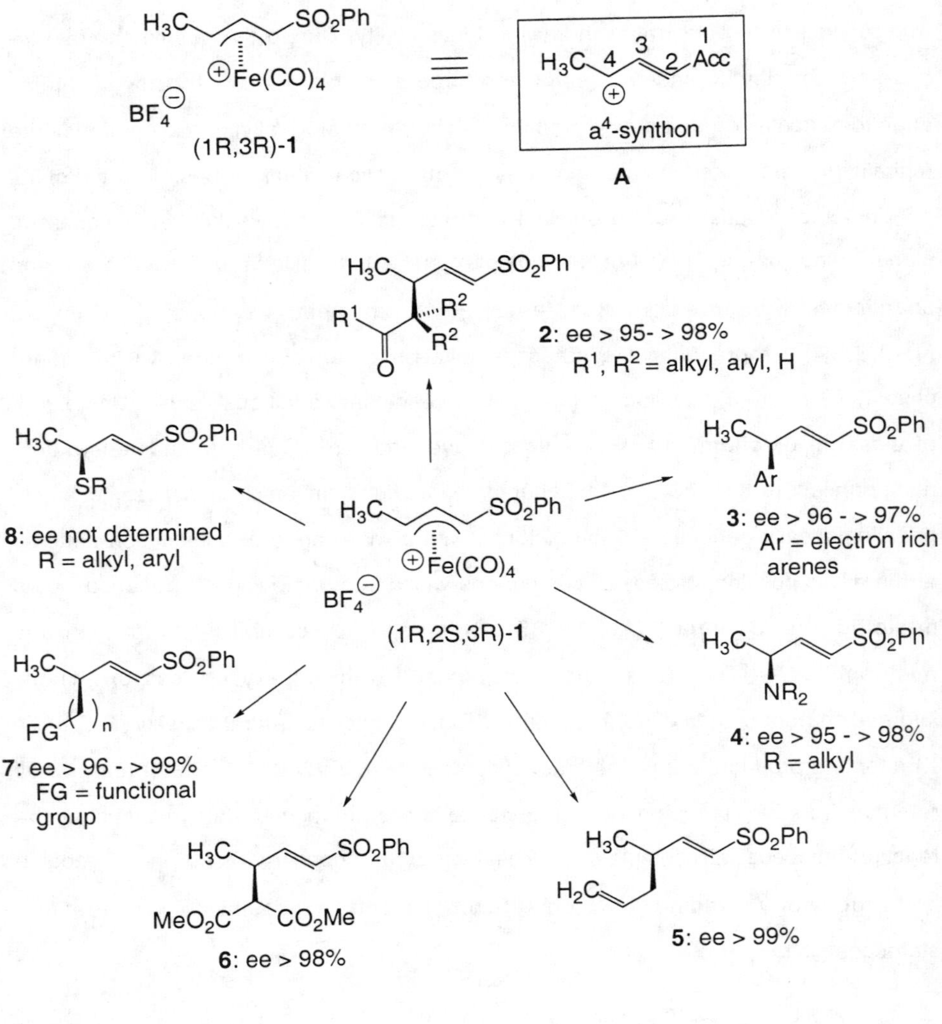

H_3C ⟍⟍ SO_2Ph
⊕ $Fe(CO)_4$
$BF_4^{\ominus}$
(1R,3R)-**1**

≡

H_3C $\overset{4}{\diagdown}\overset{3}{\diagup}\overset{2}{\diagdown}\overset{1}{Acc}$
⊕
a^4-synthon

A

H_3C ⟍⟍ SO_2Ph
R^1 ⟍ R^2
          R^2
O

2: ee > 95- > 98%
R^1, R^2 = alkyl, aryl, H

H_3C ⟍⟍ SO_2Ph
SR

8: ee not determined
R = alkyl, aryl

H_3C ⟍⟍ SO_2Ph
⊕ $Fe(CO)_4$
$BF_4^{\ominus}$
(1R,2S,3R)-**1**

H_3C ⟍⟍ SO_2Ph
Ar

3: ee > 96 - > 97%
Ar = electron rich
arenes

H_3C ⟍⟍ SO_2Ph
FG $)_n$

7: ee > 96 - > 99%
FG = functional
group

H_3C ⟍⟍ SO_2Ph
NR_2

4: ee > 95 - > 98%
R = alkyl

H_3C ⟍⟍ SO_2Ph
MeO_2C CO_2Me

6: ee > 98%

H_3C ⟍⟍ SO_2Ph
H_2C ⟍

5: ee > 99%

196

Furthermore, the planar complex (1R,2S,3R)-**1** also represents the synthetic equivalent of a d^1/a^3-butyl synthon **B** (Scheme 3). This stems from its electrophilic reactivity in the γ-position and its nucleophilic reactivity, after reductive hydrogenation and metalation, α- to the sulfonyl group, with subsequent removal of the latter. The potential bifunctionality allows a flexible sequential functionalization of the butyl-backbone of **1**. Because of their origin from enantiopure building blocks bearing methyl substituents (e.g., isoprenoids, alanine, lactic acid derivatives) many naturally occuring compounds possess methyl-branched carbon atom skeletons. Scheme 3 demonstrates some achievements made in natural product synthesis making use of complex **1**. Key steps in the syntheses of these natural products have been the nucleophilic addition of silyl enol ether or allyl-silane to **1**, respectively.[14,15,16] Starting from complex **1**, all methyl-branched natural products synthesized (**9**, **10**, **11**) have been obtained from readily accessible materials with their naturally occurring absolute configuration in excellent overall yields and in virtually enantio- and/or diastereomerically pure form (**9**, **10**: ee > 99%, **11**: ee > 99%, de > 98%).[17]

Scheme 3

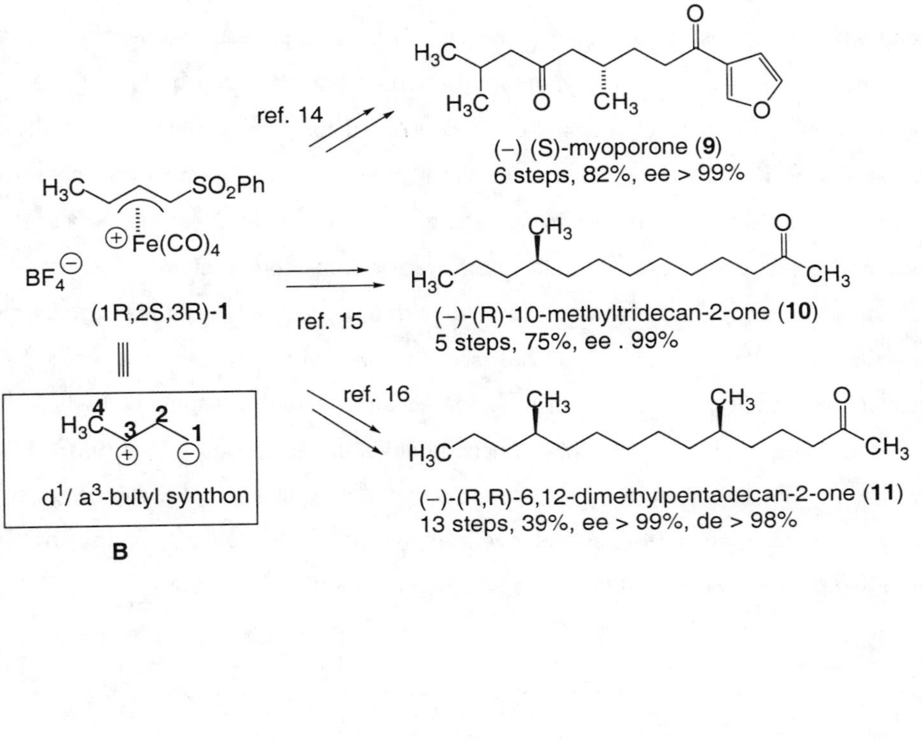

ref. 14
ref. 15
ref. 16
H3C
SO2Ph
Fe(CO)4
BF4
(1R,2S,3R)-1
d1/ a3-butyl synthon
B
(–) (S)-myoporone (9)
6 steps, 82%, ee > 99%
(–)-(R)-10-methyltridecan-2-one (10)
5 steps, 75%, ee . 99%
(–)-(R,R)-6,12-dimethylpentadecan-2-one (11)
13 steps, 39%, ee > 99%, de > 98%

1. Institut für Organische Chemie, Technical University of Aachen, Professor Pirlet-Straße 1, D-52074 Aachen, Germany. E-mail: Enders@RWTH-Aachen.de

2. New address: Symyx Technologies, 3100 Central Expressway, Santa Clara, CA 95051, USA. E-mail: bjandeleit@symyx.com.

3. For accounts and reviews see: (a) Enders, D.; Jandeleit, B.; von Berg, S. *Synlett* **1997**, 421; (b) Enders, D.; Jandeleit, B.; von Berg, S. In "Organic Synthesis via Organometallics, OSM 5"; Helmchen, G., Ed.; Vieweg: Braunschweig, 1997, 279.

4. (a) Yeh, M.-C. P.; Tau, S.-I. *J. Chem. Soc., Chem. Commun.* **1992**, 13; (b) Li, Z.; Nicholas, K. M. *J. Organomet. Chem.* **1991**, *402*, 105; (c) Dieter, J. W.; Li, Z.; Nicholas, K. M. *Tetrahedron Lett.* **1987**, *28*, 5415; (d) Silverman, G. S.; Strickland, S.; Nicholas, K. M. *Organometallics* **1986**, *5*, 2117; (e) Hafner, A.; von Philipsborn, W.; Salzer, A. *Helv. Chim. Acta* **1986**, *69*, 1757; (f) Ladoulis, S. J.; Nicholas, K. M. *J. Organomet. Chem.* **1985**, *285*, C13; (g) Salzer, A.; Hafner, A. *Helv. Chim. Acta* **1983**, *66*, 1774; (h) Dieter, J.; Nicholas, K. M. *J. Organomet. Chem.* **1981**, *212*, 107; (i) Pearson, A. J. *Tetrahedron Lett.* **1975**, 3617; (j) Whitesides, T. H.; Arhart, R. W.; Slaven, R. W. *J. Am. Chem. Soc.* **1973**, *95*, 5792.

5. (a) Enders, D.; Fey, P.; Schmitz, T.; Lohray, B. B.; Jandeleit, B. *J. Organomet. Chem.* **1996**, *514*, 227; (b) Enders, D.; Frank, U.; Fey, P.; Jandeleit, B.; Lohray, B. B. *J. Organomet. Chem.* **1996**, *519*, 147; (c) Enders, D.; Finkam, M. *Synlett* **1993**, 401.

6. (a) Zhou, T.; Green, J. R. *Tetrahedron Lett.* **1993**, *34*, 4497; (b) Gajda, C.; Green, J. R. *Synlett* **1992**, 973; (c) Green, J. R.; Carroll, M. K. *Tetrahedron Lett.* **1991**, *32*, 1141.

7. (a) Hopman, J. C. P.; Hiemstra, H.; Speckamp, W. N. *J. Chem Soc., Chem. Commun.* **1995**, 619; (b) Hopman, J. C. P.; Hiemstra, H.; Speckamp, W. N. *J.*

Chem Soc., Chem. Commun. **1995**, 617; (c) Koot, W.-J.; Hiemstra, H.; Speckamp, W. N. *J. Chem Soc., Chem. Commun.* **1993**, 156.

8. Enders, D.; Jandeleit, B.; Raabe, G. *Angew. Chem.* **1994**, *106*, 2033; *Angew. Chem., Int. Ed. Engl.* **1994**, *33*, 1949.

9. (a) Danishefsky, S. *Acc. Chem. Res.* **1979**, *12*, 66; (b) Wong, H. N. C.; Hon, M.-Y.; Tse, C.-W.; Yip, Y.-C.; Tanko, J.; Hudlicky, T. *Chem. Rev.* **1989**, *89*, 165.

10. (a) Seebach, D. *Angew. Chem.* **1979**, *91*, 259; *Angew. Chem., Int. Ed. Engl.* **1979**, *18*, 239; (b) Hase, T. A. "Umpoled Synthons: A Survey of Sources and Uses in Syntheses"; Wiley: New York, 1987.

11. (a) Simpkins, N. S. In "Sulphones in Organic Synthesis, Tetrahedron Organic Chemistry Series", 1st ed.; Baldwin, J. E.; Magnus, P. D., Eds.; Pergamon: Oxford, 1993; Vol. 10; (b) Simpkins, N. S. *Tetrahedron* **1990**, *46*, 6951; (c) "The Chemistry of Sulphones and Sulphoxides"; Patai, S.; Rappoport, Z.; Stirling, C. J., Eds.; Wiley: Chichester, 1988; (d) Fuchs, P. L.; Braish, T. F. *Chem. Rev.* **1986**, *86*, 903.

12. Enders, D.; Jandeleit, B.; von Berg, S., unpublished results.

13. (a) Enders, D.; von Berg, S.; Jandeleit, B. *Synlett* **1996**, 18; (b) Jackson R. F. W.; Turner, D.; Block, M. H. *Synlett* **1997**, 789.

14. Enders, D.; Jandeleit, B. *Synthesis* **1994**, 1327.

15. Enders, D.; Jandeleit, B. *Liebigs Ann.* **1995**, 1173.

16. Enders, D.; Jandeleit, B.; Prokopenko, O. F. *Tetrahedron* **1995**, *51*, 6273.

17. For additional natural product syntheses see: Enders, D.; Finkam, M. *Liebigs Ann. Chem.* **1993**, 551.

Appendix

Chemical Abstracts Nomenclature (Collective Index Number); (Registry Number)

(+)-(1R,2S,3R)-Tetracarbonyl[(1-3η)-1-(phenylsulfonyl)-but-2-en-1-yl]iron(1+) tetrafluoroborate: Iron(1+), tetracarbonyl[(1,2,3-η)-1-(phenylsulfonyl)-2-butenyl]-, stereoisomer, tetrafluoroborate(1-) (13); (162762-06-3)

(+)-(E,1R,3S)-Tetracarbonyl[(3-benzyloxy)-1-(phenylsulfonyl)-η²-but-1-ene]iron(0): Iron, tetracarbonyl[[[[(2-3η)-1-methyl-3-(phenylsulfonyl)-2-propenyl]oxy]methyl]-benzene]-, stereoisomer (13); (168431-28-5)

(-)-(E,S)-3-(Benzyloxy)-1-butenyl phenyl sulfone: Benzene[[[1-methyl-3-(phenylsulfonyl)-2-propenyl]oxy]methyl]-, [S-(E)]- (13); (168431-27-4)

Carbon monoxide (8,9); (630-08-0)

Nonacarbonyldiiron: Iron, tri-μ-carbonylhexacarbonyldi- (Fe-Fe) (8,9); (15321-51-4)

Fluoroboric acid: Borate(1-), tetrafluoro-, hydrogen (8,9); (16872-11-0)

PREPARATION AND DIELS-ALDER REACTION OF A 2-AMIDO SUBSTITUTED FURAN: tert-BUTYL 3a-METHYL-5-OXO-2,3,3a,4,5,6-HEXAHYDROINDOLE-1-CARBOXYLATE

(1H-Indole-1-carboxylic acid, 2,3,3a,4,5,6-hexahydro-3a-methyl-5-oxo-, 1,1-dimethylethyl ester)

Submitted by Albert Padwa,[1] Michael A. Brodney, and Stephen M. Lynch.

Checked by Sivaraman Dandapani and Dennis P. Curran.

1. Procedure

A. Furan-2-ylcarbamic acid tert-butyl ester. In a 250-mL, one-necked, round-bottomed flask equipped with a magnetic stirring bar are placed 10 g (0.07 mol) of 2-furoyl chloride (Note 1), 80 mL of tert-butyl alcohol (Note 2), and 5.1 g (0.08 mol) of sodium azide (Note 3). After the flask is stirred at 25°C for 20 hr under an argon atmosphere, it is placed behind a protective shield (Note 4) and the solution is heated at reflux for 15 hr under a constant flow of argon. The solvent is removed with a rotary evaporator at aspirator vacuum to provide a white solid that is purified by flash silica gel chromatography (10% ethyl acetate/hexane) to give 10.8 g (81%) of furan-2-ylcarbamic acid tert-butyl ester as a white solid: mp 98-99°C (Note 5).

B. 4-Bromo-2-methyl-1-butene. In a 500-mL, three-necked, round-bottomed flask equipped with a magnetic stirring bar, reflux condenser, and a dropping funnel are placed 20 g (0.23 mol) of 3-methyl-3-buten-1-ol (Note 6), 160 mL of freshly distilled dichloromethane (Note 7) and 36 mL (0.24 mol) of triethylamine (Note 8). The reaction mixture is cooled to -5°C and 14.4 g (0.24 mol) of freshly distilled methanesulfonyl chloride (Note 9) is added dropwise from the dropping funnel. After the reaction mixture is stirred at 0°C for an additional 2 hr, it is quenched with 80 mL of water and the aqueous phase is extracted three times with 40-mL portions of dichloromethane. The combined organic phase is dried over magnesium sulfate and the solvent is removed with a rotary evaporator at aspirator vacuum. The crude yellow oil is taken up in 25 mL of dry acetone and added dropwise from a dropping funnel to a slurry of 60 g (0.68 mol) of lithium bromide in 115 mL of dry acetone in a 250-mL round-bottomed flask fitted with the dropping funnel and a reflux condenser. The solution is slowly warmed to 35°C and is stirred at this temperature for 18 hr, at which time the reaction is quenched with 120 mL of water. The aqueous phase is extracted three times with 40-mL portions of ether. The combined organic phase is dried over magnesium sulfate and the solvent is removed with a rotary evaporator at aspirator vacuum. The resulting oil is distilled at aspirator pressure to provide 17.7 g (51%) of 4-bromo-2-methyl-1-butene as a colorless oil: bp 41-42°C at 34-39 torr (Note 10).

C. *tert-Butyl N-(3-methyl-3-butenyl)-N-(2-furyl)carbamate.* In a flame-dried, 500-mL, one-necked, round-bottomed flask equipped with a magnetic stirring bar and reflux condenser are placed 4.0 g (21.8 mmol) of furan-2-ylcarbamic acid tert-butyl ester and 150 mL of toluene (Note 11) under an argon atmosphere. To this solution are added 3.1 g (76.4 mmol) of freshly ground powdered sodium hydroxide, 6.04 g (43.7 mmol) of potassium carbonate, and 1.48 g (4.4 mmol) of tetrabutylammonium hydrogen sulfate (Note 12). The solution is heated at 80°C for 25 min with vigorous stirring and then 3.9 g (26.2 mmol) of freshly distilled 4-bromo-2-methyl-1-butene is added as a solution in 10 mL of toluene over a 30-min period. After being heated at 80°C for 30 min, the solution is charged with an additional 0.98 g (6.6 mmol) of 4-bromo-2-methyl-1-butene. The mixture is heated at 80°C for an additional 1 hr. After the reaction is cooled to room temperature, it is quenched by the addition of 200 mL of water and the aqueous phase is extracted three times with 100-mL portions of dichloromethane. The combined organic phase is dried over magnesium sulfate and the solvent is removed with a rotary evaporator at aspirator vacuum. The crude residue is purified by silica gel chromatography (10% ethyl acetate-hexane) to give 5.0 g (91%) of tert-butyl N-(3-methyl-3-butenyl)-N-(2-furyl)carbamate as a colorless oil (Note 13).

D. *tert-Butyl 3a-methyl-5-oxo-2,3,3a,4,5,6-hexahydroindole-1-carboxylate.* Into an oven-dried, 35-mL heavy-wall, high pressure tube (Note 14) equipped with a magnetic stirring bar and rubber septum are placed 3.7 g (14.7 mmol) of tert-butyl N-(3-methyl-3-butenyl)-N-(2-furyl)carbamate and 20 mL of toluene under an argon atmosphere. Argon is vigorously bubbled through the solution for 30 min at which time the septum is replaced with a threaded plunger valve equipped with an O-ring seal

(Note 14). The vessel is placed behind a protective shield (Note 4) and immersed into a preheated oil bath at 160°C for 14 hr. After the solution is cooled to room temperature, solvent is removed with a rotary evaporator at aspirator vacuum and the crude residue is purified by silica gel chromatography (40% ethyl acetate-hexane) to give 2.8 g (70-75%) of tert-butyl 3a-methyl-5-oxo-2,3,3a,4,5,6-hexahydroindole-1-carboxylate as a white solid: mp 112-113°C (Note 15).

2. Notes

1. 2-Furoyl choride was purchased from Aldrich Chemical Company, Inc., and used without further purification.

2. 2-Methyl-2-propanol (HPLC grade; tert-butyl alcohol) was purchased from Aldrich Chemical Company, Inc., and used without further purification.

3. Sodium azide (99%) was purchased from Aldrich Chemical Company, Inc.; a Teflon spatula was used when handling this reagent. *Caution*: avoid contact with metal and heat when using sodium azide.

4. The protective shield was purchased from Lab-Line, Inc. and was used for protection when heating at high temperatures.

5. The product has the following spectral characteristics: IR (neat) cm^{-1}: 3267, 2980, 1700, and 1546; ^{1}H NMR (CDCl$_3$, 300 MHz) δ: 1.50 (s, 9 H), 6.04 (brs, 1 H), 6.34 (m, 1 H), 6.63 (brs, 1 H), and 7.06 (m, 1 H); ^{13}C NMR (CDCl$_3$, 75 MHz) δ: 28.2, 81.3, 95.1, 111.2, 136.0, 145.4, 151.9. Anal. Calcd for C$_9$H$_{13}$NO$_3$: C, 59.00; H, 7.15; N, 7.64. Found: C, 59.09; H, 7.13; N, 7.67.

6. 3-Methyl-3-buten-1-ol was purchased from Aldrich Chemical Company, Inc., and used without further purification.

7. Dichloromethane was distilled from calcium hydride prior to use.

8. Triethylamine was purchased from Aldrich Chemical Company, Inc., and used without further purification.

9. Methanesulfonyl chloride was purchased from Aldrich Chemical Company, Inc., and was distilled before use.

10. The product has the following spectral characteristics: IR (neat) cm^{-1}: 3075, 1652 1445, and 890; ^{1}H NMR (CDCl$_3$, 400 MHz) δ: 1.75 (s, 3 H), 2.58 (t, 2 H, J = 7.4), 3.47 (t, 2 H, J = 7.4), 4.77 (s, 1 H), and 4.86 (s, 1 H); ^{13}C NMR (CDCl$_3$, 100 MHz) δ: 22.1, 31.0, 41.1, 112.9, and 142.6

11. Toluene was distilled from calcium hydride prior to use.

12. Tetrabutylammonium hydrogen sulfate (97%) was purchased from Aldrich Chemical Company, Inc., and used without further purification.

13. The product has the following spectral characteristics: IR (neat) cm^{-1}: 2975, 1711, 1606, and 1369; ^{1}H NMR (CDCl$_3$, 400 MHz) δ: 1.45 (s, 9 H), 1.74 (s, 3 H), 2.27 (t, 2 H, J = 7.2), 3.67 (dd, 2 H, J = 9.2 and 6.0), 4.71 (s, 1 H), 4.76 (s, 1 H), 6.33 (brs, 1 H), 7.14 (t, 1 H, J = 1.2), and 6.0 (brs, 1 H); ^{13}C NMR (CDCl$_3$, 100 MHz) δ: 22.2, 28.0, 36.6, 46.9, 80.7, 100.9, 110.7, 111.8, 137.8, 142.3, 148.3, and 153.5. Anal. Calcd for C$_{14}$H$_{21}$NO$_3$: C, 66.91; H, 8.42; N, 5.57. Found: C, 66.93; H, 8.38; N, 5.60. The broad resonance at δ 6.0 in the ^{1}H NMR spectrum merges into a sharp multiplet when the spectrum is recorded at 50°C.

14. The 35-mL heavy-wall high pressure tube, Teflon plug, and O-ring were purchased from Ace Glass and were oven dried prior to use.

15. The product has the following spectral characteristics: IR (KBr) cm^{-1}: 2961, 1709, and 1388; ^{1}H NMR (DMSO-d$_6$, 400 MHz) δ: 0.94 (s, 3 H), 1.42 (s, 9 H), 1.72 (m, 2 H), 2.35 (d, 1 H, J = 14.6), 2.49 (d, 1 H, J = 14.6), 2.63 (dd, 1 H, J = 14.6 and 2.8), 2.89 (dd, 1 H, J = 14.6 and 4.8), 3.51 (m, 1 H), 3.66 (m, 1 H), and 5.78 (brs, 1 H); ^{13}C NMR (DMSO-d$_6$, 100 MHz) δ: 22.8, 27.7, 35.2, 36.7, 42.5, 46.1, 51.6, 79.6, 96.4,

143.7, 151.5, and 208.3. Anal. Calcd for $C_{14}H_{21}NO_3$: C, 66.91; H, 8.42; N, 5.57. Found: C, 66.99; H, 8.38; N, 5.49.

Waste Disposal Information

All toxic materials were disposed of in accordance with "Prudent Practices in the Laboratory"; National Academy Press; Washington, DC, 1995.

3. Discussion

Heterocycles such as furan, thiophene, and pyrrole undergo Diels-Alder reactions despite their stabilized 6π-aromatic electronic configuration.[2] By far the most extensively studied five-ring heteroaromatic system for Diels-Alder cycloaddition is furan and its substituted derivatives.[3] The resultant 7-oxabicyco[2.2.1]heptanes are valuable synthetic intermediates that have been further elaborated to substituted arenes, carbohydrate derivatives, and various natural products.[4-6] A crucial synthetic transformation employing these intermediates involves the cleavage of the oxygen bridge to produce functionalized cyclohexene derivatives.[7,8] While the bimolecular Diels-Alder reaction of alkyl-substituted furans has been the subject of many reports in the literature,[9] much less is known regarding the cycloaddition behavior of furans that contain heteroatoms attached directly to the aromatic ring.[10] In this regard, we have become interested in the Diels-Alder reaction of 2-aminofurans as a method for preparing substituted aniline derivatives since these compounds are important starting materials for the preparation of various pharmaceuticals.[11] Many furan Diels-Alder reactions require high pressure or Lewis acid catalysts to give satisfactory yields of cycloadduct.[12] In contrast to this situation, 2-amino-5-carbomethoxyfuran readily reacted with several monoactivated olefins by simply heating in benzene at 80°C. The initially formed cyclohexadienol underwent a subsequent dehydration when treated with 1 equiv of boron trifluoride etherate (BF$_3$·OEt$_2$) to give the substituted aniline derivative.[13] In each case, the cycloaddition proceeded with complete regioselectivity,

with the electron-withdrawing group being located ortho to the amino group. The regiochemical results are perfectly consistent with FMO theory.[14] The most favorable FMO interaction is between the HOMO of the furanamine and the LUMO of the dienophile. The atomic coefficient at the ester carbon of the furan is larger than at the amino center, and this nicely accommodates the observed regioselectivity.

The intramolecular Diels-Alder reaction of furans, often designated as IMDAF,[15] helps to overcome the sluggishness of this heteroaromatic ring system toward [4+2]-cycloaddition. Not only do IMDAF reactions allow for the preparation of complex oxygenated polycyclic compounds, but they also often proceed at lower temperatures than their intermolecular counterparts.[9] Even more significantly, unactivated π-bonds are often suitable dienophiles for the internal cycloaddition. Indeed, the submitters discovered that the IMDAF reaction of a series of furanamide derivatives occurred

n = 1, 2

smoothly to furnish cyclized aromatic carbamates as the only isolable products in high yield.[16] When the alkenyl group possesses a substituent at the 2-position of the π-

bond, the thermal reaction furnishes a rearranged hexahydroindolinone.[17] With this system, the initially formed cycloadduct cannot aromatize. Instead, ring opening of the oxabicyclic intermediate occurs to generate a zwitterion that undergoes hydride transfer to give the rearranged ketone. The procedure described here provides a

simple and general approach for the construction of various hexahydroindolinones. This strategy can be cleanly applied toward the synthesis of more complex octahydroindole-based alkaloids.

1. Department of Chemistry, Emory University, Atlanta, GA 30322.

2. Lipshutz, B. H. *Chem. Rev.* **1986**, *86*, 795.

3. Sargent, M. V.; Dean, F. M. In "Comprehensive Heterocyclic Chemistry"; Bird, C. W.; Cheesman, G. W. H., Eds.; Pergamon Press: London, 1984; Vol. 4, pp. 599-656.

4. Vogel, P.; Fattori, D.; Gasparini, F.; Le Drian, C. *Synlett* **1990**, 173; Reymond, J.-L.; Pinkerton, A. A.; Vogel, P. *J. Org. Chem.* **1991**, *56*, 2128.

5. Renaud, P.; Vionnet, J.-P. *J. Org. Chem.* **1993**, *58*, 5895.

6. Cox, P. J.; Simpkins, N. S. *Synlett* **1991**, 321.

7. Le Drian, C.; Vieira, E.; Vogel, P. *Helv. Chem. Acta* **1989**, *72*, 338; Takahashi, T.; Iyobe, A.; Arai, Y.; Koizumi, T. *Synthesis* **1989**, 189; Guildford, A. J.; Turner, R. W. *J. Chem. Soc., Chem. Commun.* **1983**, 466; Yang, W.; Koreeda, M. *J. Org Chem.* **1992**, *57*, 3836.

8. Suami, T. *Pure Appl. Chem.* **1987**, *59*, 1509; Harwood, L. M.; Jackson, B.; Prout, K.; Witt, F. J. *Tetrahedron Lett.* **1990**, *31*, 1885; Koreeda, M.; Jung, K.-Y.; Hirota, M. *J. Am. Chem. Soc.* **1990**, *112*, 7413; Reynard, E.; Reymond, J. L.; Vogel, P. *Synlett* **1991**, 469; Ogawa, S.; Yoshikawa, M.; Taki, T. *J. Chem. Soc., Chem. Commun.* **1992**, 406; Ogawa, S.; Tsunoda, H. *Liebigs Ann. Chem.* **1992**, 637.

9. Kappe, C. O.; Murphree, S. S.; Padwa, A. *Tetrahedron* **1997**, *53*, 14179.

10. Gewald, K. *Chem. Ber.* **1966**, *99*, 1002; Boyd, G. V.; Heatherington, K. *J. Chem. Soc., Perkin Trans. 1* **1973**, 2523; Nixon, W. J., Jr.; Garland, J. T.; Blanton, C. D., Jr. *Synthesis* **1980**, 56; Aran, V. J.; Soto, J. L. *Synthesis* **1982**, 513; Semmelhack, M. F.; Park, J. *Organometallics* **1986**, *5*, 2550; Chatani, N.; Hanafusa, T. *J. Org. Chem.* **1987**, *52*, 4408; Cutler, S. J.; El-Kabbani, F. M.; Keane, C.; Fisher-Shore, S. L.; Blanton, C. D., Jr. *Heterocycles* **1990**, *31*, 651.

11. Wulfman, D. S. In "The Chemistry of Diazonium and Diazo Groups"; Patai, S., Ed.; Wiley: New York, 1978; Part 1, see pp. 286-297.

12. Matsumoto, K.; Sera, A. *Synthesis* **1985**, 999.

13. Padwa, A.; Dimitroff, M.; Waterson, A. G.; Wu, T. *J. Org. Chem.* **1997**, *62*, 4088.

14. Fleming, I. "Frontier Orbitals and Organic Chemical Reactions"; Wiley-Interscience: New York, 1976.

15. Sternbach, D. D.; Rossana, D. M.; Onan, K. D. *J. Org. Chem.* **1984**, *49*, 3427; Jung, M. E.; Gervay, J. *J. Am. Chem. Soc.* **1989**, *111*, 5469; Klein, L. L. *J. Org. Chem.* **1985**, *50*, 1770.

16. Padwa, A.; Dimitroff, M.; Waterson, A. G.; Wu, T. *J. Org. Chem.* **1998**, *63*, 3986.

17. Padwa, A.; Brodney, M. A.; Dimitroff, M. *J. Org. Chem.* **1998**, *63*, 5304.

Appendix

Chemical Abstracts Nomenclature (Collective Index Number); (Registry Number)

tert-Butyl 3a-methyl-5-oxo-2,3,3a,4,5,6-hexahydroindole-1-carboxylate: 1H-Indole-1-carboxylic acid, 2,3,3a,4,5,6-hexahydro-3a-methyl-5-oxo-, 1,1-dimethylethyl ester (14); (212560-98-0)

Furan-2-ylcarbamic acid tert-butyl ester: Carbamic acid, 2-furanyl-, 1,1-dimethylethyl ester (9); (56267-47-1)

2-Furoyl chloride (8): 2-Furancarbonyl chloride (9); (527-69-5)

tert-Butyl alcohol (8): 2-Propanol, 2-methyl- (9); (75-65-0)

Sodium azide (8,9); (26628-22-8)

4-Bromo-2-methyl-1-butene: 1-Butene, 4-bromo-2-methyl- (8,9); (20038-12-4)

3-Methyl-3-buten-1-ol: 3-Buten-1-ol, 3-methyl- (8,9); (763-32-6)

Methanesulfonyl chloride (8,9); (124-63-0)

Lithium bromide (8,9); (7550-35-8)

tert-Butyl N-(3-methyl-3-butenyl)-N-(2-furyl)carbamate: Carbamic acid, 2-furanyl(3-methyl-3-butenyl)-, 1,1-dimethylethyl ester (14); (212560-95-7)

Toluene (8); Benzene, methyl- (9); (108-88-3)

Tetrabutylammonium hydrogen sulfate: Ammonium, tetrabutyl-, sulfate (1:1) (8); 1-Butanaminium, N,N,N-tributyl-, sulfate (1:1) (9); (32503-27-8)

[4 + 3] CYCLOADDITION OF AMINOALLYL CATIONS WITH 1,3-DIENES: 11-OXATRICYCLO[4.3.1.1^{2,5}]UNDEC-3-EN-10-ONE

(11-Oxatricyclo[4.3.1.1^{2,5}]undec-3-en-10-one, (1α,2β,5β,6α)-)

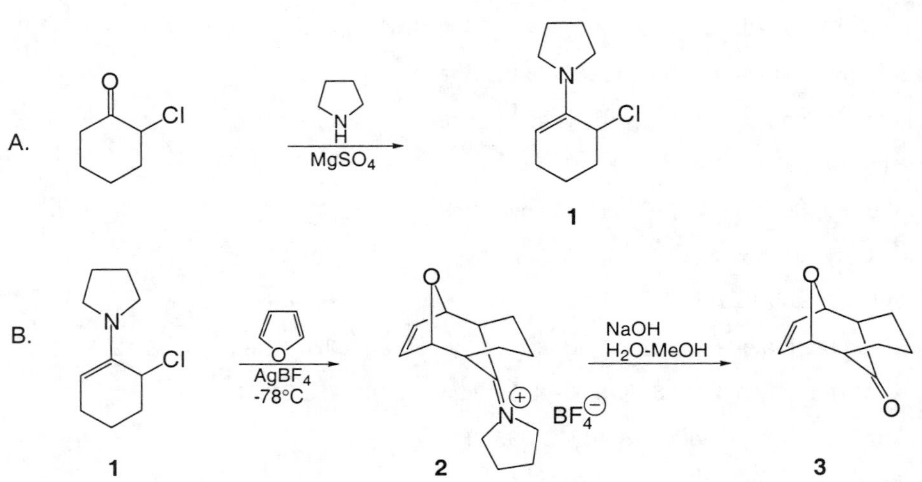

A.

1

B.

1   2   3

Submitted by Jonghoon Oh, Chewki Ziani-Cherif, Jong-Ryoo Choi, and Jin Kun Cha.[1,2]

Checked by Matthew Surman and Marvin J. Miller.

1. Procedure

A. 6-Chloro-1-pyrrolidinocyclohexene (**1**).[3] A 500-mL, round-bottomed flask equipped with a magnetic stirring bar is charged with 15 g (0.11 mol) of 2-chlorocyclohexanone (Note 1) and 170 mL of cyclohexane (Note 2). To the stirred solution under a nitrogen atmosphere is added 60 g of anhydrous magnesium sulfate in one portion. The mixture is cooled to 0°C with an ice bath, and 48 mL of pyrrolidine

(0.57 mol, 5 equiv) is added dropwise via a syringe over a 15-in period. After the reaction mixture has been stirred for an additional 30 min at 0°C, the ice bath is removed and the mixture is stirred overnight at room temperature. Magnesium sulfate is removed by filtration and rinsed thoroughly with hexane (3 x 50 mL). The combined filtrate and rinsings are concentrated under reduced pressure without heating to give crude 6-chloro-1-pyrrolidinocyclohexene (1) as an orange oil (20.76 g, 0.11 mol, 99%), which is sufficiently pure for the next step (Note 3).

    B. *11-Oxatricyclo[4.3.1.1[2,5]]undec-3-en-10-one* (3).[2,4,5] A 500-mL, two-necked, round-bottomed flask is equipped with a magnetic stirring bar and a rubber septum. The flask is quickly charged with 30.8 g (0.158 mol) of silver tetrafluoroborate (Note 4) and wrapped in aluminum foil to exclude room light. After the system has been dried under vacuum pump and purged with nitrogen, 150 mL of anhydrous methylene chloride (Note 5) is added, followed by furan (80 mL, 1.1 mol) (Note 6). The reaction mixture is cooled to -78°C with an acetone-dry ice bath and stirred under a nitrogen atmosphere. To this cooled solution is added dropwise over a period of 1 hr via cannula (using positive nitrogen pressure) a solution of crude 1 in 120 mL of methylene chloride. After the addition is completed, the cooling bath is removed, and the mixture is allowed to warm slowly to room temperature, and then stirred for an additional 12 hr. The reaction mixture is filtered through Celite, insoluble material is rinsed with methylene chloride (3 x 70 mL), and the combined filtrate and rinsings are concentrated under reduced pressure to afford the immonium salt 2 as a dark brown oil (47.4 g, 0.16 mol).

    To this crude concentrate 2, which is placed in a 1-L, one-necked, round-bottomed flask, are added 300 mL of deionized water and 250 mL of methanol, followed by 18 g (0.45 mol) of sodium hydroxide. The reaction mixture is stirred for 10 hr at room temperature and extracted with ether (300 mL, then 7 x 200 mL). The combined organic layers are washed with brine, dried over magnesium sulfate, filtered, and concentrated under reduced pressure to a dark orange oil. The crude product is purified by chromatography on silica gel (400 g) with 1:4 ethyl acetate/hexane as eluant to afford 13.0 g (72% overall from 2-chlorocyclohexanone) of the cycloadduct **3** as a white solid, mp 46-47.5°C (Note 7).

2. Notes

1. (a) 2-Chlorocyclohexanone (98%) was purchased from Aldrich Chemical Company, Inc., and used without further purification. All glassware was oven-dried (>100°C) and quickly assembled prior to use. (b) Newman, M. S.; Farbman, M. D.; Hipsher, H. *Org. Synth., Coll. Vol. III* **1955**, 188.

2. Cyclohexane was dried by distillation from calcium hydride and stored over 4Å molecular sieves. Unless stated otherwise, all solvents and reagents in this procedure were obtained commercially and used as received.

3. Most conveniently, the crude α'-chloroenamine was used immediately without further purification for the next step. Although it can be distilled under high vacuum with some loss (bp 93-94°C/0.5 mm),[3] there is little advantage to be gained from purification. Spectral data are as follows: ^{1}H NMR (300 MHz, CDCl3) δ: 1.62-2.22 (m, 10 H), 2.96-3.22 (m, 4 H), 4.40 (m, 1 H), 4.68 (m, 1 H); [13]NMR (75 MHz, CDCl3) δ: 17.0, 24.2, 24.6, 32.9, 47.1, 56.3, 97.7, 142.0.

4. Silver tetrafluoroborate (98%) was purchased from Aldrich Chemical Company, Inc.

5. Methylene chloride was dried by distillation from calcium hydride.

6. Immediately before use, furan was shaken with aqueous 5% sodium hydroxide twice, dried over sodium sulfate, filtered, and distilled from sodium hydroxide pellets under a nitrogen atmosphere.

7. Overall yields in several runs were obtained in the range of 73-86% by the submitters, and 50-72% by the checkers. Previous literature reported mp 45.5-46°C.[5a] Spectral data for **3** (R_f = 0.44 (1:4 ethyl acetate-hexane) are as follows: IR (CHCl$_3$) cm^{-1}: 1725; ^{1}H NMR (360 MHz, CDCl$_3$) δ: 1.49-1.56 (m, 1 H), 2.04-2.13 (m, 2 H), 2.27-2.33 (m, 2 H), 2.35-2.36 (m, 2 H), 2.52-2.62 (m, 1 H), 4.94 (d, 2 H, J = 1.9), 6.37 (s, 2 H); ^{13}C NMR (90 MHz, CDCl$_3$) δ: 20.8, 31.0, 53.1, 83.5, 135.4, 214.8; Mass spectrum: m/z 164 (M·+, 42), 136 (33), 107 (64), 96 (62), 81 (37), 79 (40), 68 (58), 67 (32), 55 (72), 41 (60), 39 (100). The spectral data are identical to those reported in the literature.[2,4,5]

Waste Disposal Information

All toxic materials were disposed of in accordance with "Prudent Practices in the Laboratory"; National Academy Press: Washington, DC, 1995.

3. Discussion

The [4 + 3] cycloaddition reactions of oxyallyls and related allyl cations to 1,3-dienes are general methods for the synthesis of seven-membered ring ketones and have been the subject of several excellent reviews.[6] A representative experimental procedure using acyclic oxyallyls was described by Hoffmann for the preparation of 2α,4α-dimethyl-8-oxabicyclo[3.2.1]oct-6-en-3-one.[7] Subsequent elaboration of the resulting cycloadducts provides facile preparation of seven-membered carbocycles, tetrahydrofurans, and tetrahydropyrans; these earlier investigations focused on acyclic oxyallyls. More recently, a key variant of using cyclic oxyallyls, (i.e., the oxyallyl functionality is embedded within a ring) has been developed as a convenient route to functionalized medium-sized carbocycles and heterocycles: By virtue of the spectator ring skeleton, this formal [n + 4] cycloaddition allows rapid assembly of these challenging ring systems in addition to providing a seven-membered ring.[8,9] Other key advantages of cyclic oxyallyls include (1) the geometrically constrained, rigid conformation of the oxyallyl function (especially compared to conformationally flexible acyclic oxyallyls), which results in highly diastereoselective formation of the cycloadducts via a "compact" transition state; (2) facile generation of the requisite oxyallyls, which can be attributed to the stabilizing influence of alkyl substituents (i.e., the ring carbons); and (3) well-defined diastereofacial bias present in the resulting tricyclic or tetracyclic cycloadducts made rigid by the keto bridge, which is useful for subsequent elaboration.

Among the known [4 + 3] cycloaddition protocols examined, only the Schmid[2] and Föhlisch[4,10] procedures were found to afford the [4 + 3] cycloadducts in synthetically useful yields. Schmid's rarely-used "aminoallyl" reaction appears to be most effective, particularly when sterically demanding 1,3-dienes [e.g., spiro[2.4]hepta-4,6-diene] or otherwise recalcitrant N-acylpyrroles are required for the cycloadditions. On the other hand, with sterically unencumbered 1,3-dienes and also with furans, the Föhlisch reaction has emerged as the method of choice in view of the practical aspects of simple and convenient execution.

The method described here is the direct adaptation of the original procedure developed by the late Professor Hans Schmid. As summarized in Scheme 1, the Schmid reaction has been successfully applied to [4 + 3] cycloadditions with cyclopentadiene, spiro[2.4]hepta-4,6-diene, and N-Boc-pyrrole. Additional examples can be found in references 8 and 9. Use of functionalized six-membered oxyallyls and synthetic applications of the [4 + 3] cycloadducts have also been described.[8]

Scheme 1

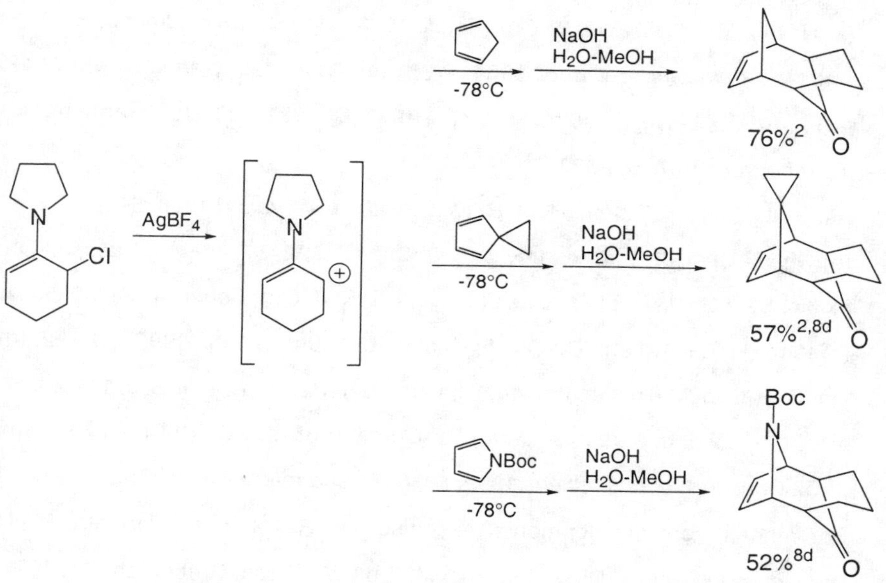

1. Department of Chemistry, University of Alabama, Tuscaloosa, AL 35487.

2. This present method is an adaptation of the original procedure of the late Professor Hans Schmid: (a) Schmid, R.; Schmid, H. *Helv. Chim. Acta* **1974**, *57*, 1883; (b) Schmid, R. Ph. D. Dissertation, University of Zürich, 1978. The submitters thank Professor Manfred Hesse for providing a copy of the Ph.D. dissertation.

3. Blazejewski, J. C.; Cantacuzene, D.; Wakselman, C. *Tetrahedron* **1973**, *29*, 4233.

4. Föhlisch, B.; Gehrlach, E.; Herter, R. *Angew. Chem., Int. Ed. Engl.* **1982**, *21*, 137.

5. (a) Noyori, R.; Baba, Y.; Makino, S.; Takaya, H. *Tetrahedron Lett.* **1973**, 1741; (b) Vinter, J. G.; Hoffmann, H. M. R. *J. Am. Chem. Soc.* **1973**, *95*, 3051.

6. (a) Noyori, R.; Hayakawa, Y. *Org. React.* **1983**, *29*, 163; (b) Hoffmann, H. M. R. *Angew. Chem., Int. Ed. Engl.* **1984**, *23*, 1; (c) Mann, J. *Tetrahedron* **1986**, *42*, 4611; (d) Rigby, J. H. *Org. React.* **1997**, *49*, 331-425; (e) Harmata, M. *Tetrahedron* **1997**, *53*, 6235.

7. Ashcroft, M. R.; Hoffmann, H. M. R. *Org. Synth., Coll. Vol. VI* **1988**, 512.

8. (a) Oh, J.; Choi, J.-R.; Cha, J. K. *J. Org. Chem.* **1992**, *57*, 6664; (b) Lee, J.; Oh, J.; Jin, S.-j.; Choi, J.-R.; Atwood, J. L.; Cha, J. K. *J. Org. Chem.* **1994**, *59*, 6955; (c) Kim, H.; Ziani-Cherif, C.; Oh, J.; Cha, J. K. *J. Org. Chem.*. **1995**, *60*, 792; (d) Jin, S.-j.; Choi, J.-R.; Oh, J.; Lee, D.; Cha, J. K. *J. Am. Chem. Soc.* **1995**, *117*, 10914; (e) Cha, J. K.; Oh, J. *Curr. Org. Chem.* **1998**, *2*, 217; (f) Cho, S. Y.; Lee, J. C.; Cha, J. K. *J. Org. Chem.* **1999**, *64*, 3394 and references therein.

9. For other recent developments, see also: (a) Ref. 6e; (b) Harmata, M. In "Advances in Cycloaddition"; Lautens, M., Ed.; JAI Press: Greenwich, CT, 1997; Vol 4, pp 41-86; (c) West, F. G. In "Advances in Cycloaddition"; Lautens, M., Ed.; JAI Press: Greenwich, CT, 1997; Vol 4, pp 1-40.

10. Sendelbach, S.; Schwetzler-Raschke, R.; Radl, A.; Kaiser, R.; Henle, G. H.; Korfant, H.; Reiner, S.; Föhlisch, B. *J. Org. Chem.* **1999**, *64*, 3398.

Appendix

Chemical Abstracts Nomenclature (Collective Index Number); (Registry Number)

11-Oxatricyclo[4.3.1.1^{2,5}]undec-3-en-10-one, (1α,2β,5β,6α)- (9); (42768-72-9)

6-Chloro-1-pyrrolidinocyclohexene: Pyrrolidine, 1-(6-chloro-1-cyclohexen-1-yl)- (9); (35307-20-1)

2-Chlorocyclohexanone: Cyclohexanone, 2-chloro- (8,9); (822-87-7)

Pyrrolidine (8,9); (123-75-1)

Silver tetrafluoroborate: Borate (1-), tetrafluoro-, silver (1+) (9); (14104-20-2)

SYNTHESIS OF AMINO ACID ESTER ISOCYANATES:
METHYL (S)-2-ISOCYANATO-3-PHENYLPROPANOATE
(Benzenepropanoic acid, α-isocyanato-, methyl ester, (S))

Submitted by James H. Tsai, Leo R. Takaoka, Noel A. Powell, and James S. Nowick.[1]

Checked by Adam Charnley and Steven Wolff.

1. Procedure

A 250-mL, three-necked, round-bottomed flask is equipped with a mechanical stirrer and charged with 100 mL of methylene chloride, 100 mL of saturated aqueous sodium bicarbonate, and 5.50 g (25.5 mmol) of L-phenylalanine methyl ester hydrochloride (Note 1). The biphasic mixture is cooled in an ice bath and stirred mechanically while 2.52 g (8.42 mmol) of triphosgene (Note 2) is added in a single portion. The reaction mixture is stirred in the ice bath for 15 min and then poured into a 250-mL separatory funnel. The organic layer is collected, and the aqueous layer is extracted with three 15-mL portions of methylene chloride. The combined organic layers are dried (MgSO₄), vacuum filtered, and concentrated at reduced pressure using a rotary evaporator to give a colorless oil. The oil is purified by Kugelrohr distillation (130°C, 0.05 mm) to afford 5.15 g (98%) of methyl (S)-2-isocyanato-3-phenylpropanoate as a colorless oil (Notes 3-6).

2. Notes

1. L-Phenylalanine methyl ester hydrochloride was purchased from Bachem California Inc.

2. Triphosgene was purchased from Aldrich Chemical Company, Inc.

3. The product has the following properties: $[\alpha]_D^{25}$ -83.8° (neat); IR (CHCl₃) cm⁻¹: 2260, 1747; ¹H NMR (400 MHz, CDCl₃) δ: 3.03 (dd, 1 H, ABX pattern, J_{AB} = 13.8, J_{BX} = 7.8), 3.16 (dd, 1 H, ABX pattern, J_{AB} = 13.6, J_{AX} = 4.8), 3.81 (s, 3 H), 4.27 (dd, 1 H, J = 7.8, 4.6), 7.18-7.21 (m, 2 H), 7.27-7.36 (m, 3 H); ¹³C NMR (125 MHz, CDCl₃) δ: 39.6, 52.8, 58.3, 126.7, 127.2, 128.4, 129.1, 135.4, 170.7. Anal. Calcd for $C_{11}H_{11}NO_3$: C, 64.38; H, 5.40; N, 6.83. Found: C, 64.18; H, 5.40; N, 6.70.

4. The yield is typically 4.97-5.15 g (95-98%).

5. The submitters previously reported an optical rotation of $[\alpha]_D^{22}$ +71.9° (neat) for methyl (S)-2-isocyanato-3-phenylpropanoate.[2] This value does not match the current value of $[\alpha]_D^{25}$ -83.8° (neat) and is in error. The origin of this discrepancy involves the path length of the polarimeter cell. With a 5-cm cell, a correct α value of -48.15° is obtained. If a 10-cm cell is used, a spurious positive α value is obtained, which gives rise to a erroneous positive value of $[\alpha]_D$.

6. The optical purity of the product was determined to be >99.5% by trapping with (S)-1-phenylethylamine and ¹H NMR analysis of the resulting urea adduct, as described in reference 2.

Waste Disposal Information

All toxic materials were disposed of in accordance with "Prudent Practices in the Laboratory"; National Academy Press; Washington, DC, 1995.

3. Discussion

This procedure provides a convenient, rapid, high yielding route to amino acid ester isocyanates. It is based upon procedures the submitters have previously reported for the preparation of both amino acid ester isocyanates[2] and peptide isocyanates.[3,4] These procedures use either triphosgene or a solution of phosgene in toluene as a one-carbon electrophile and either pyridine or aqueous sodium bicarbonate as a base. The current procedure uses triphosgene and sodium bicarbonate to minimize the hazard and toxicity of the reagents and waste products. These mild reaction conditions are superior to alternative methods for the preparation of amino acid ester isocyanates, which include refluxing the amino acid ester hydrochloride in toluene for several hours while purging with gaseous phosgene,[5] or treating the amino acid ester hydrochloride with di-tert-butyl dicarbonate and 4-dimethylaminopyridine (DMAP).[6]

Amino acid ester isocyanates are produced cleanly by this method and can often be used without purification. If desired, volatile amino acid ester isocyanates, such as the title compound, can be purified to analytical purity by Kugelrohr distillation. The amino acid ester isocyanates generated by this method are formed without detectable racemization (>99.5% ee); the enantiomeric purity of the isocyanates can be checked by trapping with (S)-1-phenylethylamine, followed by ^{1}H NMR analysis of the resulting urea adducts.[2] If this method is used to generate isocyanates of peptides, then efficient stirring is necessary to prevent epimerization of the peptide isocyanates.[3,4]

Amino acid ester isocyanates are useful synthetic building blocks, precursors to peptides and azapeptides,[7,8] chiral derivatizing agents,[9,10] and reagents for the preparation of chiral chromatographic media.[11,12] (S)-2-Isocyanato-3-phenylpropanoate (phenylalanine methyl ester isocyanate) has been used as a building block for 1,2,4-triazine azapeptides,[8] and inhibitors of thermolysin[13] and human leukocyte elastase (HLE).[14]

1. Department of Chemistry, University of California Irvine, Irvine, CA 92697-2025. This work was supported by the National Institutes of Health (Grant GM-49076). J.S.N. thanks the following agencies for support in the form of awards: The National Science Foundation (Presidential Faculty Fellow Award), the Camille and Henry Dreyfus Foundation (Teacher-Scholar Award), and the Alfred P. Sloan Foundation (Alfred P. Sloan Research Fellowship).

2. Nowick, J. S.; Powell, N. A.; Nguyen, T. M.; Noronha, G. *J. Org. Chem.* **1992**, *57*, 7364-7366.

3. Nowick, J. S.; Holmes, D. L.; Noronha, G.; Smith, E. M.; Nguyen, T. M.; Huang, S.-L. *J. Org. Chem.* **1996**, *61*, 3929-3934.

4. Nowick, J. S.; Holmes, D. L.; Noronha, G.; Smith, E. M.; Nguyen, T. M.; Huang, S.-L.; Wang, E. H. *J. Org Chem.* **1998**, *63*, 9144 (Addition and correction for Ref. 3).

5. Goldschmidt, S.; Wick, M. *Justus Liebigs Ann. Chem.* **1952**, *575*, 217-231.

6. Knölker, H.-J.; Braxmeier, T. *Synlett* **1997**, 925-928.

7. Gante, J. *Synthesis* **1989**, 405-413.

8. Gante, J.; Neunhoeffer, H.; Schmidt, A. *J. Org. Chem.* **1994**, *59*, 6487-6489.

9. Pirkle, W. H.; Hoekstra, M. S. *J. Org. Chem.* **1974**, *39*, 3904-3906.

10. Pirkle, W. H.; Simmons, K. A.; Boeder, C. W. *J. Org. Chem.* **1979**, *44*, 4891-4896.

11. Pirkle, W. H.; Hyun, M. H. *J. Chromatogr.* **1985**, *322*, 295-307.

12. Armstrong, D. W.; Chang, C. D.; Lee, S. H. *J. Chromatogr.* **1991**, *539*, 83-90.

13. Bates, S. R. E.; Guthrie, D. J. S.; Elmore, D. T. *J. Chem. Res. Synop.* **1993**, 48-49.

14. Groutas, W. C.; Brubaker, M. J.; Zandler, M. E.; Mazo-Gray, V.; Rude, S. A.; Crowley, J. P.; Castrisos, J. C.; Dunshee, D. A.; Giri, P. K. *J. Med. Chem.* **1986**, *29*, 1302-1305.

Appendix
Chemical Abstracts Nomenclature (Collective Index Number); (Registry Number)

Methyl (S)-2-isocyanato-3-phenylpropanoate: Benzenepropanoic acid, α-isocyanato-, methyl ester, (S)- (9); (40203-94-9)

L-Phenylalanine methyl ester hydrochloride: L-Phenylalanine methyl ester, hydrochloride (9); (7524-50-7)

Triphosgene: Carbonic acid, bis(trichloromethyl) ester (8,9); (32315-10-9)

IN SITU CATALYTIC EPOXIDATION OF OLEFINS WITH TETRAHYDROTHIOPYRAN-4-ONE AND OXONE:

trans-2-METHYL-2,3-DIPHENYLOXIRANE

(Oxirane, 2-methyl-2.3-diphenyl-, trans-)

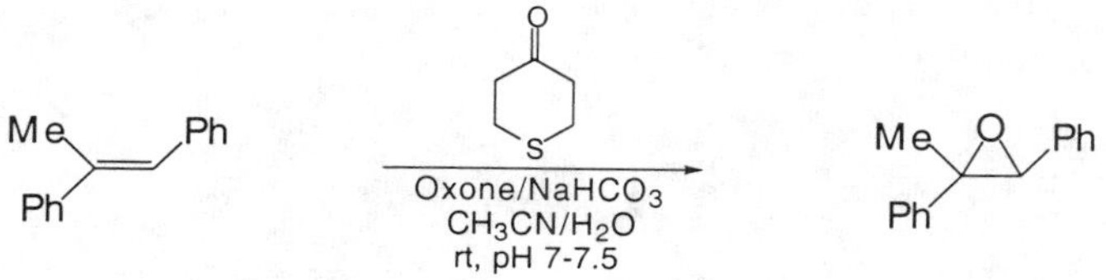

Submitted by Dan Yang, Yiu-Chung Yip, Guan-Sheng Jiao, and Man-Kin Wong.[1]
Checked by Jason M. Diffendal and Rick L. Danheiser.

1. Procedure

Caution! This procedure should be conducted in an efficient fume hood to assure the adequate removal of oxygen, a flammable gas that forms explosive mixtures with air.

A 250-mL, round-bottomed flask equipped with a Teflon-coated magnetic stirring bar is charged with 3.89 g (20.0 mmol) of trans-α-methylstilbene (Note 1), 0.12 g (1.0 mmol) of tetrahydrothiopyran-4-one (Note 2), 90 mL of acetonitrile (Note 3) and 60 mL of aqueous 4×10^{-4} M ethylenediaminetetraacetic acid, disodium salt ($Na_2 \cdot EDTA$) solution (Note 4). To this stirred mixture is added in portions a mixture of 18.4 g (30.0 mmol) of Oxone® (Note 5) and 7.8 g (93 mmol) of sodium bicarbonate over a period of 3 hr at room temperature. A slow evolution of gas bubbles is observed (Note 6). The reaction is complete in 3.5 hr as shown by TLC analysis (Note 7). The contents of the flask are poured into a 250-mL separating funnel and extracted

225

with three 50-mL portions of ethyl acetate. The combined organic layers are washed with 25 mL of saturated sodium chloride (NaCl) solution, dried over anhydrous magnesium sulfate, filtered, and concentrated under reduced pressure. The residue is purified by flash column chromatography (Note 8) to afford 4.10-4.18 g (98-99%) of 2-methyl-2,3-diphenyloxirane as a colorless oil that solidifies on standing to a white solid (Note 9).

2. Notes

1. trans-α-Methylstilbene was purchased from Aldrich Chemical Company, Inc. and used without further purification.

2. Tetrahydrothiopyran-4-one was purchased from Acros Chemical Company, Inc. and used without further purification. Oxone® rapidly converts this ketone to 1,1-dioxotetrahydrothiopyran-4-one, which functions as the catalyst for the epoxidation (Figure 1).[2]

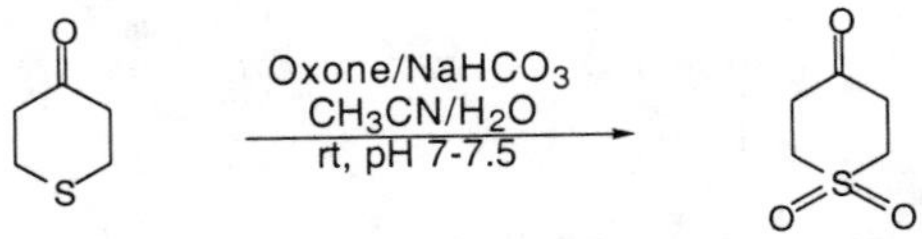

Figure 1

3. Analytical reagent grade acetonitrile was obtained from Labscan Ltd. or Mallinckrodt Inc., and was used as received.

4. An aqueous 4×10^{-4} M $Na_2 \cdot$EDTA solution is prepared by dilution of 37.2 mg of ethylenediaminetetraacetic acid disodium salt hydrate (Acros Chemical Company, Inc.) with deionized water to the mark in a 250-mL volumetric flask.

5. Oxone® $(2KHSO_5 \cdot KHSO_4 \cdot K_2SO_4)$ was purchased from Aldrich Chemical Company, Inc., and was used as received.

6. Oxygen is formed as by-product in two processes, the decomposition of Oxone® by dioxiranes and self-decomposition of Oxone®.

7. The progress of the reaction is monitored by thin layer chromatography. A 0.1-mL sample is removed and dissolved in 2 mL of hexane. The solution is spotted on a TLC plate (1 cm x 4 cm, silica gel 60 F_{254}, Merck), and the plate is developed in 1:20 ethyl acetate/hexane. Visualization by short-wavelength ultraviolet light shows the olefin at $R_f = 0.50$ and the epoxide at $R_f = 0.35$.

8. Flash column chromatography[3] is performed on Merck silica gel 60 (230-400 mesh ASTM) with 5% ethyl acetate in hexane as eluent.

9. trans-2-Methyl-2,3-diphenyloxirane exhibits the following physical and spectroscopic characteristics: mp 43.5-44.5°C; lit.[4] 46-47°C; ^{1}H NMR (500 MHz, CDCl$_3$) δ: 1.47 (s, 3 H), 3.97 (s, 1 H), 7.28-7.48 (m, 10 H); ^{13}C NMR (125 MHz, CDCl$_3$) δ: 16.9, 63.3, 67.3, 125.3, 126.7, 127.7, 127.9, 128.4, 128.7, 136.1, 142.5; EIMS (20 eV) m/z 210 (M+, 91), 209 (52), 195 (44), 181 (39), 167 (100); HRMS for $C_{15}H_{14}O$ (M+), calcd 210.1045, found 210.1047.

Waste Disposal Information

All toxic materials were disposed of in accordance with "Prudent Practices in the Laboratory"; National Academy Press; Washington, DC, 1995.

3. Discussion

Dioxiranes[5] are powerful oxidants for epoxidation of olefins under mild and neutral reaction conditions.[6] These epoxides are important and versatile intermediates in organic synthesis.[7] The most commonly used dioxiranes, i.e., dimethyldioxirane and methyl(trifluoromethyl)dioxirane, can be obtained by distillation.[8] For preparative epoxidation, an operationally simple method is to generate dioxiranes in situ from ketones and Oxone®. Compared with acetone, 1,1,1-trifluoroacetone is much more reactive for in situ epoxidation although a 10-fold excess is usually used.[9] Therefore, it will be desirable to employ commercially available ketones in low catalyst loading with a minimal amount of Oxone® for epoxidation.

The procedure described here provides a simple and convenient method for the preparation of a variety of epoxides. It uses Oxone®, an inexpensive, safe, and easily handled reagent as the terminal oxidant. The epoxidation reactions are environmentally acceptable processes as Oxone® only produces non-toxic potassium hydrogen sulfate and oxygen as the by-products.

As shown in the Table, with 5 mol% of 1,1-dioxotetrahydrothiopyran-4-one as catalyst,[10] epoxidation of various olefins (2-mmol scale) in a homogeneous acetonitrile-water solvent system with 1.5 equiv of Oxone® at room temperature can be achieved in a short period of time with excellent yields of epoxides (80-97%) isolated by flash column chromatography.[2] As the pH of the reaction is maintained at 7-7.5 by sodium bicarbonate, acid- or base-labile epoxides (entries 12-14) can be easily isolated without decomposition. More importantly, the in situ epoxidation of olefins can be performed on a large scale directly with 5 mol% of tetrahydrothiopyran-4-one, which is oxidized immediately by Oxone® to 1,1-dioxotetrahydrothiopyran-4-one during the epoxidation reactions. For example, with 5 mol% of tetrahydrothiopyran-4-one, substrates **3**, **5** (20 mmol each) and **11** (100 mmol) were epoxidized with excellent isolated yields of epoxides (91-96%).

1. Department of Chemistry, The University of Hong Kong, Pokfulam Road, Hong Kong.

2. Yang, D.; Yip, Y.-C.; Jiao, G.-S.; Wong, M.-K. *J. Org. Chem.* **1998**, *63*, 8952.

3. Still, W. C.; Kahn, M.; Mitra, A. *J. Org. Chem.* **1978**, *43*, 2923.

4. Schaap, A. P.; Siddiqui, S.; Prasad, G.; Palomino, E.; Sandison, M. *Tetrahedron* **1985**, *41*, 2229.

5. For excellent reviews on dioxirane chemistry see: (a) Adam, W.; Curci, R.; Edwards, J. O. *Acc. Chem. Res.* **1989**, *22*, 205; (b) Murray, R. W. *Chem. Rev.* **1989**, *89*, 1187; (c) Curci, R. In "Adv. Oxygenated Processes"; Baumstark, A. L., Ed.; JAI Press: Greenwich, CT, 1990; Vol. 2, p. 1; (d) Adam, W.; Hadjiarapoglou, L. P. In "Topics in Current Chemistry"; Springer-Verlag: Berlin, 1993; Vol. 164, p. 45.

6. For recent examples of synthetic applications of dioxirane epoxidations, see: (a) Deshpande, P. P.; Danishefsky, S. J. *Nature* **1997**, *387*, 164; (b) Roberge, J. Y.; Beebe, X.; Danishefsky, S. J. *J. Am. Chem. Soc.* **1998**, *120*, 3915; (c) Nicolaou, K. C.; Winssinger, N.; Pastor, J.; Ninkovic, S.; Sarabia, F.; He, Y.; Vourloumis, D.; Yang, Z.; Li, T.; Giannakakou, P.; Hamel, E. *Nature* **1997**, *387*, 268; (d) Yang, D.; Ye, X.-Y.; Xu, M.; Pang, K.-W.; Zou, N.; Letcher, R. M. *J. Org. Chem.* **1998**, *63*, 6446.

7. For reviews, see: (a) Gorzynski Smith, J. *Synthesis* **1984**, 629; (b) Besse, P.; Veschambre, H. *Tetrahedron* **1994**, *50*, 8885.

8. For an example of isolation of dimethyldioxirane by the distillation method, see: (a) Murray, R. W.; Jeyaraman, R. *J. Org. Chem.* **1985**, *50*, 2847. For examples of isolation of methyl(trifluoromethyl)dioxirane by the distillation method, see: (b) Mello, R.; Fiorentino, M.; Fusco, C.; Curci, R. *J. Am. Chem. Soc.* **1989**, *111*, 6749; (c) Adam, W.; Curci, R.; Gonzalez-Nunez, M. E.; Mello, R. *J. Am. Chem. Soc.* **1991**, *113*, 7654.

9. Yang, D.; Wong, M.-K.; Yip, Y.-C. *J. Org. Chem.* **1995**, *60*, 3887.

10. Although 1,1-dioxotetrahydrothiopyran-4-one acts as a catalyst for epoxidation and can be recovered (~80%) by column chromatography (see reference 2), it is more convenient to use tetrahydrothiopyran-4-one directly for epoxidation.

TABLE

IN SITU EPOXIDATION OF OLEFINS

Substrates:

Entry	Substrate	Reaction Time (hr)[a]	Epoxide Yield (%)[b]
1	**1**	5	95
2	**2**	4.5	87
3	**3**	1.5 (1.7[c])	95 (96[c])
4	**4**	1.5	81
5	**5**	4 (3.5[c])	97 (91[c])
6	**6**	2.5	94
7	**7**	0.5	83
8	**8**	4	95

Table (contd.)

Entry	Substrate	Reaction Time (hr)[a]	Epoxide Yield (%)[b]
9	9	2.5	85
10	10	3	96
11	11	0.5 (2[d])	96 (92[d])
12	12	4.5	80
13	13	3.5	95
14	14	2.5	92

[a]Time for epoxidation to be completed as shown by TLC or GC analysis. [b]Isolated yield. [c]20-mmol scale reaction with 5 mol% of tetrahydrothiopyran-4-one. [d]100-mmol scale reaction with 5 mol% of tetrahydrothiopyran-4-one.

Appendix
Chemical Abstracts Nomenclature (Collective Index Number); (Registry Number)

Tetrahydrothiopyran-4-one: 4H-Thiopyran-4-one, tetrahydro- (8,9); (1072-72-6)

Oxone: Peroxymonosulfuric acid, monopotassium salt, mixt. with dipotassium sulfate and potassium hydrogen sulfate (9) (37222-66-5)

trans-2-Methyl-2,3-diphenyloxirane: Oxirane, 2-methyl-2,3-diphenyl-, trans- (9); (23355-99-9)

trans-α-Methylstilbene: Stilbene, α-methyl-, (E)-; Benzene, 1,1'-(1-methyl-1,2-ethenediyl)bis-, (E)- (9); (833-81-8)

Acetonitrile (8,9); (75-05-8)

Ethylenediaminetetraacetic acid, disodium salt, dihydrate: Acetic acid, (ethylenedinitrilo)tetra-, disodium salt, dihydrate (8); Glycine, N,N'-1,2-ethanediylbis[N-(carboxymethyl)-, disodium salt, dihydrate (9); (6381-92-6)

METHYL CARBAMATE FORMATION VIA
MODIFIED HOFMANN REARRANGEMENT REACTIONS:
METHYL N-(p-METHOXYPHENYL)CARBAMATE
(Carbamic acid, (4-methoxyphenyl)-, methyl ester)

Submitted by Jeffrey W. Keillor[1] and Xicai Huang.

Checked by Scott Ceglia and Edward J. J. Grabowski.

1. Procedure

To a 1-L, round-bottomed flask equipped with a stirring bar are added p-methoxybenzamide (10 g, 66 mmol), N-bromosuccinimide (NBS) (11.9 g, 66 mmol), 1,8-diazabicyclo[5.4.0]undec-7-ene (DBU) (22 mL, 150 mmol) and methanol (300 mL) (Note 1). The solution is heated at reflux for 15 min (Note 2), at which point an additional aliquot of NBS (11.9 g, 66 mmol) is added slowly. The reaction is allowed to continue for another 30 min (Note 3). Methanol is removed by rotary evaporation and the residue is dissolved in 500 mL of ethyl acetate (EtOAc). The EtOAc solution is washed with 6 N hydrochloric acid (HCl) (2 x 100 mL), 1 N sodium hydroxide (NaOH) (2 x 100 mL) and saturated sodium chloride (NaCl), and then dried over magnesium sulfate (MgSO$_4$). The solvent is removed by rotary evaporation and the product, methyl N-(p-methoxyphenyl)carbamate, is purified by flash column chromatography [50 g of silica gel, EtOAc / hexane (1:1)] to give a pale yellow solid (11.1 g, 93%), which is further purified by recrystallization from 500 mL of hexane (Note 4). Another

1.4 g of product (total 8.8 g, 73%) is obtained from the mother liquor by recrystallization from 100 mL of hexane.

2. Notes

1. p-Methoxybenzamide, NBS and DBU were purchased from the Aldrich Chemical Company, Inc.

2. The reaction mixture is heated using an oil bath.

3. Reaction progress can be followed by thin layer chromatography using EM Science silica gel 60 F254 aluminum-backed plates, eluted with EtOAc/hexane (1:1) and visualized using a 254 nm UV lamp.

4. Methyl N-(p-methoxyphenyl)carbamate prepared by this procedure was characterized as follows: mp (uncorr.) 88.5-89.5°C (lit.[2] mp 88-89°C); ^{1}H NMR (500 MHz, CDCl$_3$) δ: 3.77 (s, 3 H), 3.78 (s, 3 H), 6.50 (bs, 1 H), 6.90 (m, 2 H), 7.26 (m, 2 H); ^{13}C NMR (100 MHz, CDCl$_3$) δ: 52.1, 55.3, 114.1, 120.6, 130.9, 154.5, 155.8; FTIR spectrum (CHCl$_3$) cm^{-1}: 3437, 3080, 2963, 1734, 1600, 1511, 1464, 1298, 1226, 1181, 1076, 1035; Calcd for C$_9$H$_{11}$NO$_3$: C, 59.7; H, 6.1; N, 7.7. Found: C, 59.6; H, 6.3; N, 7.7.

Waste Disposal Information

All toxic materials were disposed of in accordance with "Prudent Practices in the Laboratory"; National Academy Press; Washington, DC, 1995.

3. Discussion

This method offers a rapid, efficient and particularly mild preparation of methyl carbamates, and has been used with success with a variety of primary amides[3] (see Table). These carbamates are easily hydrolyzed to free amines, making this method particularly useful for making ^{15}N-labeled anilines containing electron-donating substituents, such as p-anisidine, 2,4-dimethoxyaniline and 4-dimethylaminoaniline. Although other methods are available for the preparation of methyl carbamates, including the use of NaOH/Br$_2$,[4] iodine(III) species,[5] lead tetraacetate,[6] and NBS-Hg(OAc)$_2$,[7] none of these methods are mild enough to permit the clean conversion of p-methoxybenzamide, since they cause further oxidation of the product. Currently, the best alternative method is the use of NBS/NaOMe reported earlier by this laboratory.[8]

1. Département de chimie, Université de Montréal, Montréal, Québec, Canada H3C 3J7.

2. Esch, P. M.; Hiemstra, H.; Speckamp, W. N. *Tetrahedron* **1992**, *48*, 3445-3462.

3. Huang, X.; Seid, M.; Keillor, J. W. *J. Org. Chem.* **1997**, *62*, 7495-7496.

4. Wallis, E .S.; Lane, J. F. *Org. React.* **1946**, *3*, 267-306.

5. Moriarty, R. M.; Chany II, C. J.; Vaid, R. K.; Prakash, O.; Tuladhar, S. M. *J. Org. Chem.* **1993**, *58*, 2478-2482.

6. Baumgarten, H. E.; Smith, H. L.; Staklis, A. *J. Org. Chem.* **1975**, *40*, 3554-3561.

7. Jew, S.-S.; Park, H. G.; Park, H.-J.; Park, M.-S.; Cho, Y.-S. *Tetrahedron Lett.* **1990**, *31*, 1559-1562.

8. Huang, X.; Keillor, J. W. *Tetrahedron Lett.* **1997**, *38*, 313-316.

Appendix

Chemical Abstracts Nomenclature (Collective Index Number);
(Registry Number)

Methyl N-(p-methoxyphenyl)carbamate: Carbamic acid, (4-methoxyphenyl)-, methyl ester (9); (14803-72-6)

p-Methoxybenzamide: Benzamide, 4-methoxy- (9); (3424-93-9)

N-Bromosuccinimide: Succinimide, N-bromo- (8); 2,5-Pyrrolidinedione, 1-bromo- (9); (128-08-5)

1,8-Diazabicyclo[5.4.0]undec-7-ene: DBU: Pyrimido[1,2-a]azepine, 2,3,4,6,7,8,9,10-octahydro- (8,9); (6674-22-2)

TABLE

CONVERSION OF AMIDES TO METHYL CARBAMATES WITH NBS/DBU

R[a]	Yield,[b] (%)	Obs. mp,[c] (°C)	Lit. mp, (°C)
3,4-$(MeO)_2C_6H_3$-	89	80-81	81 [d]
p-MeC_6H_4-	84	98-99	99-101[e]
C_6H_5-	95	45-46	47-48.5[e]
p-ClC_6H_4-	94	113-115	115-117[e]
p-$NO_2C_6H_4$-	70[f]	177-178	177.5-178[g]
$C_6H_5CH_2$-	95	64-65	65[h]
$CH_3(CH_2)_8$-	90	< r.t.	-
$CH_3(CH_2)_{14}$-	73	61-62	61-62[i]

[a]Prepared from the corresponding carboxylic acid or acid chloride.

[b]Refers to pure isolated and characterized product.

[c]Determined in capillary tubes and uncorrected.

[d]Brunner, O.; Wöhrl, R. Monatsh. **1933**, 63, 374-384.

[e]Fujisaki, S.; Tomiyasu, K.; Nishida, A.; Kajigaeshi, S. Bull. Chem. Soc. Jpn. **1988**, 61, 1401-1403.

[f]Overnight reflux.

[g]Hegarty, A. F.; Frost, L. N. J. Chem. Soc., Perkin Trans. 2 **1973**, 1719-1728.

[h]Chabrier de la Saulnière, P. Ann. Chim. **1942**, 17, 353-370.

[i]Reference 3.

TRIBUTYLSTANNANE (Bu₃SnH)-CATALYZED BARTON-McCOMBIE DEOXYGENATION OF ALCOHOLS: 3-DEOXY-1,2:5,6-BIS-O-(1-METHYLETHYLIDENE)-α-D-RIBO-HEXOFURANOSE

(α-D-ribo-Hexofuranose, 3-deoxy-1,2:5,6-bis-O-(1-methylethylidene)-)

Submitted by Jordi Tormo and Gregory C. Fu.[1]

Checked by Jan W. Thuring and Andrew B. Holmes.

1. Procedure

A. 1,2:5,6-Bis-O-(1-methylethylidene)-O-phenyl carbonothioate-α-D-gluco-furanose (1). All glassware is oven-dried. A 250-mL, two-necked, round-bottomed flask is fitted with a magnetic stirring bar and an argon inlet. Under a gentle flow of argon, the flask is charged with 100 mL of anhydrous dichloromethane (CH₂Cl₂) (Note 1) and 12.6 g (48.2 mmol) of 1,2:5,6-di-O-isopropylidene-D-glucose (Note 2). The flask is immersed in an ice bath, stirring is started, and 7.34 mL (53.1 mmol) of phenyl chlorothionoformate (Note 3) and 4.63 mL (57.9 mmol) of pyridine (Note 4) are added by syringe. After 30 min, the ice bath is removed, and the resulting mixture is stirred at room temperature for 14 hr. Then, in order to destroy the excess phenyl chlorothionoformate, 5 mL of anhydrous methanol (MeOH) (Note 5) is added via syringe, and the mixture is stirred at room temperature for 15 min. The resulting solution is washed with 100 mL of aqueous hydrochloric acid (HCl) (1 N) and 100 mL of brine, dried over anhydrous sodium sulfate (Na_2SO_4), filtered, and concentrated under reduced pressure. The crude product (20.5 g; yellow oil) is triturated by the addition of hexane (50 mL), scratched with a spatula and stirred for 30 min at 0°C. The cream-colored solid is collected by filtration of the cold suspension and washed with a minimal amount (5-10 mL) of hexane to give 15.2 g of **1**. Recrystallization from hexane (50 mL) affords the product as white crystals (13.9 g; 74%). The mother liquors are combined and evaporated to dryness to give a residue that is recrystallized from hexane (10 mL) to give a second crop of **1** (1.50 g; 8%) (Notes 6, 7 and 8).

B. *3-Deoxy-1,2:5,6-bis-O-(1-methylethylidene)-α-D-ribo-hexofuranose* (**2**). All glassware is oven-dried. A 250-mL, two-necked, round-bottomed flask is fitted with a magnetic stirring bar and a reflux condenser with an argon inlet. Under a gentle flow of argon, the flask is charged with 15 mL of anhydrous benzene (Note 9) and 13.0 g (32.7 mmol) of **1**. To this mixture are added via cannula a solution of 0.620 mL (1.21 mmol) of bis(tributyltin) oxide [$(Bu_3Sn)_2O$] (Note 10), 0.800 g (4.90 mmol) of 2,2'-azobis(isobutyronitrile) (AIBN) (Note 11), 9.81 g (164 mmol) of poly(methylhydrosiloxane) (PMHS) (Note 12), and 16.4 mL (180 mmol) of 1-butanol (Note 13) in 20 mL of anhydrous benzene. The resulting mixture is heated at reflux for 3 hr, after which time a solution of 0.620 mL (1.21 mmol) of $(Bu_3Sn)_2O$ and 0.800 g (4.90 mmol) of AIBN in 9 mL of anhydrous benzene is added via cannula. The reaction mixture is heated at reflux for an additional 3 hr. The solution is allowed to cool to room temperature, and then transferred to a 1-L, one-necked, round-bottomed flask. Benzene and excess 1-butanol are removed under water pump vacuum, and the resulting residue is redissolved in 100 mL of THF (Note 14). To this solution is added SLOWLY (Note 15) 400 mL of aqueous sodium hydroxide (NaOH) (2 N). The resulting mixture is stirred at room temperature for 15 hr. The layers are separated, and the aqueous layer is extracted twice with 100 mL of diethyl ether. The combined organic layers are washed with 100 mL of aqueous HCl (1 N) and 100 mL of brine, dried over anhydrous Na_2SO_4, filtered, and concentrated under reduced pressure. The resulting residue is purified by chromatography (Note 16) to give 6.2-6.4 g (76-80%) of **2** as a pale-yellow oil (Notes 17 and 18).

2. Notes

1. Laboratory grade dichloromethane was first distilled and then was further purified and dried by distillation from calcium hydride. The submitters used solvent supplied by EM Science.

2. 1,2:5,6-Di-O-isopropylidene-D-glucose (diacetone-D-glucose; Aldrich Chemical Company, Inc., 98%) was used without purification.

3. Phenyl chlorothionoformate (Aldrich Chemical Company, Inc., 99%) was used without purification.

4. Pyridine (Aldrich Chemical Company, Inc.) was purified by distillation from calcium hydride.

5. Methanol was used as supplied by Merck & Company, Inc. or Mallinckrodt Inc.

6. Both crops were pure as judged by elemental analysis. The combined yield ranged from 15.4-17.3 g (82-90%). The checkers found the recrystallization procedure to be more convenient (albeit in slightly lower yield) than isolation by chromatography. The residue can alternatively be purified by flash column chromatography with 30 g of

241

silica (Merck 9385 Kieselgel 60, 230-400 ASTM) per g of residue, eluting with a gradient of 0 → 50% EtOAc/hexane to afford 18.0 g (94%; the checkers obtained 91% on half scale) of **1** as a colorless solid, mp 108-110°C (Notes 7 and 8).

7. TLC analyses (R_f = 0.47, in EtOAc/hexane 7:3; the submitters observed R_f = 0.49 in EtOAc/hexane 4:1) were performed on 0.25-mm Merck 60 F_{254} silica gel plates (the submitters used 0.25-mm silica gel 60 plates supplied by EM Reagents) that were stained with a solution of phosphomolybdic acid in 95% ethanol.

8. Compound **1** ($[\alpha]_D^{18}$ -44.9° (CHCl$_3$, c 0.6), lit.[2] $[\alpha]_D^{18}$ -43° (CHCl$_3$)) has the following spectral data: ^1H NMR (250 MHz, CDCl$_3$) δ: 1.35 (s, 3 H), 1.38 (s, 3 H), 1.45 (s, 3 H), 1.56 (s, 3 H), 4.04-4.15 (m, 2 H), 4.31 (m, 2 H), 4.78 (d, 1 H, J = 4), 5.65 (d, 1 H, J = 2), 5.97 (d, 1 H, J = 4), 7.12 (d, 2 H, J = 8), 7.31 (t, 1 H, J = 8), 7.44 (t, 2 H, J = 7); ^{13}C NMR (62.5 MHz, CDCl$_3$) δ: 25.7, 26.7, 27.1, 27.3, 67.5, 72.7, 80.1, 83.3, 85.5, 105.4, 109.9, 112.9, 122.1, 127.2, 130.0, 153.7, 194.1; IR (KBr pellet) cm^{-1}: 2987, 1164, 1082, 948, 917. HRMS (FAB) m/z Calcd for C$_{19}$H$_{24}$O$_7$S: 396.1243. Found: 396.1243. Anal. Calcd for C$_{19}$H$_{24}$O$_7$S: C, 57.6; H, 6.1. Found: C, 57.5; H, 6.1.

9. Benzene (Aldrich Chemical Company, Inc. or EM Science) was purified by distillation from sodium benzophenone ketyl. CAUTION: Benzene is harmful as a vapor and by skin absorption, and it should always be handled in a well-ventilated hood. Chronic exposure may cause fatal blood disease.

10. (Bu$_3$Sn)$_2$O (Aldrich Chemical Company, Inc. or Gelest) was distilled [bp 140-142°C (0.1 mm)] before its use.

11. 2,2'-Azobis(isobutyronitrile) (98%; Aldrich Chemical Company, Inc.) was used without purification.

12. Poly(methylhydrosiloxane) (Fluka Chemical, Corp.) was degassed overnight under full vacuum at 25°C (~0.1 mm) prior to use.

13. Anhydrous 1-butanol (99.8%; Aldrich Chemical Company, Inc.) was used

without purification.

14. Tetrahydrofuran (Merck or EM Science) was used without purification.

15. The addition of base should be slow, as the aqueous NaOH (2 N) cleaves the siloxanes and ionizes the phenol that is generated in the reaction. The reaction is vigorous.

16. Flash column chromatography is performed with 30 g of silica (Merck 9385 Kieselgel 60, 230-400 ASTM) per gram of residue, eluting with a gradient of 0 → 50% EtOAc/hexane.

17. TLC analyses (R_f = 0.32, in hexane:EtOAc 7:3; the submitters observed R_f = 0.33 in hexane:EtOAc 4:1) were performed on 0.25-mm Merck 60 F_{254} silica gel plates (the submitters used 0.25-mm silica gel 60 plates supplied by EM Reagents) that were stained with a solution of phosphomolybdic acid in 95% ethanol.

18. Compound **2** ($[\alpha]_D^{20}$ -8.1° (CHCl$_3$, c 2.4), lit.[3] $[\alpha]_D^{18}$ -8.5° (CHCl$_3$, c 1.5)) has the following spectral data: ^{1}H NMR (250 MHz, CDCl$_3$) δ: 1.32 (s, 3 H), 1.36 (s, 3 H), 1.43 (s, 3 H), 1.51 (s, 3 H), 1.77 (m, 1 H), 2.19 (dd, 1 H, J = 4, 14), 3.82 (m, 1 H), 4.12 (m, 3 H), 4.75 (t, 1 H, J = 4), 5.81 (d, 1 H, J = 4); ^{13}C NMR (62.5 MHz, CDCl$_3$) δ: 25.5, 26.5, 26.8, 27.1, 35.6, 67.6, 77.2, 79.0, 80.8, 106.0, 110.0, 111.7; IR (neat) cm^{-1}: 1064, 957, 941. HRMS (FAB) m/z Calcd for C$_{12}$H$_{20}$O$_5$: 244.1311. Found: 244.1312. Anal. Calcd for C$_{12}$H$_{20}$O$_5$: C, 59.0; H, 8.2. Found: C, 59.0; H, 8.3.

Waste Disposal Information

All toxic materials were disposed of in accordance with "Prudent Practices in the Laboratory"; National Academy Press; Washington, DC, 1995.

3. Discussion

The Barton-McCombie protocol for the deoxygenation of alcohols[4,5] is an extremely useful method that has found widespread application in synthetic organic chemistry.[6] This radical-mediated process typically employs 1.5-3 equiv of Bu_3SnH[7] as the reducing agent. Because some tributyltin-containing compounds are toxic,[8] and product isolation from large quantities of organotin residues can be difficult,[9] the development of alternative reducing agents to Bu_3SnH has been an active area of investigation. Indeed, it has been established that silicon hydrides[10] and dialkyl phosphites[11] can serve as substitutes for Bu_3SnH in many instances. Despite these facts, Bu_3SnH continues to be the reagent most commonly used for effecting this reduction.[12]

Given this, the development of a reaction variant in which Bu_3SnH is employed as a *catalyst*, while a non-toxic second metal hydride serves as the stoichiometric reductant, has significant practical advantages. The procedure reported here uses 15 mol% Bu_3SnH [generated in situ from $(Bu_3Sn)_2O$[13]] in conjunction with poly(methylhydrosiloxane) (PMHS)[14,15] (for the proposed catalytic cycle, see the figure below).[16]

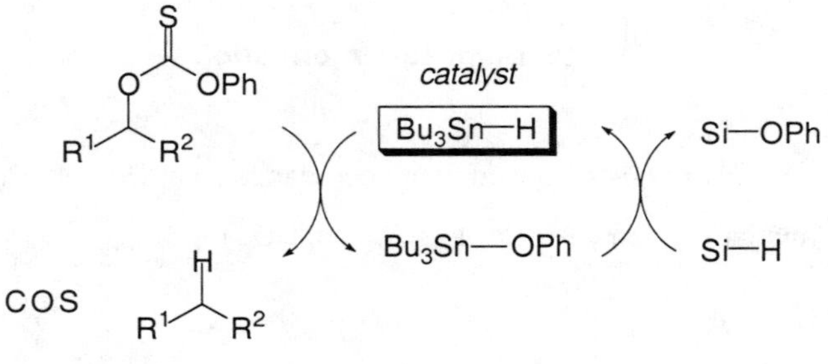

Additional applications of this method to the deoxygenation of secondary alcohols are provided in the Table.

TABLE

Substrate	Product	Isolated Yield (%)
		66
		70
		63
		68

1. Department of Chemistry, Massachusetts Institute of Technology, Cambridge, MA 02139.

2. Shasha, B. S.; Doane, W. M; Russell, C. R.; Rist, C. E. *Carbohyd. Res.* **1968**, *6*, 34-42.

3. Sano, H.; Takeda, T.; Migita, T. *Synthesis* **1988**, 402-403.

4. Barton, D. H. R.; McCombie, S. W. *J. Chem. Soc., Perkin Trans. 1* **1975**, 1574-1585.

5. The use of phenyl thionocarbonate esters was pioneered by Robins: (a) Robins, M. J.; Wilson, J. S. *J. Am. Chem. Soc.* **1981**, *103*, 932-933; (b) Robins, M. J.; Wilson, J. S.; Hansske, F. *J. Am. Chem. Soc.* **1983**, *105*, 4059-4065.

6. (a) Hartwig, W. *Tetrahedron* **1983**, *39*, 2609-2645; (b) McCombie, S. W. In "Comprehensive Organic Synthesis"; Trost, B. M., Ed.; Pergamon: New York, 1991; Vol. 8, Chapter 4.2; (c) Crich, D.; Quintero, L. *Chem. Rev.* **1989**, *89*, 1413-1432; (d) Pereyre, M.; Quintard, J.-P.; Rahm, A. "Tin in Organic Synthesis"; Butterworths: Boston, 1987; Chapter 5.

7. For reviews of the chemistry of Bu$_3$SnH, see: (a) Neumann, W. P. *Synthesis* **1987**, 665-683; (b) RajanBabu, T. V. In "Encyclopedia of Reagents for Organic Synthesis"; Paquette, L. A., Ed.; Wiley: New York, 1995; Vol. 7, pp. 5016-5023.

8. (a) De Mora, S. J. "Tributyltin: Case Study of an Environmental Contaminant"; Cambridge University Press: Cambridge, UK, 1996; (b) Boyer, I. J. *Toxicology* **1989**, *55*, 253-298.

9. For a succinct discussion, see: Crich, D.; Sun, S. *J. Org. Chem.* **1996**, *61*, 7200-7201.

10. For an overview of the use of silanes as reducing agents in the Barton-McCombie deoxygenation, see: Chatgilialoglu, C.; Ferreri, C. *Res. Chem. Intermed.* **1993**, *19*, 755-775.

11. Barton, D. H. R.; Jang, D. O.; Jaszberenyi, J. Cs. *J. Org. Chem.* **1993**, *58*, 6838-6842.

12. This statement is based on a search of the Beilstein Crossfire database.

13. Prices from Aldrich Chemical Company, Inc., per mole of tin are as follows: $(Bu_3Sn)_2O$: \$38; Bu_3SnH: \$250. Unlike Bu_3SnH, $(Bu_3Sn)_2O$ is not sensitive to light, O_2, or adventitious impurities.

14. Prices from Aldrich Chemical Company, Inc., per mole of hydride are as follows: PMHS: \$6; Bu_3SnH: \$250.

15. D_{50} of PMHS: 80 g/kg: Klyaschitskaya, A. L.; Krasovskii, G. N.; Fridlyand, S. A. *Gig. Sanit.* **1970**, *35*, 28-31; *Chem. Abs.* **1970**, *72*, 124864r.

16. Lopez, R. M.; Hays, D. S.; Fu, G. C. *J. Am. Chem. Soc.* **1997**, *119*, 6949-6950. The Bu_3SnH-catalyzed deoxygenation of dithiocarbonates requires more vigorous reaction conditions.

Appendix

Chemical Abstracts Nomenclature (Collective Index Number); (Registry Number)

Tributylstannane: Stannane, tributyl- (8,9); (688-73-3)

3-Deoxy-1,2:5,6-bis-O-(methylethylidene)-α-D-ribo-hexofuranose: D-ribo-Hexofuranose, 3-deoxy-1,2:5,6-bis-O-isopropylidene, α- (8); α-D-ribo-Hexofuranose, 3-deoxy-1,2:5,6-bis-O-(1-methylethylidene)- (9); (4613-62-1)

1,2:5,6-Bis-O-(1-methylethylidene)-, O-phenyl carbonothioate-α-D-glucofuranose: Glucofuranose, 1,2:5,6-di-O-isopropylidene, O-phenyl thiocarbonate, α-D- (8); α-D-Glucofuranose, 1,2:5,6-bis-O-(1-methylethylidene)-, O-phenyl thiocarbonate (9); (19189-62-9)

1,2:5,6-Di-O-isopropylidene-D-glucose: Glucofuranose 1,2:5,6-di-O-isopropylidene-α-D- (8); α-D-Glucofuranose, 1,2:5,6-bis-O-(1-methylethylidene)- (9); (582-52-5)

Phenyl chlorothionoformate: Formic acid, chlorothio-, O-phenyl ester (8); Carbonochloridothioic acid, O-phenyl ester (9); (1005-56-7)

Pyridine (8,9); (110-86-1)

Benzene: CANCER SUSPECT AGENT (8,9); (71-43-2)

Bis(tributyltin) oxide: Distannoxane, hexabutyl- (8,9); (56-35-9)

Azobisisobutyronitrile: Propionitrile, 2,2'-azobis[2-methyl- (8); Propanenitrile, 2,2'-azobis[2-methyl- (9); (78-67-1)

Poly(methylhydrosiloxane): PMHS: Poly[oxy(methylsilylene)] (8,9); (9004-73-3)

1-Butanol: Butyl alcohol (8); 1-Butanol (9); (71-36-3)

2-(3-OXOBUTYL)CYCLOPENTANONE-2-CARBOXYLIC
ACID ETHYL ESTER
(Cyclopentanecarboxylic acid, 2-oxo-1-(3-oxobutyl)-, ethyl ester)

Submitted by Jens Christoffers.[1]

Checked by Richard Heid and Edward J. J. Grabowski.

1. Procedure

A 50-mL, round-bottomed flask (Note 1), equipped with a magnetic stirring bar, is charged with cyclopentanone-2-carboxylic acid ethyl ester (25.0 g, 160 mmol) (Note 2) and iron(III)chloride hexahydrate (865 mg, 3.20 mmol). The flask is kept in a water bath at room temperature (external temperature) (Note 3), and methyl vinyl ketone (MVK) (15.0 mL, 12.7 g, 182 mmol) (Notes 4 and 5) is added within 1 hr using a syringe pump. The resulting mixture is stirred for 12 hr at room temperature, then all volatile materials are removed under reduced pressure from the reaction mixture (Note 6) at room temperature for 3 hr with continued stirring. Subsequently, the flask is equipped with a Claisen top and condenser and the product is distilled under high vacuum (Note 7). The distillate is collected in a single receiver flask to afford 33.3-33.7 g (91-93%) of analytically pure 2-(3-oxobutyl)cyclopentanone-2-carboxylic acid ethyl ester (Notes 8 and 9).

2. Notes

1. The reaction flask must be wide-necked to facilitate rapid distillation.

2. All starting materials were purchased from the Aldrich Chemical Company, Inc., and used without further purification.

3. A water cooling bath is required to prevent the volatile MVK from being evolved, since the reaction is slightly exothermic.

4. MVK is a hazardous and toxic material. All operations must be carried out in a hood.

5. A little excess of MVK (1.1 equiv) is required, since this starting material is very volatile. To obtain a very pure product, it is easier to remove an excess of the Michael acceptor instead of the donor.

6. An excess of MVK is removed as well as small amounts of decomposition (by hydrolysis) product cyclopentanone.

7. The bp of the product is 130°C at 1 mm. An oil bath temperature of 160-170°C is necessary to achieve rapid distillation. Temperatures above 190°C lead to decomposition, although a bath temperature of 200°C and the use of a heat gun at the end might be necessary to transfer the distillate completely into one receiver flask. Moreover, vigorous stirring during distillation is advisable, since the compound tends to delayed boiling.

8. The distillate is pure by elemental analysis and is free from solvent contamination. The physical properties are as follows: $C_{12}H_{18}O_4$ (226.27): Anal. Calcd for C, 63.70; H, 8.02. Found C, 63.48; H, 7.93; Mol. mass calcd. 226.1205, found 226.1207 (HRMS). Spectral data: IR (ATR) cm^{-1}: 2976 (m), 1748 (vs), 1717 (vs), 1448 (m), 1406 (m), 1367 (m), 1318 (m), 1260 (s), 1232 (s), 1165 (s), 1116 (m), 1029 (m), 861 (m); ^{1}H NMR (400 MHz, CDCl$_3$) δ: 1.23 (t, 3 H, J = 7.2), 1.82 - 2.03 (m, 4 H), 2.03 - 2.13 (m, 1 H), 2.12 (s, 3 H), 2.24 - 2.49 (m, 4 H), 2.69 (ddd, 1 H, J = 18, J =

9.6, J = 6.0), 4.14 (q, 2 H, J = 7.1); ^{13}C NMR (50 MHz, CDCl$_3$) δ: 13.29 (CH$_3$), 18.84 (CH$_2$), 26.24 (CH$_2$), 29.00 (CH$_3$), 33.22 (CH$_2$), 37.07 (CH$_2$), 38.01 (CH$_2$), 58.23 (C), 60.23 (OCH$_2$), 170.47 (C=O), 206.61 (C=O), 213.75 (C=O).

9. As an alternative to distillation and in accord with the observations of the submitter, the checkers have shown that the reaction mixture can be diluted with 100 mL of methyl t-butyl ether (MTBE), and filtered through a column of 150 g of silica gel with sufficient flushing by MTBE to remove all product. Concentration of the MTBE on a rotary evaporator followed by keeping the resulting oil at 1 mm/25° for 24 hr affords product of comparable purity, except for traces of MTBE, and slightly improved yield.

Waste Disposal Information

All toxic materials were disposed of in accordance with "Prudent Practices in the Laboratory"; National Academy Press; Washington, DC, 1995.

3. Discussion

Transition metal catalysis of the Michael reaction of 1,3-dicarbonyl compounds with acceptor activated alkenes has been known since the early 1980's.[2,3] It is a valuable alternative to the classic base catalysis of the reaction. Because of the mild and neutral conditions, the chemoselectivity of these reactions is superior to that provided by base catalysis, since the latter suffers from various unwanted side or subsequent reactions, such as aldol cyclizations, ester solvolyses or retro-Claisen type decompositions. A number of transition metal and lanthanide compounds have been reported to catalyze the Michael reaction, but FeCl$_3$ · 6 H$_2$O is one of the most efficient systems to date. A number of β-diketones or β-oxo esters and MVK are cleanly converted to the corresponding Michael reaction products within a few hours at room temperature, with quantitative yields being achieved in most cases.[4] No significant excess of the Michael acceptor is required, and the amount of catalyst employed can be as low as 1 mol%. Importantly, as long as the product and starting materials are liquid at room temperature, solvents are unnecessary. Moreover, the reaction can be performed without any need for anhydrous or inert conditions. Since no side reactions are observed, work-up and purification are very simple: either direct distillation of the product from the reaction mixture (as in the representative example shown here), or, if the volatility of the compound does not allow this, filtration through a short column of silica gel, which removes all iron-containing materials. There are of course also a number of other very efficient and mild systems for the catalysis of the Michael reaction.[5] However, $FeCl_3 \cdot 6 H_2O$ is the most readily available catalyst in this area, and also with respect to economical and ecological considerations, it is the transition metal compound of choice. Moreover, the procedure introduced here comes very close to an "ideal synthesis",[6] since starting materials are converted stoichiometrically and atom-economically without need of any reagents or even solvents and without generation of any stoichiometric by-product.

1.   Institut für Organische Chemie der Technischen Universität Berlin, Straße des 17. Juni 135, D-10623 Berlin, Germany.

2.   (a) Nelson, J. H.; Howells, P. N.; DeLullo, G. C.; Landen, G. L.; Henry, R. A. *J. Org. Chem.* **1980**, *45*, 1246-1249; (b) Watanabe, K.; Miyazu, K.; Irie, K. *Bull. Chem. Soc. Jpn.* **1982**, *55*, 3212-3215.

3.   Review: Christoffers, J. *Eur. J. Org. Chem.* **1998**, 1259-1266.

4.   (a) Christoffers, J. *J. Chem. Soc., Chem. Commun.* **1997**, 943-944; (b) Christoffers, J. *J. Chem. Soc., Perkin Trans. 1* **1997**, 3141-3149.

5. (a) Macquarrie, D. J. *Tetrahedron Lett.* **1998**, *39*, 4125-4128; (b) Boruah, A.; Baruah, M.; Prajapati, D.; Sandu, J. S. *Synth. Commun.* **1998**, *28*, 653-658; (c) see Ref. 3 for literature of 1997 or earlier.

6. Curran, D. P. *Angew. Chem., Int. Ed. Engl.* **1998**, *37*, 1175-1196.

Appendix
Chemical Abstracts Nomenclature (Collective Index Number); (Registry Number)

2-(3-Oxobutyl)cyclopentanone-2-carboxylic acid ethyl ester: Cyclopentanecarboxylic acid, 2-oxo-1-(3-oxobutyl)-, ethyl ester (10); (61771-81-1)

Cyclopentanone-2-carboxylic acid ethyl ester: Aldrich: Ethyl 2-oxocyclopentanecarboxylic acid: Cyclopentanecarboxylic acid, 2-oxo-, ethyl ester (8,9); (611-10-9)

Iron(III) chloride hexahydrate: Iron chloride, hexahydrate (8,9); (10025-77-1)

Methyl vinyl ketone: 3-Buten-2-one (8,9); (78-94-4)

1-OXO-2-CYCLOHEXENYL-2-CARBONITRILE

(1-Cyclohexene-1-acetonitrile, 6-oxo-)

Submitted by Fraser F. Fleming[1] and Brian C. Shook.[1]
Checked by Anne F. Vergne and Marvin J. Miller.

1. Procedure

1-Oxo-2-cyclohexenyl-2-carbonitrile. A dry, 100-mL, round-bottomed flask containing a magnetic stirring bar is fitted with an inlet adapter for ozonolysis (Note 1, Figure 1) and charged with 1-cyclopenteneacetonitrile (5.0 g, 46.7 mmol, Note 2) and 60 mL of dry dichloromethane (Note 3). A gentle stream of dry ozone is passed through the solution and the flask is immediately cooled to -78°C (Note 4). Ozonolysis is continued until the distinctive blue color of excess ozone is first observed, ozonolysis is then terminated, and the excess ozone is removed by purging with a stream of nitrogen for 5-10 min. The solution is allowed to warm to room temperature, the ozonolysis adapter is replaced with a rubber septum, and neat dimethyl sulfide (3.9 g, 62.1 mmol, Note 5) is added via syringe. The solution is allowed to stir at room temperature for 36 hr during which time the solution changes in color from a pale yellow to dark red. The resulting solution is concentrated under reduced pressure using a rotary evaporator, and the resulting thick, red syrup is diluted with 40 mL of ethyl acetate and washed with water (3 x 25 mL, Note 6). The aqueous phase is extracted with ethyl acetate (3 x 25 mL), the organic phases are combined, rinsed with brine in order to remove all DMSO, dried (MgSO$_4$), filtered, and concentrated. The residual red oil (5.55 g, 98%) contains only trace impurities and can be used without purification in most cases (Note 7). If required, further purification is achieved by rapid radial chromatography (Note 8) on a 2-mm plate using the solvent delivery tip designed for a 4-mm plate and eluting with 50% ethyl acetate-hexane (Note 9). The desired fractions are combined and concentrated to provide 1-oxo-2-cyclohexenyl-2-carbonitrile (4.8 g, 85% yield) as a pink oil (Note 10).

Figure 1

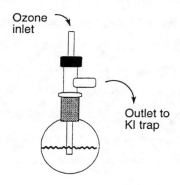

Ozone inlet

Outlet to KI trap

2. Notes

1. A short length of glass tubing (I. D. = 3 mm) is submerged (1 cm) beneath the surface of the solvent and the outlet tubing is immersed in a saturated solution of potassium iodide.[2]

2. 1-Cyclopenteneacetonitrile was purchased from Oakwood Products and purified by Kugelrhor distillation (50-60°C at 5 mm) prior to ozonolysis. 1-Cyclopenteneacetonitrile from other suppliers (Aldrich and Acros) was treated similarly.

3. Dichloromethane was distilled from calcium hydride.

4. The ozone is dried by passing the gas through a trap containing concentrated sulfuric acid. The ozonolysis is initiated prior to cooling to -78°C to prevent a vacuum from forming that would otherwise cause the potassium iodide solution to be drawn into the reaction flask.

5. The dimethyl sulfide was purchased from Aldrich Chemical Company, Inc., and used without purification.

6. This extraction procedure ensures removal of the dimethyl sulfoxide that is produced in the reaction.

7. The crude material reacts conjugately with phenylmagnesium bromide affording the addition product in 49% yield compared to 55% obtained using chromatographically pure 1-oxo-2-cyclohexenyl-2-carbonitrile.[3]

8. Rapid radial chromatography is essential since column chromatography results in significant irreversible adsorption of 1-oxo-2-cyclohexenyl-2-carbonitrile. For example, column chromatography of a relatively pure 3.0-g sample afforded only 0.5 g of pure 1-oxo-2-cyclohexenyl-2-carbonitrile. However, the checkers observed that a sample, purified as described by radial chromatography, survived flushing through a plug of silica gel.

9. The crude oxonitrile is dissolved in 10 mL of 50% ethyl acetate-hexane solution to afford a homogeneous solution. Incomplete removal of DMSO (Note 6) results in a two-phase mixture that, if loaded directly onto the silica plate, results in a diminished yield through partial absorption of the oxonitrile on the silica gel.

10. The product solidifies on standing at -4°C and can be stored neat at this temperature for several weeks or as a solution in benzene for at least two months. The spectral data are as follows: IR (film) cm^{-1}: 2233, 1698, 1615; ^{1}H NMR δ: 2.10 (br quintet, 2 H, J = 6), 2.53-2.61 (m, 4 H), 7.75 (t, 1 H, J = 4.2); ^{13}C NMR δ: 21.2 (t), 26.3 (t), 36.9 (t), 114.0 (s), 117.3 (s), 163.4 (d), 192.0 (s). MS m/e 122 (M + H$^+$).

Waste Disposal Information

All toxic materials were disposed of in accordance with "Prudent Practices in the Laboratory"; National Academy Press; Washington, DC, 1995.

3. Discussion

Highly electron-deficient alkenes are valuable reactants for both cycloaddition and conjugate addition reactions. The demand for highly reactive cycloalkenones has resulted in several syntheses of cycloalkenones containing an additional electron-withdrawing group on the α-carbon.[4] Oxonitriles represent an ideal compromise between high reactivity and stability toward storage and chromatography.[5]

The conversion of cyclopentenacetonitrile to 1-oxo-2-cyclohexenyl-2-carbonitrile proceeds by a domino ozonolysis-aldol sequence (Scheme 1).[6] Isolation and characterization of the ozonide[7] preclude the direct cyclization of the intermediate carbonyl oxide and establish that the cyclization occurs after the addition of dimethyl sulfide. Subsequent formation of the bis-oxonitrile (^{1}H NMR analysis) occurs rapidly and is followed by a slower cyclization-dehydration to afford 1-oxo-2-cyclohexenyl-2-carbonitrile.

Scheme 1

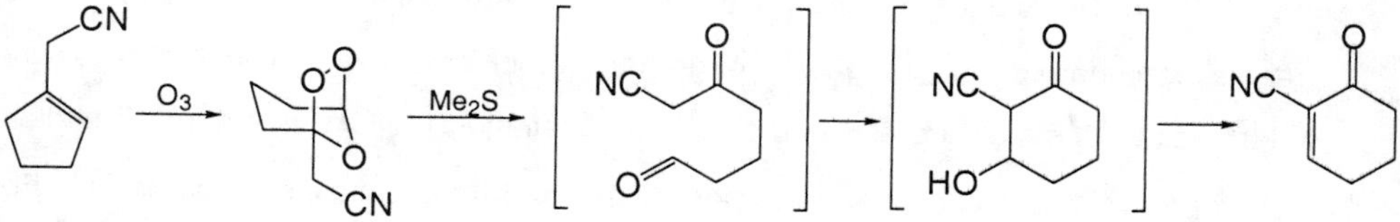

The domino ozonolysis-aldol sequence represents a general method for preparing 5-, 6-, and 7-membered cyclic oxonitriles (Scheme 2).[6] Cyclization to 6-membered ring oxonitriles proceeds as for the parent system, 1-oxo-2-cyclohexenyl-2-carbonitrile, whereas the corresponding 5- and 7-membered oxonitriles are less prone to cyclization. Cyclization of the 5- and 7-membered oxonitriles requires exposure of the aldehyde intermediates to acid or base in order to promote the geometrically more challenging cyclization.[6]

Scheme 2

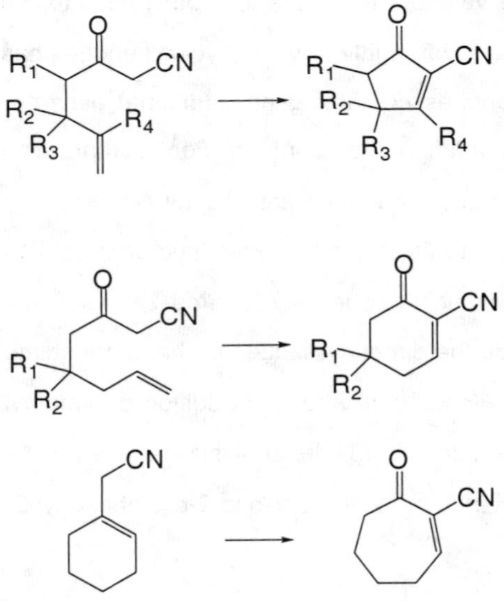

R_1 = Me, $R_2 = R_3 = R_4$ = H	83%
R_1 = H, $R_2 = R_3$ = Me, R_4 = H	91%
$R_1 = R_2 = R_3$ = H, R_4 = Me	91%

R_1 = H, R_2 = OTBDMS	85%
R_1 = Me, R_2 = OH.	53%
R_1 = Me, R_2 = OTMS	57%

58%

1-Oxo-2-cyclohexenyl-2-carbonitrile is an exceptional Michael acceptor that reacts conjugately with Grignard reagents without catalysis.[3] In most cases the intermediate enolates are silylated to afford substituted β-siloxy unsaturated nitriles, several of which are excellent precursors to cis- and trans-decalins (Scheme 3).[8] For example, unmasking the latent ketone enolate of siloxy unsaturated nitriles provides the cis-decalin in excellent yield while the trans-decalin is obtained from the same precursor by cyclizing the corresponding nitrile anion.

Scheme 3

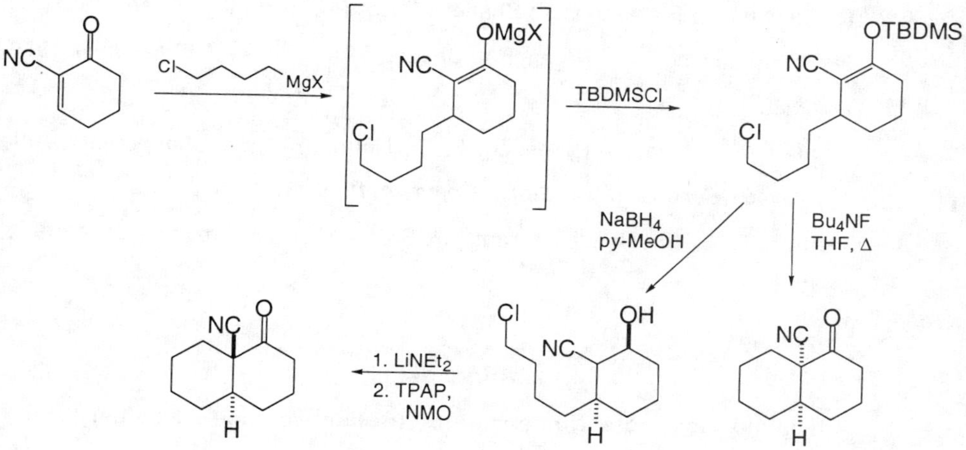

1. Department of Chemistry and Biochemistry, Duquesne University, Pittsburgh, PA 15282.

2. Belew, J. S. In "Oxidation: Techniques and Applications in Organic Synthesis"; Augustine, R. L., Ed.; Marcel Dekker, Inc.: New York, 1969; Vol. 1, p 294.

3. Fleming, F. F.; Pu, Y.; Tercek, F. *J. Org. Chem.* **1997**, *62*, 4883.

4. (a) Crimmins, M. T.; Huang, S.; Guise, L. E.; Lacy, D. B. *Tetrahedron Lett.* **1995**, *36*, 7061; (b) Funk, R. L.; Fitzgerald, J. F.; Olmstead, T. A.; Para, K. S.; Wos, J. A. *J. Am. Chem. Soc.* **1993**, *115*, 8849; (c) Schultz, A. G.; Harrington, R. E. *J. Am. Chem. Soc.* **1991**, *113*, 4926; (d) Posner, G. H.; Mallamo, J. P.; Hulce, M.; Frye, L. L. *J. Am. Chem. Soc.* **1982**, *104*, 4180; (e) Taber, D. F.; Amedio, Jr., J. C.; Sherrill, R. G. *J. Org. Chem.* **1986**, *51*, 3382.

5. The corresponding 5- and 6-membered ring esters are unstable to silica gel chromatography: (a) Liu, H. J.; Ngooi, T. K.; Browne, E. N. C. *Can. J. Chem.* **1988**, *66*, 3143; (b) Marx, J. N.; Cox, J. H.; Norman, L. R. *J. Org. Chem.* **1972**, *37*, 4489.

6. (a) Fleming, F. F.; Huang, A.; Sharief, V. A.; Pu, Y. *J. Org. Chem.* **1997**, 62, 3036; (b) Fleming, F. F.; Huang, A.; Sharief, V. A.; Pu, Y. *J. Org. Chem.* **1999**, *64*, 2830.

7. Tzou, J.–R.; Huang, A.; Fleming, F. F.; Norman, R. E.; Chang, S.-C. *Acta Crystallogr. Sect. C: Cryst. Struct. Commun.* **1996**, *C52*, 1012.

8. Fleming, F. F.; Shook, B. C.; Jiang, T.; Steward, O. W. *Organic Lett.* **1999**, *1*, 1547.

Appendix
Chemical Abstracts Nomenclature (Collective Index Number); (Registry Number)

1-Oxo-2-cyclohexenyl-2-carbonitrile: 1-Cyclohexene-1-carbonitrile, 6-oxo- (11); (91624-93-0)

1-Cyclopenteneacetonitrile: 1-Cyclopentene-1-acetonitrile (9); (22734-04-9)

Ozone (8,9); (10028-15-6)

Dimethyl sulfide: Methyl sulfide (8); Methane, thiobis- (9); (75-18-3)

Unchecked Procedures

Accepted for checking during the period September 1, 1999
through September 1, 2000. An asterisk (*) indicates that
the procedure has been subsequently checked.

Previously, *Organic Syntheses* has supplied these procedures upon request. However, because of the potential liability associated with procedures which have not been tested, we shall continue to list such procedures but requests for them should be directed to the submitters listed.

2878R (Z)- and (E)-Bis(phenylsulfonyl)ethylene.
S. Cossu and O. De Lucchi, Dipartimento di Chimica, Università Ca' Foscari di Venezia, Dorsoduro 2137, I-30123 Venezia, Italy.

2879* Preparation of Secondary Amines from Primary Amines via 2-Nitrobenzenesulfonamides: N-(4-Methoxybenzyl)-3-phenylpropylamine.
W. Kurosawa, T. Kan, and T. Fukuyama, Graduate School of Pharmaceutical Sciences, The University of Tokyo, 7-3-1 Hongo, Bunkyo-ku, Tokyo 113-0033, Japan.

2892 1-tert-Butyloxycarbonyl-4-((9-fluorenylmethyloxycarbonyl)amino)-piperidine-4-carboxylic Acid. A Convenient Preparation of an Orthogonally Protected α,α-Disubstituted Amino Acid Analog of Lysine.
L. G. J. Hammarström and M. L. McLaughlin, Department of Chemistry, Louisiana State University, Baton Rouge, LA 70803.

2893 Synthesis of tris(2-Perfluorohexylethyl)tin Hydride [Stannane, tris-(3,3,4,4,5,5,6,6,7,7,8,8,8-tridecafluorooctyl)-]: A Highly Fluorinated Tin Hydride with Advantageous Features of Easy Purification.
A. Crombie, S.-Y. Kim, S. Hadida, and D. P. Curran, Department of Chemistry, University of Pittsburgh, Pittsburgh, PA 15260.

2898 (S)-3-(tert-Butyloxycarbonylamino)-4-phenylbutanoic Acid.
M. R. Linder, S. Steurer, and J. Podlech, Institut für Organische Chemie und Isotopenforschung, der Universität Stuttgart, Pfaffenwaldring 55, D-70569 Stuttgart, Germany.

2902R Synthesis of Indoles by Palladium Catalyzed Reductive N-Heteroannulation of 2-Nitrostyrenes: Methyl Indole-4-carboxylate.
B. C. Söderberg and J. A. Shriver, Department of Chemistry, P.O. Box 6045, West Virginia University, Morgantown, WV 26506-6045.

2903A Synthesis of 1,2:4,5-Di-O-Isopropylidene-D-erythro-2,3-hexodiulo-2,6-pyranose. A Highly Enantioselective Ketone Catalyst for Epoxidation.
Y. Tu, M. Frohn, Z.-X. Wang, and Y. Shi, Department of Chemistry, Colorado State University, Fort Collins, CO 80523.

2903B Asymmetric Epoxidation of trans-β-Methylstyrene and 1-Phenylcyclo-hexene Using a D-Fructose-Derived Ketone.
Z.-X. Wang, L. Shu, M. Frohn, Y. Tu, and Y. Shi, Department of Chemistry, Colorado State University, Fort Collins, CO 80523.

2904 Tetrabenzyl Pyrophosphate.
T. D. Nelson, J. D. Rosen, M. Bhupathy, J. McNamara, M. J. Sowa, C. Rush, L. S. Crocker, Merck Research Laboratories, Merck & Co., Inc., P.O. Box 2000, R801-205, Rahway, NJ 07065-0900.

2906R Asymmetric Reductive Ring Opening of an Oxabenzonorbornadiene. Synthesis of (R)- (1H,2H)Dihydronaphthalen-1-ol.
M. Lautens, G. Bouchain, and T. Rovis, Department of Chemistry, Lash Miller Chemical Laboratories, University of Toronto, Toronto, Ontario M5S 3H6, Canada.

2907R [4+3] Cycloaddition in Water: Synthesis of 2,4-Dimethyl-8-oxabicyclo[3.2.1]oct-6-en-3-one silane.
M. Lautens and G. Bouchain, Department of Chemistry,
Lash Miller Chemical Laboratories, University of Toronto,
Toronto, Ontario M5S 3H6, Canada.

2909 (5S)-(d-Menthyloxy-2(5H)-furanone.
O. M. Moradei and L. A. Paquette, Department of Chemistry,
The Ohio State University, Columbus, OH 43210.

2911 Cholest-5-en-3-one.
J. M. Harris, Y. Liu, and J. C. Vederas, Department of Chemistry,
University of Alberta. Edmonton, Alberta. Canada, T6G 2G2.

2912 Preparation of Thioamides by Addition of Sodium Hydrosulfide to Activated Amides.
A. B. Charette and M. Grenon, Département de Chimie, Université de Montréal, P.O. Box 6128, Station Downtown, Montréal, Québec, Canada H3C 3J7.

2914* 2-Amino-3-fluorobenzoic Acid.
M. Kollmar, R. Parlitz, S. Oevers, G. Helmchen, Organisch-Chemisches Institut, der Universität Heidelberg, Im Neuenheimer Feld 270, D-69120 Heidelberg, Germany.

2915R Synthesis of Acetylated Glycals under Aprotic Conditions.
A. H. Franz and P. H. Gross, Department of Chemistry, University of the Pacific, 3601 Pacific Ave., Stockton, CA 95211.

2917 Preparation of α-Acetoxy Ethers by the Reductive Acetylation of Esters: (endo)-1-Bornoxyethyl Acetate.
D. J. Kopecky and S. D. Rychnovsky, Department of Chemistry,
University of California, Irvine, CA 92697-2025.

2920* N-Hydroxy-4-(p-chlorophenyl)thiazole-2(3H)-thione.
J. Hartung and M. Schwarz, Institut für Organische Chemie, der Universität Würzburg, Am Hubland, D-97074 Würzburg, Germany.

2923R Mild and Selective Oxidation of Sulfur Compounds in Trifluoroethanol: Diphenyl Disulfide and Methyl Phenyl Sulfoxide.
V. Kesavan, K. S. Ravikumar, D. Bonnet-Delpon, and J.-P. Bégué, BIOCIS CNRS, Faculté de Pharmacie de Chatenay-Malabry, Rue J.B. Clément, 92296 Chatenay-Malabry, France.

2925 Ruthenium-Catalyzed Addition of an Aromatic Ketone at the Ortho C-H Bond to an Olefin: 8-[2-(Triethoxysilyl)ethyl]-α-tetralone.
F. Kakiuchi and S. Murai, Department of Applied Chemistry,
Osaka University, Suita, Osaka 565-0871, Japan.

2926R Preparation of 1-[N-Benzyloxycarbonyl-(1S)-1-amino-2-oxoethyl]-4-methyl-2,6,7-trioxabicyclo[2.2.2]octane.
N. G. W. Rose, M. A. Blaskovich, G. Evindar, S. Wilkinson, Y. Luo, C. Reid, and G. A. Lajoie, Department of Chemistry. University of Waterloo, Waterloo, ON, N2L 3G1, Canada.

2927 Preparation of 1-n-Butyl-3-methyl Imidazolium Based Room Temperature Ionic Liquids.
J. Dupont, C. S. Consorti, P. A. Z. Suarez, and R. F. deSouza, Laboratory of Molecular Catalysis, Institute of Chemistry -UFRGS, Av. Bento Goncalves, 9500 Porto Alegre 91510-970 RS, Brazil.

2930 4-Carbomethoxy-2-methyl-1,3-oxazole.
J. D. White, C. L. Kranemann, and P. Kuntiyong, Department of Chemistry, Oregon State University, Corvallis, OR 97331.

2931 Asymmetric Synthesis of (M)-2-Hydroxymethyl-1-(2-hydroxy-4,6-dimethylphenyl)naphthalene via a Configurationally Unstable Biaryl Lactone.
G. Bringmann, M.Breuning, P. Henschel, and J. Hinrichs, Institut für Organische Chemie, Universität Würzburg, Am Hubland, D-97074 Würzburg, Germany.

2932 (R)-3-Phenylcyclohexanone.
T. Hayashi, M. Takahashi, Y. Takaya, and M. Ogasawara, Department of Chemistry, Faculty of Science, Kyoto University, Sakyo, Kyoto 606-8502, Japan.

2933 Synthesis of (R,R)-4,6-Dibenzofurandiyl-2,2'-bis(4-phenyloxazoline) (DBFOX/PH) - A Novel Tridentate Ligand.
U. Iserloh, Y. Oderaotoshi, S. Kanemasa, and D. P. Curran, Department of Chemistry, University of Pittsburgh, Pittsburgh, PA 15260.

2941 Ethyl 3-(p-Cyanophenyl)propionate from Ethyl 3-Iodopropionate and p-Cyanophenylzinc Bromide.
A. E. Jensen, F. Kneisel, and P. Knochel, Department of Chemistry, Ludwig-Maximilians-Universität München, Butenandtstr. 5-13, Bldg. F, D-81377 München, Germany.

CUMULATIVE AUTHOR INDEX
FOR VOLUMES 75, 76, 77, AND 78

This index comprises the names of contributors to Volume **75, 76, 77, and 78** only. For authors to previous volumes, see either indices in Collective Volumes I through IX or the single volume entitled *Organic Syntheses, Collective Volumes I-VIII, Cumulative Indices*, edited by J. P. Freeman.

Acquaah, S. O., **75**, 201
Ahern, C., **77**, 220
Ahiko, T-a., **77**, 176
Akiba, T., **75**, 45
Alexakis, A., **76**, 23
Alvernhe, G., **76**, 159
Amouroux, R., **77**, 91
Andrews, D. M., **76**,37
Aujard, I., **76**, 23

Baek, D. N., **76**, 294
Bailey, W. F., **75**, 177
Baine, N. H., **77**, 198
Baker, T. J., **78**, 91
Barton, D. H. R., **75**, 124
Beck, A. K., **76**, 12
Bégué, J.-P., **75**, 153
Behrens, C., **75**, 106
Bender, D. R., **76**, 6
Bennett, G. D., **77**, 107
Beresis, R. T., **75**, 78
Bethell, D., **76**, 37
Bhatia, A. V., **75**, 184
Bierer, D. E., **77**, 186
Binger, P., **77**, 254
Bittman, R., **77**, 225
Blanchot, B., **76**, 239

Boeckman, Jr., R. K., **77**, 141
Bomben, A., **76**, 169
Bonnet-Delpon, D., **75**, 153
Boussaguet, P., **76**, 221
Boyle, R. W., **76**, 287
Brady, B., **77**, 220
Braun, M. P., **75**, 69
Brinker, U. H., **77**, 254
Brodney, M. A., **78**, 202
Brook, C. S., **75**, 189
Bruckner, C., **76**, 287
Buchwald, S. L., **78**, 23
Buck, J. R., **77**, 153, 162
Buono, G., **78**, 135
Burns, D. M., **77**, 50
Byun, H.-S., **77**, 225

Cahiez, G., **76**, 239
Capdevielle, P., **76**, 133
Carson, M. W., **75**, 177
Castaldi, M. J., **78**, 63
Cha, J. K., **78**, 212
Charette, A. B., **76**, 86
Chase, C. E., **75**, 161
Chau, F., **76**, 239
Chaudhary, S. K., **75**, 184
Chemla, F., **78**, 99

Chen, C., **77**, 12
Chen, C-y., **78**, 36
Chen, H., **77**, 29
Chen, K., **75**, 189
Choi, J.-R., **78**, 212
Collington, E. W., **76**, 37
Constantieux, T., **78**, 135
Corley, E. G., **77**, 231
Crimmins, M. T., **77**, 114

Dailey, W. P., **75**, 89, 98
Dallaire, C.. **78**, 42
Darcy, R., **77**, 220
Davis, F. A., **77**, 50
de Meijere, A., **78**, 142
Deprés, J.-P., **75**, 195
Devine, P. N., **76**, 101
Dolphin, D., **76**, 287
Dondoni, A., **77**, 64, 78
Drewes, M. W., **76**, 110
Dubenko, L. G., **77**, 186
Durand, P., **76**, 123

Ehrler, R., **77**, 236
Ejjiyar, S., **77**, 91
Ernet, T., **76**, 159
Enders, D., **78**, 169, 177, 189

Fales, K. R., **77**, 121
Fanelli, D. L., **77**, 50
Fengler-Veith, M., **78**, 123
Fleming, F. F., **78**, 254
Fraser, C. L, **78**, 51, 82
Fu, G. C., **78**, 239
Fujiwara, Y., **78**, 104
Fürstner, A., **76**, 142

Garofalo, A. W., **77**, 98
Gerber, R. E., **77**, 186
Gibson, D. T., **76**, 77
Gingras, M., **78**, 42
Gleason, J. L., **76**, 57
Goj, O., **76**, 159
Goodson, F. E., **75**, 61
Grabowski, E. J. J., **75**, 31
Greene, A. E., **75**, 195
Griesgraber, G., **77**, 1
Griesser, H., **77**, 236
Gysi, P., **76**, 12

Hara, S., **75**, 129
Hart, H., **75**, 201
Hartner, Jr., F. W., **75**, 31
Hasbun, C., **77**, 186
Haufe, G., **76**, 159
He,.S., **78**, 152, 160
Heer, J. P., **76**, 37
Hernandez, O., **75**, 184
Hill, P. D., **78**, 63
Hinkle, K. W., **77**, 98
Homsi, F., **77**, 206
Huang, X., **78**, 234
Huc, V., **76**, 221
Hudlicky, T., **76**, 77
Huff, B. E., **75**, 53
Hughes, D. L., **76**, 1, 6
Huntington, M., **77**, 231
Hupperts, A., **76**, 142
Hutchison, D. R., **75**, 223

Imamoto, T., **76**, 228
Ishiyama, T., **77**, 176

Jacobsen, E. N., **75**, 1; **76**, 46
Jäger, V., **77**, 236; **78**, 123
Jain, N. F., **75**, 78
James, B. R., **76**, 287
Jandeleit, B., **78**, 169, 177, 189
Jiao, G.-S., **78**, 225
Johnson, C. R., **75**, 69
Jones, B. P., **78**, 63.

Kanger, T., **76**, 23
Karoyan, P., **78**, 99
Kaszynski, P., **77**, 249
Kautz, U., **78**, 123
Kawatsura, M. **78**, 1
Keck, G. E., **75**, 12
Keillor, J. W. , **78**, 234
Khau, V. V., **75**, 223
King, M. F., **77**, 186
Kitamura, T., **78**, 104
Klumpp, D. A., **76**, 294

Koenig, T. M., **75**, 53
Kolber, I., **78**, 42
Kornilov, A., **75**, 153
Kozhushkov, S. I., **78**, 142
Kozmin, S. A., **78**, 152, 160
Krämer, B., **78**, 123
Krishnamurthy, D., **75**, 12
Kröger, S., **76**, 159
Kubo, K., **75**, 210

Lamba, J. J. S., **78**, 51
Lang, M., **78**, 113
Lang-Fugmann, S., **78**, 113
Larchevêque, M., **75**, 37

Larrow, J. F., **75**, 1; **76**, 46
Larsen, R. D., **78**, 36
Laurent, A., **76**, 159
La Vecchia, L., **76**, 12
Lebel, H., **76**, 86
Le Goffic, F., **76**, 123
Levin, M. D., **77**, 249
Ley, S. V., **75**, 170; **77**, 212
Liebeskind, L. S., **77**, 135
Lipshutz, B. H., **76**, 252
Liu, H., **76**, 189
Liu, Y.-Z., **76**, 151
Love, J. C., **78**, 51
Luo, F.-T., **75**, 146
Lynam, N., **77**, 220
Lynch, K. M., **75**, 89, 98
Lynch, S. M., **78**, 202

MacKinnon, J., **75**, 124
Mack, R. A., **77**, 45
Macor, J., **77**, 45
Makowski, T. W., **78**, 63
Mangeney, P., **76**, 23
Mann, J., **75**, 139
Marshall, J. A., **76**, 263; **77**, 98
Martinelli, M. J., **75**, 223
Maryanoff, B. E., **75**, 215
Maryanoff, C. A., **75**, 215
Maumy, M., **76**, 133
Mazerolles, P., **76**, 221
McComsey, D. F., **75**, 215
Meffre, P., **76**, 123
Mellinger, M., **77**, 198
Michl, J., **77**, 249
Mitchell, D., **75**, 53
Miyaura, N., **77**, 176

Moore, H. W., **76**, 189
Moore, J. R., **77**, 12
Mori, A., **75**, 210
Mullins, J. J., **77**, 141
Murata, M., **77**, 176
Myers. A. G., **76**, 57, 178; **77**, 22, 29

Nakai, T., **76**, 151
Nakajima, A., **76**, 199
Nayyar, N. K., **75**, 223
Neyer, G., **76**, 294
Novak, B. M., **75**, 61
Nowick, J. S., **78**, 220

Oh, J., **78**, 212
Oh, T., **76**, 101
Öhrlein, R., **77**, 236
Okabe, M., **76**, 275
Olah, G. A., **76**, 294
Osborn, H. M. I., **75**, 170; **77**, 212
O'Sullivan, T., **77**, 220

Padwa, A., **78**, 202
Page, P. C. B., **76**, 37
Panek, J. S., **75**, 78
Park, M., **77**, 153, 162
Paquette, L. A., **75**, 106; **77**, 107
Payack, J. F., **76**, 6;
Peña-Cabrera, E., **77**, 135
Perchet, R. N., **75**, 124
Perrone, D., **77**, 64, 78
Peterson, B. C., **75**, 223
Petit , Y., **75**, 37
Phillips, B. W., **75**, 19
Pierce, M. E., **77**, 12
Posakony, J., **76**, 287

Powell, N. A. **78**, 220
Powers, J. P., **77**, 1
Priepke, H. W. M., **75**, 170
Prudhomme, D. R., **77**, 153, 162

Qian, C.-P., **76**, 151

Ragan, J. A. , **78**, 63
Ravikumar, V. T., **76**, 271
Rawal, V. H., **78**, 152, 160
Reddy, G. V., **77**, 50
Reed, D. P., **78**, 73
Reetz, M. T., **76**, 110
Reider, P. J., **76**, 1, 6, 46
Rigby, J. H., **77**, 121
Rizzo, C. J., **77**153, 162
Robbins, M. A., **76**, 101
Roberts, E., **76**, 46
Robin, S., **77**, 206
Ross, B., **76**, 271
Rousseau, G., **77**, 206
Rousselet, G., **76**, 133
Ruel, F. S., **75**, 69
Ryan, K. M., **76**, 46
Rychnovsky, S. D., **77**, 1

Saluzzo, C., **77**, 91
Sampognaro, A. J., **77**, 45
Sasai, H., **78**, 14
Sattler, A., **76**, 159
Savage, S. A., **78**, 51
Schick, H., **75**, 116
Schwab, W., **77**, 236
Schwardt, O., **78**, 123
Schwickardi, R., **76**, 110
Seebach, D., **76**, 12

Sehon, C. A., **76**, 263
Seidel, G., **76**, 142
Selva, M., **76**, 169
Senanayake, C. H., **76**, 46
Shahlai, K., **75**, 201
Shao, P., **77**, 141
Sheldrake, P. W., **77**, 198
Shibasaki, M., **78**, 14
Shook, B. C., **78**, 254
Sisko, J., **77**, 198
Smith, III, A. B., **75**, 19, 189
Smith, A. P., **78**, 51, 82
Solomon, J. S., **75**, 78
Sorgi, K. L., **75**, 215
Späth, T., **78**, 142
Stabile, M. R., **76**, 77
Staszak, M. A., **75**, 53
Steglich, W., **78**, 113
Suffert, J., **76**, 214
Sullivan, K. A., **75**, 223
Sullivan, R. W., **76**, 189
Suzuki, A., **75**, 129
Suzuki, T., **78**, 14
Szewczyk, J. M., **77**, 50

Takaoka, L. R., **78**, 220
Takeda, K., **76**, 199
Takeda, M., **76**, 199
Takeda, N., **76**, 228
Takeshita, H., **75**, 210
Tamura, O., **75**, 45
Tan, L., **77**, 12
Taylor, C. M., **75**, 19
Terashima, S., **75**, 45
Thompson, A. S., **75**, 31; **77**, 231
Tillyer, R., **77**, 12

Tirado, R., **76**, 252
Todaka, M., **78**, 104
Tomioka, M. **78**, 91
Tomooka, C. S., **76**, 189
Tomooka, K., **76**, 151
Tormo, J., **78**, 239
Toussaint, D., **76**, 214
Tsai, J. H., **78**, 220
Tse, C.-L., **75**, 124
Tundo, P., **76**, 169

Uozumi, Y. **78**, 1

Venkatraman, S., **75**, 161
Verhoest, P. R., **77**, 45
Verhoeven, T. R., **76**, 1, 6, 46
von Berg, S., **78**, 169, 177, 189

Wallow, T. I., **75**, 61
Wang, M.-W., **75**, 146
Wang, R.-T., **75**, 146
Wang, Q., **76**, 294
Wang, Z., **77**, 153, 162
Warriner, S. L., **75**, 170
Washburn, D. G., **77**, 114
Watanabe, S., **78**, 14
Wedemann, P., **77**, 254
Wedler, C., **75**, 116
Weinreb, S. M., **75**, 161
Weisman, G. R., **78**, 73
Weymouth-Wilson, A. C., **75**, 139
Whited, G. M., **76**, 77
Wipf, P., **75**, 161
Wolfe, J. P., **78**, 23
Wong, M.-K., **78**, 225
Wood, M. R., **76**, 252

Xiong, Y., **76**, 189
Xu, F., **77**, 12
Xu, S. L., **76**, 189

Yager, K. M., **75**, 19
Yang, B. H., **77**, 22, 29
Yang, D., **78**, 225
Yang, M. G., **75**, 78
Yerxa, B. R., **76**, 189
Yip, Y.-C., **78**, 212
Yoshii, E., **76**, 199

Zarcone, L. M. J., **75**, 177
Zawacki, F. J., **77**, 114
Zhong, N., **77**, 225
Zhang, Y., **77**, 50
Zhao, D., **77**, 12
Zheng, B., **76**, 178
Ziani-Cherif, C., **78**, 212

CUMULATIVE SUBJECT INDEX
FOR VOLUMES 75, 76, 77 AND 78

This index comprises subject matter for Volumes **75**, **76**, **77**, and **78**. For subjects in previous volumes, see either the indices in Collective Volumes I through IX or the single volume entitled *Organic Syntheses, Collective Volumes I-VIII, Cumulative Indices*, edited by J. P. Freeman.

The index lists the names of compounds in two forms. The first is the name used commonly in procedures. The second is the systematic name according to **Chemical Abstracts** nomenclature, accompanied by its registry number in parentheses. While the systematic name is indexed separately, it also accompanies the common name. Also included are general terms for classes of compounds, types of reactions, special apparatus, and unfamiliar methods.

Most chemicals used in the procedure will appear in the index as written in the text. There generally will be entries for all starting materials, reagents, intermediates, important by-products, and final products. Entries in capital letters indicate compounds, reactions, or methods appearing in the title of the preparation.

ABNORMAL REACTIONS, SUPPRESSION OF, **76**, 228

Acetaldehyde; (75-07-0), **75**, 106

Acetaldehyde dimethyl acetal: Ethane, 1,1-dimethoxy-; (534-15-6), **75**, 46

Acetic acid, glacial; (64-19-7), **75**, 2, 225; **77**, 142

Acetic anhydride: Acetic acid anhydride; (108-24-7), **76**, 70; **77**, 45, 142

Acetone: 2-Propanone; (67-64-1), **76**, 13

Acetonitrile: TOXIC: (75-05-8), **75**, 146; **76**, 24, 47, 67, 191; **77**, 121; **78**, 83

Acetonylacetone: 2,5-Hexanedione; (110-13-4), **78**, 64

7α-Acetoxy-(1Hβ, 6Hβ)-bicyclo[4.4.1]undeca-2,4,8-triene: Bicyclo[4.4.1]undeca-3,7,9-trien-2-ol, acetate, endo- (±)-; (129000-83-5), **77**, 121

(E)-1-Acetoxy-1,3-butadiene: 1,3-Butadien-1-ol, acetate, (E)-; (35694-20-3), **77**, 122

1-Acetoxy-3-(methoxymethoxy)butane: 1-Butanol, 3-(methoxymethoxy)-, acetate; (167563-42-0), **75**, 177

Acetylacetaldehyde dimethyl acetal: 2-Butanone, 4,4-dimethoxy-; (5436-21-5), **78**, 152

Acetylacetone: 2,4-Pentanedione; (123-54-6), **78**, 135

ACETYLALLENE: see 3,4-PENTA-1,2-DIEN-2-ONE

9-Acetylanthracene: Ethanone, 1-(9-anthracenyl)-; (784-04-3), **76**, 276

3-Acetyl-6-butoxy-2H-pyran-2,4(3H)-dione: 2H-Pyran-2,4(3H)-dione, 6-butoxydihydro-3-(1-hydroxyethylidene)-; (182616-30-4), **77**, 115

Acetyl chloride; (75-36-5), **75**, 177; **77**, 65, 114; **78**, 99

1-Acetyl-1-cyclohexene: Ethanone, 1-(1-cyclohexen-1-yl)-; (932-66-1), **76**, 203

Acetylenes, terminal, cyanation of, **75**, 148

Acetyl Meldrum's acid: See 5-(1-hydroxyethylidene)-1,3-dioxane-4,6-dione

2-(4'-Acetylphenyl)thiophene, **77**, 135

Acrolein: 2-Propenal; (107-02-8), **77**, 237

Agar plate preparation, 76, 79

Aliquat 336: Ammonium, methyltrioctyl-, chloride; 1-Octanaminium, N-methyl-N,N-dioctyl-, chloride; (5137-55-3), **76**, 40

Alkyllithium solutions, to titrate, **76**, 68

Allene: 1,2-Propadiene; (463-49-0), **75**, 129

Allenes, stereodefined synthesis of, **76**, 185

ALLYLATION, CATALYTIC ASYMMETRIC, OF ALDEHYDES, **75**, 12

      table, **75**, 17

Allyl bromide: 1-Propene, 3-bromo-; (106-95-6), **76**, 60, 221

Allyl chloride: 1-Propene, 3-chloro-; (107-05-1), **77**, 254

L-ALLYLGLYCINE: 4-PENTENOIC ACID, 2-AMINO-, (R)-; (54594-06-8), **76**, 57

Allylic alcohols, enantioselective cyclopropanation, **76**, 97

ALLYLINDATION, **77**, 107

Allylmagnesium bromide: Magnesium, bromo-2-propenyl-; (1730-25-2), **76**, 221

Allylmagnesium bromide (ethereal complex solution), **76**, 221

Allylmagnesium bromide (THF complex solution), **76**, 222

Allyltributylstannane: Stannane, tributyl-2-propenyl-; (24850-33-7), **75**, 12

Aluminum chloride; (7446-70-0), **77**, 2

2-Amidofurans as Diels-Alder dienes, **78**, 207

Aminals, **76**, 30

Amination, palladium-catalyzed, of aryl halides and triflates, **78**, 23

      tables, **78**, 32, 34

Amine protecting (blocking) group, **78**, 69

Amino acid ester isocyanates, **78**, 220

α-Amino aldehydes:

      Boc-protected, **76**, 117

      N,N-Dibenzyl-protected, **76**, 115

Aminoallyl cations in Diels-Alder reactions, **78**, 212

α-Aminocarboxylic acids, **75**, 25

2-Amino-5-chlorobenzophenone: Methanone, (2-amino-5-chlorophenyl)phenyl-; (719-59-5), **76**, 142

(1S,2R)-(+)-2-Amino-1,2-diphenylethanol: Benzeneethanol, β-amino-α-phenyl-, [S-(R*,S*)]-; (23364-44-5), **75**, 45

(1S,2R)-1-AMINOINDAN-2-OL: 1H-INDEN-2-OL, 1-AMINO-2,3-DIHYDRO-, (1S-cis); (126456-43-7), **76**, 46

2-Amino-5-methylpyridine: 2-Pyridinamine, 5-methyl-; (1603-41-4), **78**, 52

(R)-(-)-1-Amino-1-phenyl-2-methoxyethane: Benzenemethanamine, α-(methoxymethyl)-, (R)-; (64715-85-1), **75**, 19

(S)-2-Amino-3-phenylpropanol: Benzenepropanol, 2-amino-, (S)-; (3182-95-4), **76**, 113

α-Aminophosphonates, **75**, 25, 30

table, **75**, 30

α–Aminophosphonic acids and esters, **75**, 25

3-AMINOPROPYL CARBANION EQUIVALENT, **75**, 215

1-Amino-3-siloxy-1,3-butadienes, highly reactive dienes for Diels-Alder reactions **78**, 155, 164

table, **78**, 157; of cycloadditions, **78**, 166

Ammonia, **76**, 66

Ammonium chloride; (12125-02-9), solid acid catalyst, **75**, 47

Ammonium hydroxide; (1336-21-6), **76**, 24, 63

Ammonium molybdate(VI) tetrahydrate: Molybdic acid, hexaammonium salt, tetrahydrate; (12027-67-7), **76**, 78

Ammonium sulfate, **76**, 79

Anti-inflammatory drugs, **76**, 173

L-Arginine hydrochloride: L-Arginine, monohydrochloride; (1119-34-2), **76**, 77

2-Aryl-2-cyclohexenones,

table, **75**, 75

2-ARYLPROPIONIC ACIDS, PURE, SYNTHESIS OF, **76**, 169

Aseptic transfer, **76**, 80

Asymmetric reactions, catalytic, **76**, 10

Asymmetric

    aldol reaction, **78**, 14

        table, **78**, 22

    alkylation, of carboxylic acid amides, **77**, 27

    allylation, of aldehyes, **75**, 12

        table, **75**, 17

    SYNTHESIS

        of α-AMINO ACIDS, **76**, 57

        of α-AMINOPHOSPONATES, **75**, 19

            table, **75**, 30

        of methyl (R)-(+)-β-phenylalanate, **77**, 50

Azide, asymmetric introduction of, **75**, 34

Azeotropic drying, **76**, 67

Azobisisobutyronitrile: Propanenitrile, 2,2'-azobis[2-methyl-; (78-67-1), **78**, 240

1,1'-(Azodicarbonyl)dipiperidine [ADD]: Piperidine, 1,1'-(azodicarbonyl)bis-; (10465-81-3), **77**, 99

Bacto-Agar, **76**, 79

Barton esters, **75**, 124

Barton-McCombie deoxygenation,

    of alcohols, **78**, 239

    of esters, **77**, 162

1,4-BENZADIYNE EQUIVALENT, **75**, 201

Benzaldehyde; (100-52-7), **76**, 24, 288; **77**, 52, 199

Benzene: CANCER SUSPECT AGENT (8,9); (71-43-2), **77**, 37, 163

Benzeneboronic acid: Boronic acid, phenyl-; (98-80-6), **75**, 53

Benzenethiol: See Thiophenol, **75**, 100

Benzophenone: Methanone, diphenyl-; (119-61-9), **75**, 141, 224

    as photosensitizer, **75**, 141

Benzopinacol: 1,2-Ethanediol, 1,1,2,2-tetraphenyl-; (464-72-2), **75**, 141; **76**, 294

Benzoquinolizines, **76**, 36

Benzoyl chloride; (98-88-4), **75**, 225

N-(2-Benzoyl-4-chlorophenyl)oxanilic acid ethyl ester: Acetic acid, [(2-benzoyl-4-chlorophenyl)amino]oxo-, ethyl ester; (19144-20-8), **76**, 142

5'-O-Benzoyl-3'-deoxythymidine: Thymidine, 3'-deoxy-, 5'-benzoate; (122621-07-2), **77**, 165

4a(S),8a(R)-2-Benzoyl-1,3,4,4a,5,8a-hexahydro-6(2H)-isoquinolinone: 6(2H)-Isoquinolone, 2-benzoyl-1,3,4,4a,5,8a-hexahydro-, (4aS-cis)-; (52346-14-2), **75**, 225

N-Benzoyl meroquinene tert-butyl ester: 4-Piperidineacetic acid, 1-benzoyl-3-ethenyl-, 1,1-dimethylethyl ester, (3R-cis)-; (52346-13-1), **75**, 224

4a(S),8a(R)-2-BENZOYLOCTAHYDRO-6(2H)-ISOQUINOLINONE: 6(2H)ISOQUINOLINONE, 2-BENZOYLOCTAHYDRO-, (4aS-cis)-; (52390-26-8), **75**, 223

3-Benzoyl-N-vinylpyrrolidin-2-one: 2-Pyrrolidinone, 3-benzoyl-1-ethenyl-, (±)-; (125330-80-5), **75**, 215

Benzyl alcohol: Benzenemethanol; (100-51-6), **76**, 111; **78**, 180

Benzylamine: Benzenemethanamine ; (100-46-9), **75**, 107; **78**, 91

N-BENZYL-2,3-AZETIDINEDIONE: 2,3-AZETIDINEDIONE, 1-(PHENYLMETHYL)-; (75986-07-1), **75**, 106

N-Benzylbenzamide: Benzamide, N-(phenylmethyl)-; (1485-70-7), **78**, 55, 84

Benzyl bromide: Benzene, bromomethyl-; (100-39-0), **76**, 111, 113, 240; **77**, 23

N-Benzylcinchonidinium chloride: Cinchonanium, 9-hydroxy-1-(phenylmethyl)-, chloride, (9S)-; (69221-14-3), **76**, 1

Benzyl (S)-2-(N,N-dibenzylamino)-3-phenylpropanoate: L-Phenylalanine, N,N-bis(phenylmethyl)-, phenylmethyl ester; (111138-83-1), **76**, 110

N-Benzyl-3-(Z/E)-ethylideneazetidin-2-one: 2-Azetidinone, 3-ethylidene-1-(phenylmethyl)-; (115870-02-5), **75**, 108

N-Benzyl-3-(1-hydroxyethyl)azetidin-2-one: 2-Azetidinone, 3-(1-hydroxyethyl)-1-(phenylmethyl)-, (R*,R*)-; (89368-08-1); (R*,S*)-; (89368-09-2), **75**, 107

N-Benzylidenebenzylamine: Benzylamine, N-benzylidene-; Benzenemethanamine,N-(phenylmethylene)-; (780-25-6), **76**, 182

(S)-(+)-N-(Benzylidene)-p-toluenesulfinamide: Benzenesulfinamide, 4-methyl-N-(phenylmethylene)-, [S-(E)]-; (153277-49-7), **77**, 51

2-Benzyl-2-methylcyclohexanone: Cyclohexanone, 2-benzyl-2-methyl-; Cyclohexanone, 2-methyl-2-(phenylmethyl)-; (1206-21-9), **76**, 244

2-BENZYL-6-METHYLCYCLOHEXANONE: CYCLOHEXANONE, 2-METHYL-6-(PHENYLMETHYL)-; (24785-76-0), **76**, 239

Benzyloxyacetaldehyde: Acetaldehyde, (phenylmethoxy)-; (60656-87-3), **75**, 12

(-)-(E,S)-3-(BENZYLOXY)-1-BUTENYL PHENYL SULFONE: BENZENE, [[[1-METHYL-3-(PHENYLSULFONYL)-2-PROPENYL]OXY]METHYL]-, [S-(E)]-; (168431-27-4), **78**, 177, 189

(-)-(S)-2-(Benzyloxy)propanal: Propanal, 2-(phenylmethoxy)-, (S)-; (81445-44-5), **78**, 177

O-Benzyl-2,2,2-trichloroacetimidate: Ethanimidic acid, 2,2,2-trichloro-, phenylmethyl ester; (81927-55-1), **78**, 178

    use in acid-catalyzed benzylation, **78**, 177

Benzyltriethylammonium chloride: Ammonium, benzyltriethyl-, chloride; Benzenemethanaminium, N,N,N-triethyl-, chloride; (56-37-1), **75**, 38

Benzyne precursors, **75**, 207; **78**, 104

BIARYLS, UNSYMMETRICAL, SYNTHESIS OF, **75**, 53, 63

    tables, **75**, 60, 67

BICYCLO[1.1.1]PENTANE-1,3-DICARBOXYLIC ACID; (56842-95-6), **77**, 249

Bicyclo[1.1.1]pentyl phenyl sulfide: Bicyclo[1.1.1]pentane, 1-(phenylthio)- (11); (98585-81-0), **75**, 99

BICYCLOPROPYLIDENE: CYCLOPROPANE, CYCLOPROPYLIDENE-; (27567-82-4), **78**, 142

6,6'-Bi(3,4-dihydro-2H-pyran) [Bis-DHP]: 6,6'-Bi-2H-pyran, 3,3',4,4'-tetrahydro-; (109669-49-0), **77**, 212

(R)-BINAL-H: (R)-2,2'-Dihydroxy-1,1'-binaphthol-lithium aluminum hydride, **77**, 99

(S)-(-)-BINAP: Phosphine, [1,1'-binaphthalene]-2,2'-diylbis(diphenyl-, (S)-; (76189-56-5), **77**, 3

(±)-1,1'-Bi-2-naphthol: (1,1'-Binaphthalene)-2.2'-diol; (602-09-5), **76**, 1, 6

(R)-1,1'-Bi-2-naphthol: [1,1'-Binaphthalene]-2,2'-diol, (R)-; (18531-94-7), **76**, 1, 6; **77**, 99; **78**, 2, 15

(S)-(-) [1,1'-BINAPHTHALENE]-2,2'-DIOL, (S)-: (S-BINOL); (18531-99-2), **75**, 12; **76**, 1, 6

(±)-1,1'-Bi-2-naphthyl ditriflate: Methanesulfonic acid, trifluoro-, (1,1'-binaphthalene)-2,2'-diyl ester, (±)-; (128575-34-8), **76**, 6

Biocatalytic transformations, **76**, xviii

4-BIPHENYLCARBOXALDEHYDE: [1,1'-BIPHENYL]-4-CARBOXALDEHYDE; (3218-36-8), **75**, 53

(R,R)-N,N'-Bis-(3,5-di-tert-butylsalicylidene)-1,2-cyclohexanediamine: Phenol, 2,2'-[1,2-cyclohexanediylbis(nitrilomethylidyne)]bis[4,6-bis(1,1-dimethylethyl)-[1R-(1α(E),2β(E)]]-; (135616-40-9), **75**, 4

4,4'-BIS(CHLOROMETHYL)-2,2'-BIPYRIDINE: 2,2'-BIPYRIDINE,4,4'-BIS(CHLOROMETHYL)-; (138219-98-4), **78**, 82

(R,R)-N,N'-BIS-(3,5-DI-tert-BUTYLSALICYLIDENE)1,2-CYCLOHEXANEDIAMINOMANGANESE(III) CHLORIDE: MANGANESE, CHLORO[[2,2'-[1,2-CYCLOHEXANEDIYLBIS(NITRILOMETHYLIDYNE)]BIS[4,6-BIS(1,1-DIMETHYLETHYL)PHENOLATO]](2-)-N,N',O,O']-[SP-5-13-(1R-trans)];(138124-32-0), **75**, 1; **76**, 47

Bis(1,5-dichloro-2,4-pentanedione) copper (II) complex: Copper, bis(1,5-dichloro-2,4-pentanedionato-O,O')-; (135943-96-3), **77**, 1

2,7-Bis(diethylcarbamoyloxy)naphthalene: Carbamic acid, diethyl-, 2,7-naphthalene diyl ester, **78**, 42

(R)-(+)- AND (S)-(-)-2,2'-BIS(DIPHENYLPHOSPHINO)-1,1'-BINAPHTHYL: PHOSPHINE, [1,1'-BINAPHTHALENE]-2,2'-DIYLBIS[DIPHENYL-, (R)-; (76189-55-4); (S)-; (76189-56-5), **76**, 6

rac-2,2'-Bis(diphenylphosphino)-1,1'-binaphthyl-: rac BINAP: Phosphine, [1,1'-binaphthalene]-2,2'-diylbis(diphenyl)-; (98327-87-8), **78**, 23

[1,4-Bis(diphenylphosphino)butane (dppb): Phosphine, 1,4-butanediylbis[diphenyl-; (7688-25-7), **78**, 3

[1,2-Bis(diphenylphosphino)ethane]nickel(II) chloride: Nickel, dichloro[ethanediylbis[diphenylphosphine]-P,P']-, (SP-4-2)-; (14647-23-5), **76**, 7

[1,3-Bis(diphenylphosphino)propane]nickel(II) chloride : Nickel, dichloro[1,3-propanediylbis(diphenylphosphine)-P,P']-; (15629-92-2), **77**, 154; **78**, 43

Bis(η-divinyltetramethyldisiloxane)tri-tert-butylphosphineplatinum(0): Platinum, [1,3-bis(η^2-ethenyl)-1,1,3,3-tetramethyldisiloxane][tris(1,1-dimethylethyl)phosphine]-; (104602-18-8), **75**, 79; preparation, **75**, 81

1,2:5,6-Bis-O-(1-methylethylidene)-, O-phenyl carbonothioate-)-α-D-glucofuranose: α-D-Glucofuranose, 1,2:5,6-bis-O-(1-methylethylidene)-, O-phenyl thiocarbonate; (19189-62-9), **78**, 239

BIS(PINACOLATO)DIBORON: 2,2'-BI-1,3,2-DIOXABOROLANE, 4,4,4',4',5,5,5',5'-OCTAMETHYL-; (73183-34-3), **77**, 176

Bis(ruthenium dichloride-S-BINAP)-triethylamine catalyst: Ruthenium, bis[[1,1'binaphthalene]-2,2'diylbis[diphenylphosphine]-P,P']di-μ-chlorodichloro(N,N-diethylethanamine)di-; (114717-51-0), **77**, 3

Bis(tributyltin) oxide: Distannoxane, hexabutyl-; (56-35-9), **78**, 240

Bis(trichloromethyl) carbonate: See Triphosgene, **75**, 46

(R)-2,2'-Bis(trifluoromethanesulfonyloxy)-1,1'-binaphthyl: Methanesulfonic acid, trifluoro-, [1,1'-binaphthalene]-2.2'-diyl ester, (R)-; (126613-06-7), **78**, 2

BIS(2,4,6-TRIMETHYLPYRIDINE)BROMINE(I) HEXAFLUOROPHOSPHATE: BROMINE(1+), BIS(2,4,6-TRIMETHYLPYRIDINE)-, HEXAFLUOROPHOSPHATE(1-); (188944-77-6), **77**, 206

BIS(2,4,6-TRIMETHYLPYRIDINE)IODINE(I) HEXAFLUOROPHOSPHATE: IODINE(1+), BIS(2,4,6-TRIMETHYLPYRIDINE)-, HEXAFLUOROPHOSPHATE(1-); (113119-46-3), **77**, 206

BIs(2,4,6-trimethylpyridine)silver(I) hexafluorophosphate, **77**, 207

1,2-Bis(trimethylsilyl)benzene: Silane, 1,2-phenylenebis[trimethyl-; (17151-09-6), **78**, 104

4,4'-Bis[(trimethylsilyl)methyl]-2,2'-bipyridine: 2,2'-Bipyridine, 4,4'-bis(trimethylsilyl)methyl]-; (199282-52-5), **78**, 82

2,4-Bis(trimethylsilyloxy)-5-methylpyrimidine: Pyrimidine, 5-methyl-2,4-bis[(trimethysilyl)oxy]-; (7288-28-0), **77**, 163

Borane-ammonia complex: Borane, monoammoniate; (13774-81-7), **77**, 31

Boron tribromide: Borane, tribromo-; (10294-33-4), **75**, 129; **77**, 177

Boron trifluoride etherate: Ethane, 1,1'-oxybis-, compd. with trifluoroborane (1:1); (109-63-7), **75**, 189; **76**, 13; **77**, 66

Boronic acids, alkyl-, **76**, 96

Boroxines, (anhydrides), **76**, 92, 96

BROMIDES, EFFICIENT SYNTHESIS OF, **75**, 124

Bromine; (7726-95-6), **75**, 210; **77**, 142, 208, 250; **78**, 135, 143

(2-Bromoallyl)diisopropoxyborane: Boronic acid, (2-bromo-2-propenyl)-, bis(1-methylethyl) ester; (158556-14-0), **75**, 129, 133

4-Bromobenzaldehyde: Benzaldehyde, 4-bromo-; (1122-91-4), **75**, 53

Bromobis(dimethylamino)borane: Boranediamine, 1-bromo-N,N,N',N'-tetramethyl-; (6990-27-8), **77**, 177

1-Bromobutane: Butane, 1-bromo-; (109-65-9), **76**, 87

1-Bromo-5-chloropentane: Pentane, 1-bromo-5-chloro-; (54512-75-3), **76**, 222

1-Bromo-3-chloropropane: Propane, 1-bromo-3-chloro-; (109-70-6), **76**, 222

1-Bromo-1-cyclopropylcyclopropane: 1,1'-Bicyclopropyl, 1-bromo-; (60629-95-0), **78**, 143

Bromofluorination of alkenes, **76**, 159

   Table, **76**, 165-168

1-BROMO-2-FLUORO-2-PHENYLPROPANE: BENZENE, (2-BROMO-1-FLUORO-1-METHYLETHYL)-; (59974-27-5), **76**, 159

Bromoform: Methane, tribromo- (8,9); (75-25-2), **75**, 98

6-Bromo-1-hexene: 1-Hexene, 6-bromo-; (2695-47-8), **76**, 223

(S)-(-)-2-Bromo-3-hydroxypropanoic acid: Propanoic acid, 2-bromo-3-hydroxy-, (S)-; (70671-46-4), **75**, 37

Bromomalononitrile: Propanedinitrile, bromo-; (1885-22-9), **75**, 210

   preparation, **75**, 211

4-Bromo-3-methylanisole: Benzene, 1-bromo-4-methoxy-2-methyl-; (27060-75-9), **78**, 23

4-Bromo-2-methyl-1-butene: 1-Butene, 4-bromo-2-methyl-; (20038-12-4), **78**, 203

2-Bromonaphthalene: Naphthalene, 2-bromo-; (580-13-2), **76**, 13

8-Bromo-1-octene: 1-Octene, 8-bromo; (2695-48-9), **76**, 224

2-Bromo-1-octen-3-ol: 1-Octen-3-ol, 1-bromo-, (E)-; (52418-90-3), **76**, 263

4-Bromo-3-pentenone (in situ), **78**, 135

1-Bromo-3-phenylpropane: Benzene, (3-bromopropyl)-; (637-59-2), **75**, 155

(Z/E)-1-BROMO-1-PROPENE: 1-PROPENE, 1-BROMO-; (590-14-7), **76**, 214

280

4-(2-BROMO-2-PROPENYL)-4-METHYL-γ-BUTYROLACTONE: 2(3H)-FURANONE, 5-(2-BROMO-2-PROPENYL)DIHYDRO-5-METHYL-; (138416-14-5), **75**, 129

2-Bromopyridine: Pyridine, 2-bromo-; (109-04-6), **78**, 53

N-Bromosuccinimide: 2,5-Pyrrolidinedione, 1-bromo-; (128-08-5), **77**, 107; **78**, 234

Bromotrichloromethane: Methane, bromotrichloro; (75-62-7), **75**, 125

Butane, **77**, 231

2,3-Butanedione; (431-03-8), **77**, 249

1-Butanol, **76**, 48; **78**, 240

2-Butanol; (78-92-2), **76**, 68

N-Boc-L-ALLYLGLYCINE: 4-PENTENOIC ACID, 2-[[(1,1-DIMETHYLETHOXY-CARBONYL]AMINO]-, (R)-; (170899-08-8), **76**, 57

(S)-2-[(4S)-N-tert-BUTOXYCARBONYL-2,2-DIMETHYL-1,3-OXAZOLIDINYL]-2-tert-BUTYLDIMETHYLSILOXYETHANAL: 3-OXAZOLIDINECARBOXYLIC ACID, (4-[[[(1,1-DIMETHYLETHYL)DIMETHYLSILYL]OXY-2-OXOETHYL]-2,2-DIMETHYL-,1,1-DIMETHYLETHYL ESTER, [S-(R*,R*)]-; (168326-01-0), **77** , 78

(S)-2-[[(4S)-N-tert-Butoxycarbonyl-2,2-dimethyl-1,3-oxazolidinyl]-tert-butyldimethylsiloxy]-1,3-thiazole: 3-Oxazolidinecarboxylic acid, 4-[[[(1,1-dimethylethyl)dimethylsilyl]oxy]-2-thiazolylmethyl]2,2-dimethyl-, 1,1-dimethylethyl ester, [S-(R*,R*)]-; (168326-00-9), **77**, 80

(S)-2-{[(4S)-N-tert-Butoxycarbonyl-2,2-dimethyl-1,3-oxazolidin-4-yl]hydroxymethyl}-1,3-thiazole: 3-Oxazolidinecarboxylic acid, 4-(hydroxy-2-thiazolylmethyl)-2,2-dimethyl-, 1,1-dimethylethyl ester, [(S-(R*,R*)]-; (115822-48-5), **77**, 79

tert-Butyl alcohol: 2-Propanol, 2-methyl-; (75-65-0), **75**, 225; **78**, 114, 203

Butylboronic acid: Boronic acid, butyl-; (4426-47-5), **76**, 87

tert-Butyl chloride: Propane, 2-chloro-2-methyl-; (507-20-0), **75**, 107

(5S)-(5-O-tert-BUTYLDIMETHYLSILOXYMETHYL)FURAN-2(5H)-ONE: 2(5H)-FURANONE, 5-[[[(1,1-DIMETHYLETHYL)DIMETHYLSILYL]OXY]METHYL]-, (S)-; (105122-15-4), **75**, 140

(1R*,6S*,7S*)-4-tert-BUTYLDIMETHYLSILOXY)-6-(TRIMETHYLSILYL)BICYCLO[5.4.0]UNDEC-4-EN-2-ONE, **76**, 199

(tert-BUTYLDIMETHYLSILYL)ALLENE: SILANE, (1,1-DIMETHYLETHYL)DIMETHYL-1,2-PROPADIENYL-; (176545-76-9), **76**, 178 tert-Butyldimethylsilyl chloride: Silane, chloro(1,1-dimethylethyl)dimethyl-; (18162-48-6), **75**, 140; **76**, 179, 201

1-(tert-Butyldimethylsilyl)-1-(1-ethoxyethoxy)-1,2-propadiene: Silane, (1,1-dimethylethyl)dimethyl-1-(1-ethoxyethoxy)-1,2-propadienyl-; (86486-46-6), **76**, 201

1-(tert-Butyldimethylsilyl)-1-(1-ethoxyethoxy)-3-trimethylsilyl-1,2-propadiene (crude), **76**, 180

3-(tert-Butyldimethylsilyl)-2-propyn-1-ol: 2-Propyn-1-ol, 3-[(1,1-dimethylethyl)dimethylsilyl]-; (120789-51-7), **76**, 179

tert-Butyldimethylsilyl trifluoromethanesulfonate: Methanesulfonic acid, trifluoro-, (1,1-dimethylethyl)dimethylsilyl ester; (69739-34-0), **77**, 80

(E)-1-(tert-Butyldimethylsilyl)-3-trimethylsilyl-2-propen-1-one: Silane, (1,1-dimethylethyl)dimethyl[1-oxo-3-(trimethylsilyl)-2-propenyl]-, (E)-; (83578-66-9), **76**, 202; (Z)-, **76**, 207

2-BUTYL-6-ETHENYL-5-METHOXY-1,4-BENZOQUINONE: 2,5-CYCLOHEXADIENE-1,4-DIONE, 5-BUTYL-3-ETHENYL-2-METHOXY-; (134863-12-0), **76**, 189

Butyllithium: Lithium, butyl-; (109-72-8), **75**, 22, 116, 201; **76**, 37, 59, 151, 179, 191, 201, 202, 203, 214, 230, 239; **77**, 23, 31, 32, 98, 231; **78**, 15, 82, 114

Butylmagnesium bromide: Magnesium, bromobutyl-; (693-03-8), **76**, 87

tert-Butyllithium: Lithium, (1,1-dimethylethyl)-; (594-19-4), **77**, 212; **78**, 53

tert-Butyl N-(3-methyl-3-butenyl)-N-(2-furyl)carbamate: Carbamic acid, 2-furanyl(3-methyl-3-butenyl)-, 1,1-dimethylethyl ester; (212560-95-7), **78**, 204

tert-Butyl methyl ether: Ether, tert-butyl methyl; Propane, 2-methoxy-2-methyl-; (1634-04-4), **76**, 276; **77**, 199

tert-BUTYL 3a-METHYL-5-OXO-2,3,3a,4,5,6-HEXAHYDROINDOLE-1-CARBOXYLATE: 1H-INDOLE-1-CARBOXYLIC ACID, 2,3,3a,4,5,6-HEXAHYDRO-3a-METHYL-5-OXO-, 1,1-DIMETHYLETHYL ESTER; (212560-98-0), **78**, 202

tert-Butyl perbenzoate: Benzenecarboperoxoic acid, 1,1-dimethylethyl ester; (614-45-9), **75**, 161

1-BUTYL-1,2,3,4-TETRAHYDRO-1-NAPHTHOL: 1-NAPHTHALENOL, 1-BUTYL-1,2,3,4-TETRAHYDRO-; (240802-91-9), **76**, 228

(4R-trans)-2-Butyl-N,N,N',N'-tetramethyl[1,3,2]dioxaborolane-4,5-dicarboxamide: 1,3,2-Dioxaborolane-4,5-dicarboxamide, 2-butyl-N,N,N',N'-tetramethyl-, (4R-trans)-; (161344-85-0), **76**, 88

Butyl vinyl ether: Butane, 1-(ethenyloxy)-; (111-34-2), **77**, 115

3-Butyn-2-ol: 3-Butyn-2-ol, (±)-; (65337-13-5), **75**, 79

^{13}C-labelled compounds, **78**, 113

Camphorsulfonic acid monohydrate (CSA): Bicyclo[2.2.1]heptane-1-methanesulfonic acid, 7,7-dimethyl-2-oxo-, (±)-; (5872-08-2), **75**, 46, 171; **76**, 178; **77**, 213

4-Carbomethoxy-3-dimethylamino-1-tert-butyldimethylsiloxy-1-cyclohexene:cis-2-(Dimethylamino)-4-[[(1,1-dimethylethyl)dimethylsilyl]oxy]-3-cyclohexene-1-carboxylic acid, methyl ester [194233-86-8]; trans-2-(Dimethylamino)-4- [[(1,1-dimethylethyl)dimethylsilyl]oxy]-3-cyclohexene-1-carboxylic acid, methyl ester; (194233-84-6), **78**, 161

Carbon-carbon bond formation, **77**, 111

Carbon monoxide; (630-08-0), **78**, 189

Carbon tetrachloride, CANCER SUSPECT AGENT; Methane, tetrachloro-; (56-23-5), **76**, 124

CARBONYL ADDITIONS, CERIUM(III) CHLORIDE-PROMOTED, **76**, 237

Carbonyl compounds, reactions with organolithiums or Grignard reagents, **76**, 228

      Table, **76**, 237

Carboxylic acid amides, **77**, 27

Catalytic epoxidation with Oxone, **78**, 225

      table, **78**, 231-232

Cells, storage of, **76**, 80

Centrifugation, **76**, 78

Cerium(III) chloride, anhydrous: Cerium chloride; (7790-86-5), **76**, 229

Cerium(III) chloride heptahydrate: Cerium chloride (CeCl$_3$), heptahydrate; (18618-55-8), **76**, 228

Cerium(III) chloride monohydrate: Cerium chloride (CeCl$_3$), monohydrate; (64332-99-6), **76**, 228

Cerium(III) molybdate: Molybdic acid, cerium salt; (53986-44-0), **77**, 39

Cesium carbonate: Carbonic acid, dicesium salt; (534-17-8), **78**, 24

Cesium fluoride; (13400-13-0) **78**, 83

Chiral :

    acetate, **77**, 47

    auxiliaries, **76**, 18, 73

    diamines, **76**, 30

    dioxaborolane ligands, **76**, 97

    Lewis acids, **76**, 18

    ligands, **76**, xvii, 53, 97

    mediator, **77**, 17, 27

    monodentate phosphine ligands, **78**, 8

    NON-RACEMIC DIOLS, SYNTHESIS OF, **76**, 101

        table, **76**, 107

    reagents:

        for enantioselective syntheses, **76**, xviii

        from amino acids, **76**, 128

    sulfoxides, **76**, 42

    synthons, **76**, 82

    titanium catalysts, **76**, 19

Chirald: Benzenemethanol, α-[2-(dimethylamino)-1-methylethyl]-α-phenyl-, [S-(R,R*)]-; (38345-66-3), **77**, 101

Chloroacetyl chloride: Acetyl chloride, chloro-; (79-04-9), **77**, 2

Chlorobenzene: Benzene, chloro-; (108-90-7), **76**, 77

3-Chloro-2,2-bis(chloromethyl)propanoic acid: Propionic acid, 3-chloro-2,2-bis(chloromethyl)-; (17831-70-8), **75**, 90

3-Chloro-2,2-bis(chloromethyl)propan-1-ol: 1-Propanol, 3-chloro-2,2-bis(chloromethyl)- ; (813-99-0), **75**, 89

3-Chloro-2-(chloromethyl)-1-propene: 1-propene, 3-chloro-2-(chloromethyl)-; (1871-57-4), **75**, 89

2-Chlorocyclohexanone: Cyclohexanone, 2-chloro-; (822-87-7), **78**, 212

1-CHLORO-(2S,3S)-DIHYDROXYCYCLOHEXA-4,6-DIENE: 3,5-CYCLOHEXADIENE-1,2-DIOL, 3-CHLORO-, (1S-cis)-; (65986-73-4), **76**, 77

6-CHLORO-1-HEXENE: 1-HEXENE, 6-CHLORO-; (928-89-2), **76**, 221

Chloromethyl-4-hydroxytetrahydrofuran, **77**, 7

Chloromethyl methyl ether: Methane, chloromethoxy-; (107-30-2), **77**, 100

Chloromethyl phenyl sulfide: Benzene, [(chloromethyl)thio]-; (7205-91-6), **78**, 169

8-CHLORO-1-OCTENE: 1-OCTENE, 8-CHLORO-; (871-90-9), **76**, 221

5-Chloro-1-pentyne: 1-Pentyne, 5-chloro-; (14267-92-6), **77**, 231

m-Chloroperoxybenzoic acid: Benzocarboperoxoic acid, 3-chloro-; (937-14-4), **75**, 154

2-Chlorophenol: Phenol, 2-chloro; (95-57-8), **76**, 271

4-Chlorophenol: Phenol, 4-chloro-; (106-48-9), **78**, 27

2-CHLOROPHENYL PHOSPHORODICHLORIDOTHIOATE: PHOSPHORODICHLORIDOTHIOIC ACID, O-(2-CHLOROPHENYL) ESTER; (68591-34-4), **76**, 271

4-Chlorophenyl trifluoromethanesulfonate: Methanesulfonic acid, trifluoro-, 4-chlorophenyl ester; (29540-84-9), **78**, 27

Chloroplatinic acid hexahydrate: Platinate (2-), hexachloro-, dihydrogen, (OC-6-11)-; (16941-12-1), **75**, 81

3-Chloropropionitrile: Propanenitrile, 3-chloro-; (542-76-7), **77**, 186

6-Chloro-1-pyrrolidinocyclohexene: Pyrrolidine, 1-(6-chloro-1-cyclohexen-1-yl)-; (35307-20-1), **78**, 212

N-Chlorosuccinimide: 2,5-Pyrrolidinedione, 1-chloro-; (128-09-6), **76**, 124

Chlorotrimethylsilane: Silane, chlorotrimethyl-; (75-77-4), **75**, 146; **76**, 24, 252; **77**, 26, 91,163, 199; **78**, 83, 99, 105

Chlorotriphenylmethane: Benzene, 1,1',1"-(chloromethylidyne)tris-; (76-83-5), **75**, 184

Chromium hexacarbonyl: Chromium carbonyl (OC-6-11)-; (13007-92-6), **77**, 121

trans-Cinnamaldehyde: 2-Propenal, 3-phenyl-, (E)-; (14371-10-9), **76**, 214

Cinnamyl alcohol: 2-propen-1-ol, 3-phenyl-; (104-54-1), **76**, 89

Citric acid monohydrate: 1, 2, 3-Propanetricarboxylic acid, 2-hydroxy-, monohydrate; (5949-29-1), **75**, 31

Clathrate, **76**, 14

Cobalt nitrate hexahydrate: Nitric acid, cobalt (2+ salt), hexahydrate; (10026-22-9), **76**, 79

2,4,6-Collidine: Pyridine, 2,4,6-trimethyl-; (108-75-8), **77**, 207

Copper; (7440-50-8), **76**, 94, 255

COPPER-CATALYZED CONJUGATE ADDITION OF ORGANOZINC REAGENTS TO α,β-UNSATURATED KETONES, **76**, 252

    table, **74**, 262

Copper(I) chloride: Copper chloride; (7758-89-6), **76**, 133; **78**, 64

Copper(II) chloride: Copper chloride; (7447-39-4), **77**, 213

Copper(II) chloride dihydrate: Copper chloride, dihydrate; (10125-13-0), **77**, 81

Copper cyanide: See Cuprous cyanide, **75**, 146

Copper(I) iodide; (7681-65-4), **77**, 135

Copper(II) oxide: Copper oxide; (1317-38-0), **77**, 81

Copper sulfate, **76**, 79

Crotonaldehyde: 2-Butenal, (E)-; (123-73-9), **77**, 99, 125

(E)-CROTYLSILANES, CHIRAL, **75**, 78

Crown ethers, as phase transfer catalysts, **75**, 103

Cupric acetate monohydrate: Acetic acid, copper(2+) salt, monohydrate; (6046-93-1), **77**, 2, 125

Cuprous cyanide: Copper cyanide; (544-92-3), **75**, 146; **76**, 253

Cuprous iodide: Copper iodide; (7681-65-4), **76**, 103, 264

CYANOALKYNES, SYNTHESIS OF, **75**, 146

    table, **75**, 150

2-CYANOETHANETHIOL, **77**, 186

2-Cyanoethylthiouronium hydrochloride: Carbamimidothioic acid, 2-cyanoethyl ester, monohydrochloride salt; (6634-40-8), **77**, 186

Cyclen (1,4,7,10-tetraazacyclododecane), **78**, 77

Cyclizations:

dehydrative, **76**, 296

reductive, with low-valent titanium reagents, **76**, 149

CYCLOADDITION, CHROMIUM-MEDIATED, **77**, 121

[4 + 3] Cycloadditions, **78**, 212

Cycloalkenones, α-substituted, **75**, 73

Cyclobutenediones, synthesis of, table, **76**, 196

β-Cyclodextrin hydrate: β-Cyclodextrin, hydrate; (68168-23-0), **77**, 220, 226

β-Cyclodextrins, functionalized, **77**, 229

Cycloheptatriene: 1,3,5-Cycloheptatriene; (544-25-2), **77**, 122

Cycloheptatrienylium tetrafluoroborate: Aldrich: Tropylium tetrafluoroborate: Cycloheptatrienylium, tetrafluoroborate (1-); (27081-10-3), **75**, 210

Cycloheptenones, all-cis, bicyclic, **76**, 209

table, **76**, 213

Cycloheptenones, synthesis, using [3+4] annulation, **76**, 213

1,2-Cyclohexanedione; (765-87-7), **75**, 170

Cyclohexanone; (108-94-1), **75**, 116, 119; **78**, 36

Cyclohexanone, 2-methyl-2-(phenylmethyl)-; (1206-21-9), **76**, 244

2-Cyclohexen-1-one, HIGHLY TOXIC; (930-68-7), **75**, 69; **76**, 253

Cyclohexenones with electron-attracting substituents, **78**, 254

Cyclohexenone synthesis, **78**, 164

table, **78**, 167

1,5-Cyclooctadieneruthenium(II) chloride: Ruthenium, dichloro[(1,2,5,6-η)-1,5-cyclooctadiene]-; (50982-12-2), **77**, 3

Cyclopentadiene: 1,3-Cyclopentadiene; (542-92-7), **77**, 255

1,3-Cyclopentanedione; (3859-41-4), **75**, 189

Cyclopentanone-2-carboxylic acid ethyl ester: Cyclopentanecarboxylic acid, 2-oxo-, ethyl ester; (611-10-9), **78**, 249

1-Cyclopenteneacetonitrile: 1-Cyclopentene-1-acetonitrile; (22734-04-9) **78**, 254

3-CYCLOPENTENE-1-CARBOXYLIC ACID; (7686-77-3), **75**, 195

3-Cyclopentene-1,1-dicarboxylic acid; (88326-51-6), **75**, 195

Cyclopropanations, enantioselective, **76**, 97

CYCLOPROPENE; (2781-85-3), **77**, 254

CYCLOPROPYLACETYLENE: CYCLOPROPANE, ETHYNYL-; (6746-94-7), **77**, 231

1-Cyclopropylcyclopropanol: [1,1'-Bicyclopropyl]-1-ol; (54251-80-8), **78**, 142

DABCO: See 1,4-Diazabicyclo[2.2.2]octane, **75**, 107; **76**, 7

DBU: See 1,8-Diazabicyclo[5.4.0]undec-7-ene, **75**, 31, 108

DEC-9-ENYL BROMIDE: 1-DECENE, 10-BROMO-; (62871-09-4), **75**, 124

3-DEOXY-1,2:5,6-BIS-O-(METHYLETHYLIDENE)-α-D-RIBOHEXOFURANOSE: α-D-RIBO-HEXOFURANOSE, 3-DEOXY-1,2:5,6-BIS-O-(METHYLETHYLIDENE)-; (4613-62-1), **78**, 239

Deoxygenation, of thionocarbonyls, **77**, 156

    of other esters, table, **77**, 160

2'-DEOXYRIBONUCLEOSIDES, **77**, 162

    table, **77**, 174-175

DESS-MARTIN PERIODINANE, **77**, 141

    table, **77**, 150

(Diacetoxyiodo)benzene: Iodine, bis(aceto-O)phenyl-; (3240-34-4), **78**, 105

1,3-Diacetylbicyclo[1.1.1]pentane: Ethanone, 1,1'(bicyclo[1.1.1]pentane-1,3-diyl)bis-; (115913-30-9), **77**, 249

(±)-trans-1,2-Diaminocyclohexane: 1,2-Cyclohexanediamine, trans-; (1121-22-8), **75**, 2

(R,R)-1,2-Diammoniumcyclohexane mono-(+)-tartrate: 1,2-Cyclohexanediamine, (1R-trans)-, [R-(R*,R*)-2,3-dihydroxybutanedioate (1:1); (39961-95-0), **75**, 2

Diaryne equivalents, **75**, 204

Diastereomeric purity, determination of, **76**, 70; **77,** 26

Diastereoselective

addition, of chiral acetates, **77**, 47

alkylation, of pseudoephedrine amides, **77**, 22

synthesis, of protected vicinal amino alcohols, **77**, 78

1,4-Diazabicyclo[2.2.2]octane [DABCO]; (280-57-9), **75**, 106; **78**, 36;

1,8-Diazabicyclo[5.4.0]undec-7-ene [DBU]: Pyrimido[1,2-a]azepine, 2,3,4,6,7,8,9,10-octahydro-; (6674-22-2), **75**, 31, 108; **78**, 234

Diazene, propargylic, **76**, 185

Dibenzo-18-crown-6: Dibenzo[b,k][1,4,7,10,13,16]hexaoxacyclooctadecin, 6,7,9,10,17,18,20,21-; (14187-32-7), **75**, 98

β-3',5'-DI-O-BENZOYLTHYMIDINE: THYMIDINE, 3',5'-DIBENZOATE; (35898-30-7), **77**, 162

3'5'-Di-O-benzoyl-2'-O[(3-trifluoromethyl)benzoyl]-5-methyluridine: Uridine, 5-methyl-, 3',5'-dibenzoate 2'-[3-(trifluoromethyl)benzoate]; (182004-59-7), **77**, 164

(S)-2-(N,N-DIBENZYLAMINO)-3-PHENYLPROPANAL: BENZENEPROPANAL, α-[BIS(PHENYLMETHYL)AMINO]-, (S)-; (111060-64-1), **76**, 110

(S)-2-(N,N-Dibenzylamino)-3-phenyl-1-propanol: Benzenepropanol, β-[bis(phenylmethyl)amino]-, (S)-; (111060-52-7), **76**, 111

Diborons, tetraalkoxo-, **77**, 181

1,1-Dibromo-2,2-bis(chloromethyl)cyclopropane: Cyclopropane, 1,1-dibromo-2,2-bis(chloromethyl)- (11); (98577-44-7), **75**, 98; **77**, 249

1,4-Dibromobutane: Butane, 1,4-dibromo- (110-52-1), **77**, 12

3,6-Dibromocarbazole: 9H-Carbazole, 3,6-dibromo-; (6825-20-3), **77**, 153

1,2-Dibromoethane: Ethane, 1,2-dibromo-; (106-93-4), **76**, 24, 252; **78**, 99

3,6-Dibromo-9-ethylcarbazole: 9H-Carbazole, 3,6-dibromo-9-ethyl-; (33255-13-9), **77**, 153

Dibromotriphenylphosphorane: Phosphorane, dibromotriphenyl-; (1034-39-5) **78**, 135

N,N'-Di-Boc-N"-benzylguanidine : [1,3-Bis(tert-butoxycarbonyl)-N"-benzylguanidine]: Carbamic acid, [[(phenylmethyl)imino]methylene]bis-, bis(1,1-dimethylethyl) ester; (145013-06-5), **78**, 92

N.N'-Di-Boc-guanidine [1,3-Bis(tert-butoxycarbonyl)guanidine]: Carbamic acid, carbonimidoylbis-, bis(1,1-dimethylethyl) ester; (154476-57-0), **78**, 91

N,N'-DI-Boc-N"-TRIFLYLGUANIDINE [N,N' BIS(tert-BUTOXYCARBONYL)-N"-TRIFLUOROMETHANESULFONYLGUANIDINE]: CARBAMIC ACID, [[(TRIFLUOROMETHYL)SULFONYL]CARBONIMIDOYL]BIS-, BIS(1,1-DIMETHYLETHYL) ESTER; (207857-15-6), **78**, 91

Di-tert-butyl dicarbonate: Dicarbonic acid, bis(1,1-dimethylethyl) ester; (24424-99-5), **76**, 65; **77**, 65; **78**, 92

2,4-Di-tert-butylphenol: Phenol, 2,4-bis(1,1-dimethylethyl)-; (96-76-4), **75**, 3

3,5-Di-tert-butylsalicylaldehyde: Benzaldehyde,3,5-bis(1,1-dimethylethyl)-2-hydroxy-; (37942-07-7), **75**, 3

1,2-Dichlorobenzene: Benzene, 1,2-dichloro-; (95-50-1), **78**, 105

Dichlorobis(triphenylphosphine)palladium(II): Palladium, dichlorobis(triphenylphosphine)-; (13965-03-2), **76**, 264

cis-1,4-Dichlorobut-2-ene: 2-Butene, 1,4-dichloro-, (Z)-; (1476-11-5), **75**, 195

(2R,4R)-1,5-Dichloro-2,4-pentanediol: 2,4-Pentanediol, 1,5-dichloro-, [R-(R*,R*)]-; (136030-28-9), **77**, 3

(2R,4R)-1,5-Dichloro-2,4-pentanediol bis-Mosher ester, **77**, 7

1,5-Dichloro-2,4-pentanedione: 2,4-Pentanedione, 1,5-dichloro-; (40630-12-4), **77**, 3

8,8-DICYANOFULVENE: PROPANEDINITRILE, 2,4,6-CYCLOHEPTATRIEN-1-YLIDENE-; (2860-54-0), **75**, 210

Dicyclohexylcarbodiimide: HIGHLY TOXIC. Cyclohexanamine, N,N'-methanetetraylbis-; (538-75-0), **75**, 124; **77**, 72

DIELS-ALDER REACTION, **77**, 254

(S,S)-1,2,3,4-DIEPOXYBUTANE: 2,2'-BIOXIRANE, [S-(R*,R*)]-; (30031-64-2), **76**, 101

(R,R)-1,2:4,5-DIEPOXYPENTANE: D-threo-PENTITOL, 1,2:4,5-DIANHYDRO-3-DEOXY-; (109905-51-3), **77**, 1

Diethanolamine: Ethanol, 2,2'-iminobis-; (111-42-2), **76**, 88

     Butylboronate complex, **76**, 88

Diethylamine: Ethanamine, N-ethyl-; (109-89-7), **76**, 264

DIETHYL (R)-(-)-(1-AMINO-3-METHYLBUTYL)PHOSPHONATE: PHOSPHONIC ACID,(1-AMINO-3-METHYLBUTYL)-, DIETHYL ESTER, (R)-; (159171-46-7), **75**, 19

N,N-Diethylaniline: Benzenamine, N,N-diethyl-; (91-66-7), **76**, 29

Diethyl azodicarboxylate: Diazenedicarboxylic acid, diethyl ester; (1972-28-7), **76**, 180

Diethylcarbamoyl chloride: Carbamic chloride, diethyl-; (88-10-8), **78**, 42

Diethyl [2-^{13}C]malonate: Propanedioic –2-^{13}C acid, diethyl ester; (67035-94-3) **78**, 113

Diethyl (R)-(-)-[1-((N-(R)-(1-phenyl-2-methoxyethyl)amino)-3-methylbutyl)]phosphonate: Phosphonic acid, [1-(2-methoxy-1-phenylethyl)amino]-3-methylbutyl]-, diethyl ester, [R-(R*,R*)]-; (159117-09-6), **75**, 20

DIETHYL [(PHENYLSULFONYL)METHYL]PHOSPHONATE: PHOSPHONIC ACID, [(PHENYLSULFONYL)METHYL, DIETHYL ESTER; (56069-39-7), **78**, 169, 179

as a precursor of unsaturated sulfones, 78, 177

Diethyl [(phenylthio)methyl]phosphonate: Phosphonic acid, [(phenylthio)methyl]-, diethyl ester; (38066-16-9), **78**, 170

Diethyl phosphite: Phosphonic acid, diethyl ester; (762-04-9), **75**, 22; **78**, 5

Diethyl sulfate: Sulfuric acid, diethyl ester; (64-67-5), **77**, 153

Diethyl tartrate, **76**, 93

Diethylzinc: Zinc, diethyl-; (557-20-0), **76**, 89

6,7-DIHYDROCYCLOPENTA-1,3-DIOXIN-5(4H)-ONE: CYCLOPENTA-1,3-DIOXIN-5(4H)-ONE,6,7-DIHYDRO-; (102306-78-5), **75**, 189

1,4-Dihydronaphthalene 1,4-oxide: 1,4-Epoxynaphthalene, 1,4-dihydro-; (573-57-9), **78**, 106

3,4-Dihydro-2H-pyran: 2H-Pyran, 3,4-dihydro-; (110-87-2), **76**, 178; **77**, 212

(2S,3S)-DIHYDROXY-1,4-DIPHENYLBUTANE: 2,3-BUTANEDIOL, 1,4-DIPHENYL-, [S-(R*,R*)]-; (133644-99-2), **76**, 101

2,7-Dihydroxynaphthalene: 2,7-Naphthalenediol; (582-17-2), **78**, 42

1,3-Diiodobicyclo[1.1.1]pentane: Bicyclo[1.1.1]pentane,1,3-diiodo- (1); (105542-98-1), **75**, 100

Diiodomethane: Methane, diiodo; (75-11-6), **76**, 89

Diisobutylaluminum hydride: Aluminum, hydrobis(2-methylpropyl)-; (1191-15-7), **78**, 74, 178

Diisopropylamine: 2-Propanamine, N-(1-methylethyl)-; (108-18-9), **75**, 116; **76**, 59, 203, 239; **77**, 23, 31, 32, 98; **78**, 82

Diisopropyl ether: Propane, 2,2'-oxybis-; (108-20-3), **75**, 130

N,N-Diisopropylethylamine: 2-Propanamine, N-ethyl-N-(1-methylethyl)-;
(7087-68-5), **75**, 177; **77**, 68, 100; **78**, 3

1,2:5,6-Di-O-isopropylidene-D-glucose: α-D-Glucofuranose, 1,2:5,6-bis-O-
(1-methylethylidene)-; (582-52-5), **78**, 239

Dilithium tetrachloromanganate: Manganate (2–), tetrachloro-, dilithium, (I-4)-;
(57384-24-4), **76**, 240

(+)-[(8,8-Dimethoxycamphoryl)sulfonyl]imine: 3H-3a,6-Methano-2,1-
benzisothiazole,4,5,6,7-tetrahydro-7,7-dimethoxy-8,8-dimethyl-, 2,2-dioxide,
[3aS]-; (131863-80-4), **76**, 40

(+)-[(8,8-Dimethoxycamphoryl)sulfonyl]oxaziridine: 4H-4a,7-Methanoxazirino[3,2-
i][2,1]benzisothiazole, tetrahydro-8,8-dimethoxy-9,9-dimethyl-, 3,3-dioxide, [2R-
(2α,4aα,7α,8aS*)]-; (131863-82-6), **76**, 38, 39

Dimethoxyethane: Ethane, 1,2-dimethoxy-; (110-71-4), **76**, 89

(3,4-Dimethoxyphenyl)acetonitrile: Benzeneacetonitrile, 3,4-dimethoxy-;
(93-17-4), **76**, 133

2-(3,4-Dimethoxyphenyl)-N,N-dimethylacetamidine; (240797-77-7), **76**, 133

2,2-Dimethoxypropane: Propane, 2,2-dimethoxy-; (77-76-9), **77**, 66

Dimethylamine: Methanamine, N-methyl-; (124-40-3), **76**, 93, 133; **77**, 176;
78, 152

(E)-4-Dimethylamino-3-buten-2-one: 3-Buten-2-one, 4-(dimethylamino)-,
(E)-; (2802-08-6), **78**, 152

1-DIMETHYLAMINO-3-tert-BUTYLDIMETHYLSILOXY-1,3-BUTADIENE: 1,3-
BUTADIEN-1-AMINE, 3-[[(1,1-DIMETHYLETHYL)DIMETHYLSILYL]OXY]-N,N-
DIMETHYL-, (E)-; (194233-66-4), **78**, 152, 161

1-(3-Dimethylamino)propyl-3-ethylcarbodiimide hydrochloride: 1,3-
Propanediamine, N'-(ethylcarbonimidoyl)-N,N-dimethyl-, monohydrochloride;
(25952-53-8), **77**, 35

4-Dimethylaminopyridine: HIGHLY TOXIC: 4-Pyridinamine, N,N-dimethyl-;
(1122-58-3), **75**, 184; **76**, 70; **77**, 7, 38, 72, 80

4-DIMETHYLAMINO-N-TRIPHENYLMETHYLPYRIDINIUM CHLORIDE:
PYRIDINIUM, 4-(DIMETHYLAMINO)-1-(TRIPHENYLMETHYL)-, CHLORIDE;
(78646-25-0), **75**, 184

4,4'-Dimethyl-2,2'-bipyridine: 2,2'-Bipyridine, 4,4'-dimethyl-; (1134-35-6), **78**, 82

(1R,2R)-(+)-N,N'-Dimethyl-1,2-bis(3-trifluoromethyl)phenyl-1,2-ethanediamine:;
(120263-19-6), **76**, 127

Dimethyl carbonate: Carbonic acid, dimethyl ester; (616-38-6), **76**, 170

Dimethyl 3-cyclopentene-1,1-dicarboxylate; 3-Cyclopentene-1,1-dicarboxylic acid, dimethyl ester; (84646-68-4), **75**, 197

N,N'-DIMETHYL-1,2-DIPHENYLETHYLENEDIAMINE: 1,2-ETHANEDIAMINE, N,N'-DIMETHYL-1,2-DIPHENYL, (R*,S*)-; (60509-62-8); [R-(R*,R*)-; (118628-68-5); [S-(R*,R*)]; (70749-06-3), **76**, 23

N-[(1,1-Dimethylethoxy)carbonyl]-N,O-isopropylidene-L-serinol: 3-Oxazolidinecarboxylic acid, 4-(hydroxymethyl)-2,2-dimethyl-, 1,1-dimethylethyl ester, (R)-; (108149-63-9), **77**, 66

N-[(1,1-Dimethylethoxy)carbonyl]-L-serine methyl ester: L-Serine, N-[(1,1-dimethylethoxy)carbonyl]-, methyl ester; (2766-43-0), **77**, 65

1,1-DIMETHYLETHYL 2,2-DIMETHYL-(S)-4-FORMYLOXAZOLIDINE-3-CARBOXYLATE: 3-OXAZOLIDINECARBOXYLIC ACID, 4-FORMYL-2,2-DIMETHYL-, 1,1-DIMETHYLETHYL ESTER, (S)-; (102308-32-7),**77**, 64, 79, 89

3-(1,1-Dimethylethyl) 4-methyl (S)-2,2-dimethyl-3,4-oxazolidine-dicarboxylate: 3,4-Oxazolidinedicarboxylic acid, 2,2-dimethyl-, 3-(1,1-dimethylethyl) 4-methyl ester, (S)-; (108149-60-6), **77**, 66

N,N-Dimethylformamide: CANCER SUSPECT AGENT: Formamide, N,N-dimethyl-(8,9); (68-12-2), **75**, 162; **77**, 35, 37, 221; **78**, 36, 64

N,N-DIMETHYLHOMOVERATRYLAMINE: BENZENEETHANAMINE, 3,4-DIMETHOXY-N,N-DIMETHYL-; (3490-05-9), **76**, 133

(R,R)-Dimethyl O,O-isopropylidenetartrate: 1,3-Dioxolane-4,5-dicarboxylic acid, 2,2-dimethyl-, dimethyl ester, (4R-trans)-; (37031-29-1), **76**, 13

Dimethyl malonate: Propanedioic acid, dimethyl ester; (108-59-8), **75**, 195

2,7-DIMETHYLNAPHTHALENE: NAPHTHALENE, 2,7-DIMETHYL-; (582-16-1), **78**, 42

3,3-DIMETHYL-1-OXASPIRO[3.5]NONAN-2-ONE: 1-OXASPIRO[3.5]NONAN-2-ONE, 3,3-DIMETHYL-; (22741-15-7), **75**, 116

Dimethylphenylsilane: Silane, dimethylphenyl-; (766-77-8), **75**, 79

(±)-1-(Dimethylphenylsilyl)-1-buten-3-ol: 3-Buten-2-ol, 1-(dimethylphenylsilyl)-, (E)-(±)-; (137120-08-2), **75**, 78

(3R)-1-(Dimethylphenylsilyl)-1-buten-3-ol: 3-Buten-2-ol, 1-(dimethylphenylsilyl)-, [R-(E)]-; (133398-25-1), **75**, 79

(3S)-1-(Dimethylphenylsilyl)-1-buten-3-ol: 3-Buten-2-ol, 4-(dimethylphenylsilyl)-, [S-(E)]-; (133398-24-0), **75**, 79

(3R)-1-(Dimethylphenylsilyl)-1-buten-3-ol acetate: 3-Buten-2-ol, 4-(dimethylphenylsilyl)-, acetate, [R-(E)]-; (129921-47-7), **75**, 79

2-(2,2-Dimethylpropanoyl)-1,3-dithiane: 1-Propanone, 1-(1,3-dithian-2-yl)-2,2-dimethyl-; (73119-31-0), **76**, 37

(1S)-(2,2-Dimethylpropanoyl)-1,3-dithiane 1-oxide: 1-Propanone, 2,2-dimethyl-1-(1-oxido-1,3-dithian-2-yl)-, (1S-trans)-; (160496-17-3); (1S-cis)-, **76**, 38

N,N'-Dimethylpropyleneurea [DMPU]: 2(1H)-Pyrimidinone, tetrahydro-1,3-dimethyl-; (7226-23-5), **75**, 195

2,5-Dimethylpyrrole as amine-protecting (blocking) group, **78**, 69

DIMETHYL SQUARATE: 3-CYCLOBUTENE-1,2-DIONE, 3,4-DIMETHOXY-; (5222-73-1), **76**, 189

Dimethyl sulfide: Methane, thiobis-; (75-18-3), **77**, 107; **78**, 254

Dimethyl sulfoxide: Methyl sulfoxide; Methane, sulfinylbis-; (67-68-5), **75**, 146; **76**, 112; **77**, 67; **78**, 3, 144

(R,R)-Dimethyl tartrate: Butanedioic acid, 2,3-dihydroxy-, [R-(R*,R*)]-dimethyl ester; (608-68-4), **76**, 13

(4R,5R)-2,2-DIMETHYL-α,α,α',α'-TETRA(NAPHTH-2-YL)-1,3-DIOXOLANE-4,5-DIMETHANOL: [1,3-DIOXOLANE-4,5-DIMETHANOL, 2,2-DIMETHYL-α,α,α',α'-TETRA-2-NAPHTHALENYL-, (4R-trans)-]; (137365-09-4), **76**, 12

3,5-Dinitrobenzoyl chloride: Benzoyl chloride, 3,5-dinitro-; (99-33-2), **76**, 277

2,4-Dinitrofluorobenzene, **76**, 52

Diol metabolites, **76**, 82

Dioxaborolane ligand, **76**, 89

p-Dioxane, CANCER SUSPECT AGENT: 1,4-Dioxane; (123-91-1), **76**, 65; **78**, 3

1,3-Dioxin vinylogous esters, **75**, 192

Diphenylacetic acid, **77**, 36

1,3-Diphenylacetone p-tosylhydrazone: p-Toluenesulfonic acid, (α-benzylphenethylidene)hydrazide; Benzenesulfonic acid, 4-methyl-, [2-phenyl-1-(phenylmethyl)ethylidene]hydrazide; (19816-88-7), **76**, 206

(4R,5S)-4-Diphenylhydroxymethyl-5-tert-butyldimethylsiloxymethylfuran-2(5H)-one, **75**, 45

(4R,5S)-4,5-Diphenyl-2-oxazolidinone: 2-Oxazolidinone, 4,5-diphenyl-, (4R-cis)-; (86286-50-2), **75**, 45

9,10-DIPHENYLPHENANTHRENE: PHENANTHRENE, 9,10-DIPHENYL-; (602-15-3), **76**, 294

Diphenylphosphine: Phosphine, diphenyl-; (829-85-6), **76**, 7

Diphenylphosphine oxide: Phosphine oxide, diphenyl-; (4559-70-0), **78**, 3, 5

(R)-2-DIPHENYLPHOSPHINO-2'-METHOXY-1,1'-BINAPHTHYL: PHOSPHINE, (2'-METHOXY[1,1'-BINAPHTHALENE]-2-YL)DIPHENYL-, (R)-; (145964-33-6), **78**, 1

(R)-(-)-2-Diphenylphosphinyl-2'-hydroxy-1,1'-binaphthyl: [1,1'-Binaphthalene]-2-ol, 2'-(diphenylphosphinyl)-, (R)-; (132548-91-5), **78**, 3

(R)-(+)-2-Diphenylphosphinyl-2'-methoxy-1,1'-binaphthyl: Phosphine oxide, (2'-methoxy[1,1'-binaphthalene]-2-yl)diphenyl-, (R)-; (172897-73-3) **78**, 4

(R)-(+)-2-Diphenylphosphinyl-2'-trifluoromethanesulfonyloxy-1,1'-binaphthyl: Methanesulfonic acid, trifluoro-, 2'-(diphenylphosphinyl)[1,1'binaphthalen]-2-yl ester, (R)-; (132532-04-8) **78**, 2

Diphenylphosphoryl azide: Phosphorazidic acid, diphenyl ester; (26386-88-9), **75**, 31

5,15-DIPHENYLPORPHYRIN: PORPHINE, 5,15-DIPHENYL-; (22112-89-6), **76**, 287

(4R,5S)-4,5-DIPHENYL-3-VINYL-2-OXAZOLIDINONE: 2-OXAZOLIDINONE, 3-ETHENYL-4,5-DIPHENYL-, (4R-cis)-; (143059-81-8), **75**, 45

(4S,5R)-4,5-Diphenyl-3-vinyl-2-oxazolidinone: 2-Oxazolidinone, 3-ethenyl-4,5-diphenyl-, (4S-cis)-; (128947-27-3), **75**, 45

Diphosgene, **75**, 48

2.2'-Dipyridyl: 2,2'-Bipyridine; (366-18-7), **76**, 68

Disodium hydrogen phosphate: Phosphoric acid, disodium salt; (7558-79-4), **76**, 78

Dispiroketal (dispoke), **77**, 215

1,3-Dithiane; (505-23-7), **76**, 37

(1S)-(−)-1,3-DITHIANE 1-OXIDE: 1,3-DITHIANE, 1-OXIDE, (S)-; (63865-78-1), **76**, 37

3,3'-Dithiobispropionitrile: Propanenitrile, 3,3'-dithiobis-; (42841-31-6), **77**, 191

Dithiooxamide: Ethanedithioamide; (79-40-3) **78**, 74

DiTOX (1,3-Dithiane 1-oxide), **76**, 42

1,3-Divinyltetramethyldisiloxane: Disiloxane, 1,3-diethenyl-1,1,3,3-tetramethyl-; (2627-95-4); **75**, 81

DMPU: See N,N'-Dimethylpropyleneurea, **75**, 195

Duff reaction, adaptation of, **75**, 6, 9

Electron-transfer deoxygenation, photoinduced, **77**, 160, 170

Enantiomeric

excess (ee), determination of, **75**, 2, 33, 82, 83; **76**, 29, 30, 51, 95; **77**, 34, 37; **78**, 17

purity, determination, **77**, 71

ENANTIOMERICALLY PURE α-N,N-DIBENZYLAMINO ALDEHYDES, SYNTHESIS OF, **76**, 110

Enantiomerically pure products, syntheses, **76**, xviii

Enantioselective addition, **77**, 17

Epoxidation, **75**, 153; **76**, 50, 53

asymmetric, **75**, 9

EPOXIDATION CATALYST, ENANTIOSELECTIVE, **75**, 1

Epoxides, optically acitive, **76**, 107

Ergosterol: Ergosta-5,7,22-trien-3-ol, (3β)-; (57-87-4), **76**, 276

3-ETHENYL-4-METHOXYCYCLOBUTENE-1,2-DIONE: 3-CYCLOBUTENE-1,2-DIONE, 3-ETHENYL-4-METHOXY-; (124022-02-2), **76**, 189

1-(1-Ethoxyethoxy)-1,2-propadiene: 1,2-Propadiene, 1-(1-ethoxyethoxy)-; (20524-89-4), **76**, 200

1-(1-Ethoxyethoxy)-1-propyne: 1-Propyne, 3-(1-ethoxyethoxy)-; (18669-04-0), **76**, 200

Ethyl acetate: Acetic acid, ethyl ester; (141-78-6), **77**, 30

Ethyl [2-^{13}C] acetate: Acetic-2-^{13}C acid, ethyl ester; (58735-82-3), **78**, 114

ETHYL (R)-2-AZIDOPROPIONATE: PROPANOIC ACID, 2-AZIDO-, ETHYL ESTER, (R)-; (124988-44-9), **75**, 31

Ethyl benzoate: Benzoic acid, ethyl ester; (93-89-0), **75**, 215

(-)-(S)-Ethyl 2-(benzyloxy)propanoate: Propanoic acid, 2-(phenylmethoxy)-, ethyl ester,(S)-; (54783-72-1), **78**, 178

Ethyl bromide: Bromoethane; (74-96-4), **75**, 38

Ethyl 5-bromovalerate: Pentanoic acid, 5-bromo-, ethyl ester; (14660-52-7), **76**, 255

Ethyl chloroformate: Carbonochloridic acid, ethyl ester; (541-41-3), **78**, 114

ETHYL 5-CHLORO-3-PHENYLINDOLE-2-CARBOXYLATE: 1H-INDOLE-2-CARBOXYLIC ACID, 5-CHLORO-3-PHENYL-, ETHYL ESTER; (212139-32-2), **76**, 142

Ethyl p-dimethylaminobenzoate: Benzoic acid, 4-dimethylamino-, ethyl ester; (10287-53-3), **76**, 276

9-ETHYL-3,6-DIMETHYLCARBAZOLE: CARBAZOLE, 9-ETHYL-3,6-DIMETHYL-; (51545-42-7), **77**, 153

   as a deoxygenation catalyst, table, **77**, 160

Ethyl 2,2-dimethylpropanoate: Propanoic acid, 2,2-dimethyl-, ethyl ester; (3938-95-2), **76**, 37

ETHYL (E)-(-)-4.6-O-ETHYLIDENE-(4S,5R,1'R)-4,5,6-TRIHYDROXY-2-HEXENOATE: D-erythro-HEX-2-ENONIC ACID, 2,3-DIDEOXY-4,6-O-ETHYLIDENE-, ETHYL ESTER [2E,4(R)]-; (125567-87-5); [2Z,4(S)]-; (125567-86-4), **78**, 123

Ethylene: Ethene; (74-85-1), **76**, 24

Ethylenediaminetetraacetic acid, disodium salt: Glycine, N,N'-1,2-ethanediylbis[N-(carboxymethyl)-, disodium salt, dihydrate; (6381-92-6), **78**, 53, 225

Ethylenediaminetetraacetic acid, tetrasodium salt: Glycine, N,N'-1,2-ethanediylbis[N-carboxymethyl)-, tetrasodium salt, trihydrate; (67401-50-7), **76**, 78

Ethylene glycol, **77**, 213

Ethylene glycol dimethyl ether: Ethane, 1,2-dimethoxy-; (110-71-4), **76**, 143

ETHYL (R)-(+)-2,3-EPOXYPROPANOATE: See ETHYL GLYCIDATE, **75**, 37

ETHYL GLYCIDATE (ETHYL (R)-(+)-2,3-EPOXYPROPANOATE): OXIRANECARBOXYLIC ACID, ETHYL ESTER, (R)-; (111058-33-4), **75**, 37

Ethyl 4-hydroxy[1-^{13}C]benzoate: Benzoic-1-^{13}C acid, 4-hydroxy-, ethyl ester; (211519-29-8) **78**, 114

(S)-Ethyl hydroxypropanoate: Propanoic acid, 2-hydroxy-, ethyl ester, (S)-; (687-47-8), **78**, 178

(1'R)-(-)-2,4-O-ETHYLIDENE-D-ERYTHROSE: 1, 3-DIOXANE-4-CARBOXALDEHYDE, 5-HYDROXY-2-METHYL-, [2R-(2α,4α,5β)]-; (70377-89-8), **78**, 123

(1'R)-(-)-4,6-O-Ethylidene-D-glucose: D-Glucopyranose, 4,6-O-ethylidene-; (18465-50-4), **78**, 124

Ethyl 5-iodovalerate: Pentanoic acid, 5-iodo-, ethyl ester; (41302-32-3), **76**, 252

Ethyl (S)-(-)-lactate: Propanoic acid, 2-hydroxy-, ethyl ester, (S)-; (687-47-8), **75**, 31

Ethyl levulinate: Pentanoic acid, 4-oxo-, ethyl ester; (539-88-8), **75**, 130

Ethylmagnesium bromide: Magnesium, bromoethyl-; (925-90-6), **78**, 142

Ethyl oxalyl chloride: Acetic acid, chlorooxo-, ethyl ester; (4755-77-5), **76**, 142

(±)-3-Ethyl-1-oxaspiro[3.5]nonan-2-one: 1-Oxaspiro[3.5]nonan-2-one, 3-ethyl-; (160890-30-2), **75**, 119

ETHYL 5-(3-OXOCYCLOHEXYL)PENTANOATE: PENTANOIC ACID, 5-(3-OXOCYCLOHEXYL)-, ETHYL ESTER; (167107-91-7), **76**, 252

Ethyl trifluoroacetate: Acetic acid, trifluoro-, ethyl ester; (383-63-1), **75**, 153

Ethyl trimethylacetate: Propanoic acid, 2,2-dimethyl-, ethyl ester; (3938-95-2), **76**, 37

Ethyl vinyl ether: Ethene, ethoxy-; (109-92-2), **76**, 200

Ferrous sulfate, **76**, 78

Fluorinated compounds, **76**, 154, 161

Fluoroboric acid: Borate(1-), tetrafluoro-, hydrogen; (16872-11-0), **78**, 190

2-Fluoro-4-iodoaniline: Aniline, 2-fluoro-4-iodo-; (29632-74-4), **78**, 64

1-(2-Fluoro-4-iodophenyl)-2,5-dimethyl-1H-pyrrole: 1H-Pyrrole, 1-(2-fluoro-4-iodophenyl)-2,5-dimethyl-; (217314-30-2), **78**, 64

2-FLUORO-4-METHOXYANILINE: BENZENAMINE, 2-FLUORO-4-METHOXY-; (458-52-6), **78**, 63

1-(2-Fluoro-4-methoxyphenyl)-2,5-dimethyl-1H-pyrrole: 1H-Pyrrole, 1-(2-fluoro-4-methoxyphenyl)-2,5-dimethyl-; (217314-30-2), **78**, 64

Formaldehyde; (50-00-0), **77**, 108

Formamide; (75-12-7), **77**, 199

Furan; (110-00-9), **75**, 201; **78**, 106, 213

Furan-2-ylcarbamic acid tert-butyl ester: Carbamic acid, 2-furanyl-, 1,1-dimethylethyl ester; (56267-47-1), **78**, 203

2-Furoyl chloride: 2-Furancarbonyl chloride; (527-69-5), **78**, 203

Garner aldehyde, **77**, 73

Glove box, nitrogen, **77**, 4

D-Glucose: α-D-Glucopyranose; (492-62-6), **78**, 124

L-(S)-Glyceraldehyde acetonide (2,3-O-Isopropylidene-L-glyceraldehyde): 1,3-Dioxolane-4-carboxaldehyde, (S)-; (22323-80-4), **75**, 139

Glycine methyl ester; (616-34-20), **76**, 58

Glycine methyl ester hydrochloride: Glycine methyl ester, hydrochloride; (5680-79-5), **76**, 66

Green Chemistry, **76**, 174

GRIGNARD REAGENTS, **76**, 87, 221, 222, 228

Guanidine hydrochloride: Guanidine, monohydrochloride; (50-01-1), **78**, 92

Guanidinylation of amines, **78**, 95

Haloalkenes, as synthons, **76**, 224

HALOBORATION, of 1-alkynes, **75**, 134

OF ALLENE, **75**, 129

Hanovia mercury lamp, **76**, 276

Hexachloroethane: Ethane, hexachloro-; (67-72-1), **78**, 83

Hexafluoroisopropyl alcohol: 2-Propanol, 1,1,1,3,3,3-hexafluoro-; (920-66-1), **76**, 151

Hexafluorophosphate salts, vs. perchlorate salts, **77**, 210

2,3,5,6,8,9-Hexahydrodiimidazo[1,2-a:2',1'-c]pyrazine: Diimidazo[1,2-a:2',1'-c]pyrazine, 2,3,5,6,8,9-hexahydro-; (180588-23-2), **78**, 73

Hexahydroindolinones, preparation, **78**, 202

Hexamethyldisilane, **75**, 155

Hexamethyldisilazane: Silanamine, 1,1,1-trimethyl-N-(trimethylsilyl)-; (999-97-3) **78**, 113

Hexamethylenetetramine: 1,3,5,7-Tetraazatricyclo[3.3.1.1^{3,7}]decane; (100-97-0), **75**, 3

Hexamethylphosphoric triamide, HIGHLY TOXIC: Phosphoric triamide, hexamethyl-; (680-31-9), **76**, 201; **78**, 105

Hexylamine: 1-Hexanamine; (111-26-2), **78**, 24

N-HEXYL-2-METHYL-4-METHOXYANILINE, **78**, 23

1-Hexyne; (693-02-7), **76**, 191

Hunsdiecker reaction, **75**, 127

Hydratropic acids, syntheses, **76**, 173

Hydrazine monohydrate, HIGHLY TOXIC. CANCER SUSPECT AGENT: Hydrazine; (302-01-2), **76**, 183

Hydrofluoric acid; (7664-39-3), **78**, 162

Hydrogen;(1333-58-3), **77**, 4

Hydrogen bromide: Hydrobromic acid; (10035-10-6), **75**, 37; **76**, 264

Hydrogenation, **75**, 21; **77**, 3

Hydrogen chloride, **77**, 178, 240

Hydrogen chloride trap, **77**, 2

Hydrogen peroxide; (7722-84-1), **76**, 40; **78**, 171

Hydrogen sulfide; (7783-06-4), **78**, 73

Hydrosilation, **75**, 85

1-Hydroxy-1,2-benziodoxol-3(1H)-one 1-oxide: 1,2-Benziodoxol-3(1H)-one, 1-hydroxy-, 1-oxide; (61717-82-6), **77**, 141

4-HYDROXY[1-^{13}C]BENZOIC ACID: BENZOIC-1-^{13}C ACID, 4-HYDROXY-; (211519-30-1) **78**, 113

1-Hydroxybenzotriazole hydrate: 1H-Benzotriazole, 1-hydroxy-, hydrate; (12333-53-9), **77**, 35

5-(1-Hydroxyethylidene)-2,2-dimethyl-1,3-dioxane-4,6-dione: 1,3-Dioxane-4,6-dione, 5-(1-hydroxyethylidene)-2,2-dimethyl; (85920-63-4), **77**, 114

Hydroxylamine hydrochloride: Hydroxylamine, hydrochloride; (5470-11-1), **78**, 65

(4R,5S)-4-HYDROXYMETHYL-(5-O-tertBUTYLDIMETHYLSILOXYMETHYL)-
FURAN-2(5H)-ONE: D-erythro-PENTONIC ACID, 2,3-DIDEOXY-5-O-[(1,1-
DIMETHYLETHYL)DIMETHYLSILYL]-3-HYDROXYMETHYL)-, γ-LACTONE;
(164848-06-0), **75**, 139

4-HYDROXYMETHYL-2-CYCLOHEXEN-1-ONE: 2-CYCLOHEXEN-1-ONE,4-
(HYDROXYMETHYL)-; (224578-91-0), **78**, 160

4-Hydroxymethyl-3-dimethylamino-1-tert-butyldimethylsiloxy-1-cyclohexene,;
78, 161

(S)-5-Hydroxymethylfuran-2(5H)-one: 2(5H)-Furanone, 5-(hydroxymethyl)-, (S)-;
(78508-96-0), **75**, 140

(1S,2S)-N-(2-Hydroxy-1-methyl-2-phenylethyl)-N-methylpropionamide:
Propanamide, N-(2-hydroxy-1-methyl-2-phenylethyl)-N-methyl- [R-(R*,R*)]-
(192060-67-6); [S-(R*,R*)]-; (159213-03-3), **77**, 22

[1S(R)2S]-N-(2-HYDROXY-1-METHYL-2-PHENYLETHYL)-N,2-
DIMETHYLBENZENEPROPIONAMIDE OR (1S,2S)- PSEUDOEPHEDRINE-(R)-
2-METHYLHYDROCINNAMIDE: BENZENEPROPANAMIDE, N-(2-HYDROXY-1-
METHYL-2-PHENYLETHYL)-N,α-DIMETHYL-, [1S-[1R*(R*),2R*]]
(159345-08-1); [1S-[1R*(S*),2R]]-; (159345-06-9), **77**, 23, 30

2-Hydroxy-4-methylpyridine: 2(1H)-Pyridinone, 4-methyl-; (13466-41-6), **78**, 52

2-Hydroxy-5-methylpyridine: 2(1H)-Pyridinone, 5-methyl-; (1003-68-5), **78**, 51

2-Hydroxy-6-methylpyridine: 2(1H)-Pyridinone, 6-methyl-; (3279-76-3), **78**, 52

4-HYDROXY-1,1,1,3,3-PENTAFLUORO-2-HEXANONE HYDRATE: 2,2,4-
HEXANETRIOL,1,1,1,3,3-PENTAFLUORO-; (119333-90-3), **76**, 151

3-Hydroxyquinuclidine: 1-Azabicyclo[2.2.2]octan-3-ol; (1619-34-7), **77**, 109

N-Hydroxythiopyridone: 2(1H)-Pyridinethione, 1-hydroxy-; (1121-30-8), **75**, 124

(R)-(+)-2-Hydroxy-1,1,2-triphenyl-1,2-ethanediol: 1,2-Ethanediol, 1,1,2-triphenyl-,
(R)-; (95061-246-4), **77**, 45

(R)-(+)-2-HYDROXY-1,2,2-TRIPHENYLETHYL ACETATE [(R)-HYTRA]: 1,2-
ETHANEDIOL, 1,1,2-TRIPHENYL-, 2-ACETATE, (R)-; (95061-47-5), **77**, 45

(R)- and (S)-HYTRA (2-Hydroxy-1,2,2-triphenylethyl acetate) esters, **77**, 45

Imidazole: 1H-Imidazole; (288-32-4), **75**, 140; **77**, 225

[(2-)-N,O,O'[2,2'-Iminobis[ethanolato]]]-2-butylboron, **76**, 88

Indene: 1H-Indene; (95-13-6), **76**, 47

(1S,2R)-Indene oxide: 6H-Indeno[1,2-b]oxirene, 1a,6a-dihydro-; (768-22-9), **76**, 47

Indigo test, **76**, 80

Indinavir (Crixivan®), **76**, 52

Indium; (7440-74-6), **77**, 108

Indole: 1H-Indole; (120-72-9), **76**, 80

Indole synthesis, **78**, 36

Indoloquinolizines, **76**, 36

Intramolecular Diels-Alder reactions, **78**, 202

Iodine; (7553-56-2), **75**, 69, 100; **76**, 13; **77**, 207; **78**, 105

4-Iodoacetophenone: Ethanone, 1-(4-iodophenyl)-; (13329-40-3), **77**, 135

o-Iodoaniline: Benzenamine, 2-iodo-; (615-43-0), **78**, 36

4-Iodoanisole: Benzene, 1-iodo-4-methoxy-; (696-62-8), **75**, 61

2-Iodobenzoic acid: Benzoic acid, 2-iodo-; (88-67-5), **77**, 141

2-IODO-2-CYCLOHEXEN-1-ONE: 2-CYCLOHEXEN-1-ONE, 2-IODO-; (33948-36-6), **75**, 69

Iron(III) chloride hexahydrate: Iron chloride, hexahydrate; (10025-77-1), **78**, 249

Irradiation, **76**, 276

Irradiation apparatus, **75**, 141

Isobutyryl chloride: See 2-Methylpropanoyl chloride, **75**, 118

Isobutyraldehyde: Propanal, 2-methyl-; (78-84-2), **77**, 109

ISOMERIZATION OF β-ALKYNYL ALLYLIC ALCOHOLS TO FURANS CATALYZED BY SILVER NITRATE ON SILICA GEL, **76**, 263

Isomerization, of a meso-diamine to the dl-isomer, **76**, 25

Isoprene: 1,3-Butadiene, 2-methyl-; (78-79-5), **76**, 25

Isopropenyl acetate: 1-Propen-2-ol, acetate; (108-22-5), **77**, 125

O^4,O^5-Isopropylidene-3,6-anhydro-1-deoxy-1-iodo-D-glucitol, **77**, 91

O^4,O^5-ISOPROPYLIDENE-1,2:3,6-DIANHYDRO-D-GLUCITOL, **77**, 91

2,3-O-Isopropylidene-L-glyceraldehyde: 1,3-Dioxolane-4-carboxaldehyde, 2,2-dimethyl-, L-; (22323-80-4), **75**, 139

2,3-O-Isopropylidene-L-threitol: 1,3-Dioxolane-4,5-dimethanol, 2,2-dimethyl-, (4S-trans)-; (50622-09-8), **76**, 102

2,3-O-Isopropylidene-L-threitol 1,4-bismethanesulfonate, **76**, 101

(3R*,4R*)- and (3R*,4S*)-4-Isopropyl-4-methyl-3-octyl-2-oxetanone, **75**, 119

Isosorbide: D-Glucitol, 1,4:3,6-dianhydro-; (652-67-5), **77**, 91

Isovaleraldehyde: Butanal, 3-methyl-; (590-86-3), **75**, 20

Karl Fischer titration, **76**, 68

KETONES, ALDOLIZATION OF, **75**, 116

Kinetic resolution, **75**, 79

β-LACTONES, decarboxylation of, **75**, 120

      regioselective fission of, **75**, 120

      stereoselective reactions of, **75**, 120

      SYNTHESIS OF, **75**, 116

LANTHANUM-LITHIUM-(R)-BINOL COMPLEX (LLB): **78**, 14

Lanthanum trichloride heptahydrate: Lanthanum chloride, heptahydrate; (10025-84-0), **78**, 15

LIGANDLESS PALLADIUM CATALYST, **75**, 61

Lipase Amano AK, **75**, 79

Lithium aluminum hydride: Aluminate (1-), tetrahydro-, lithium, (T-4); (16853-85-3), **75**, 80; **76**, 111; **77**, 29, 39, 66, 99; **78**, 161

Lithium amidotrihydroborate: Borate (1-), amidotrihydro-, lithium, I-4)-; (99144-67-9), **77**, 31; **78**, 161

Lithium bis(trimethylsilyl)amide: Silanamine, 1,1,1-trimethyl-N-(trimethylsilyl)-, lithium salt; (4039-32-1), **77**, 51

Lithium bromide; (7550-35-8), **78**, 179, 203

Lithium chloride; (7447-41-8), **76**, 58, 60, 241; **77**, 23; **78**, 53

Lithium diethyl phosphite, **75**, 20

Lithium diisopropylamide: 2-Propanamine, N-(1-methylethyl)-, lithium salt; (4111-54-0), **75**, 116; **76**, 59, 203, 239; **77**, 23, 31, 32, 98

Lithium dimethylcyanocuprate, **76**, 253

Lithium hydride; (7580-67-8), **75**, 195

Lithium hydroxide monohydrate; Lithium hydroxide, monohydrate; (1310-66-3), **75**, 195

Lithium methoxide: Methanol, lithium salt; (865-34-9), **76**, 58

LITHIUM PENTAFLUOROPROPEN-2-OLATE: 1-PROPEN-2-OL, 1,1,3,3,3-PENTAFLUORO-, LITHIUM SALT; (116019-90-0), **76**, 151

   reaction with electrophiles, table, **76**, 157-158

Lithium triethoxyaluminum hydride: Aluminate (1-), triethoxyhydro-, lithium (I-4); (17250-30-5), **77**, 30

Lithium wire; (7439-93-2), **76**, 25

2,6-Lutidine: Pyridine, 2,6-dimethyl-; (108-48-5), **77**, 163

Magnesium; (7439-95-4), **75**, 107; **76**, 13, 87, 221; **78**, 104

Magnesium perchlorate hexahydrate: Perchloric acid, magnesium salt, hexahydrate; (13446-19-0), **77**, 164

Malononitrile: HIGHLY TOXIC. Propanedinitrile; (109-77-3), **75**, 210

Manganese acetate tetrahydrate: Acetic acid, manganese (2+ salt), tetrahydrate; (6156-78-1), **75**, 4

Manganese(II) chloride tetrahydrate, **76**, 242

Manganese(II) chloride: Manganese chloride; (7773-01-5), **76**, 241, 242

Manganese(II) sulfate, **76**, 79

McMurry olefin synthesis, two extensions to, **76**, 145

Meldrum's acid: 1,3-Dioxane-4,6-dione, 2,2-dimethyl-; (2033-24-1), **77**, 114

l-Menthol, **76**, 29

(1R,2S,5R)-(-)-Menthyl (S)-p-toluenesulfinate: Benzenesulfinic acid, 4-methyl-, 5-methyl-2-(1-methylethyl)cyclohexyl ester, [1R-[1α(S*),2β,5α]];(188447-91-8), **77**, 51

β-MERCAPTOPROPIONITRILE: PROPANENITRILE, 3-MERCAPTO-; (1001-58-7), **77**, 186

Meroquinene tert-butyl ester: 4-Piperidineacetic acid, 3-ethenyl-, 1,1-dimethylethyl ester, (3R-cis)-; (52346-11-9), **75**, 225

Meroquinene esters, **75**, 231

Metal complexes, **76**, 18

Metallation conditions, **76**, 73

"Metals 44 solution", **76**, 78

Methanesulfonic acid; (75-75-2), **76**, 102

Methanesulfonyl chloride; (124-63-0), **75**, 108; **76**, 102; **78**, 99, 203

L-Methionine methyl ester hydrochloride: L-Methionine, methyl ester, hydrochloride; (2491-18-1), **76**, 123

p-Methoxybenzamide: Benzamide, 4-methoxy-; (3424-93-9) **78**, 234

(4R,5S)-3-(1-Methoxyethyl)-4,5-diphenyl-2-oxazolidinone: 2-Oxazolidinone, 3-(1-methoxyethyl)-4,5-diphenyl-, [4R-[3(R*),4α,5α]]-; (142977-52-4), **75**, 46

3-(METHOXYMETHOXY)-1-BUTANOL: 1-BUTANOL, 3-(METHOXYMETH-OXY)-;(60405-27-8), **75**, 177

(S,E)-1-(METHOXYMETHOXY)-1-TRIBUTYLSTANNYL-2-BUTENE: STANNANE, TRIBUTYL[1-(METHOXYMETHOXY)-2-BUTENYL]-,[S-(E)]-; (131433-64-2), **77**, 98

4-METHOXY-2'-METHYLBIPHENYL: 1,1'-BIPHENYL, 4'-METHOXY-2-METHYL-; (92495-54-0), **75**, 61

4-Methoxyphenylboronic acid: Boronic acid, (4-methoxyphenyl)-; (5720-07-0), **75**, 70

2-(4-METHOXYPHENYL)-2-CYCLOHEXEN-1-ONE: 2-CYCLOHEXEN-1-ONE, 2-(4-METHOXYPHENYL); (63828-70-6), **75**, 69

(R)-(+)- and (S)-(-)-α-Methoxy-α-(trifluoromethyl)phenylacetic acid: Benzeneacetic acid, α-methoxy-β,β,β-(trifluoromethyl)-, (R)-; (20445-31-2), **77**, 37, 72; (S)-; (17257-71-5)

(R)-(-)-α-Methoxy-α-(trifluoromethyl)phenylacetyl chloride: Benzeneacetyl chloride, α–methoxy-β,β,β-(trifluoromethyl)-, (R)-, (20445-33-4), **77**, 7, 40

Methyl acetate: Acetic acid, methyl ester; (79-20-9), **77**, 52

Methyl acrylate: 2-Propenoic acid, methyl ester; (96-33-3), **75**, 106; **77**, 109; **78**, 161

Methylamine: Methanamine; (74-89-5), **76**, 24

N-Methylaniline: Benzenamine, N-methyl-; (100-61-8), **78**, 24

(R)-α-METHYLBENZENEPROPANAL: BENZENEPROPANAL, α-METHYL- (R)-; (42307-9-5), **77**, 29

(R)-α-Methylbenzenepropanoic acid: Benzenepropanoic acid, α-methyl-(R)-; (14367-67-0), **77**, 35

(R)-β-METHYLBENZENEPROPANOL: BENZENEPROPANOL, β-METHYL-, (R)-; (77493-96-5), **77**, 31

N-Methylbenzimine: Methanamine, N-(phenylmethylene)-; (622-29-7), **76**, 24

(R)-(+)-α-Methylbenzylamine: Benzenemethanamine, α-methyl-, (R)-; (3886-69-9), **77**, 35

Methyl 2-(benzylamino)methyl-3-hydroxybutanoate: Butanoic acid, 3-hydroxy-2-[[(phenylmethyl)amino]methyl]-, methyl ester; (R*,R*)- (118559-03-8); (R*,S*)-(118558-99-9), **75**, 107

(R)-N-(α-Methylbenzyl)-α-methylbenzenepropanamide, **77**, 35

Methyl 2,2'-bipyridines, **78**, 59

4-Methyl-2,2'-bipyridine: 2,2'-Bipyridine, 4-methyl-; (56100-19-7), **78**, 51

5-METHYL-2,2'-BIPYRIDINE: 2,2'-BIPYRIDINE, 5-METHYL-; (56100-20-0), **78**, 51

6-Methyl-2,2'-bipyridine: 2,2'-Bipyridine, 6-methyl-; (56100-22-2), **78**, 51

Methyl (Z)-2-(bromomethyl)-4-methylpent-2-enoate: 2-Pentenoic acid, 2-(bromomethyl)-4-methyl-, methyl ester, (Z)-; (137104-29-3), **77**, 107

3-Methylbutan-2-one; (563-80-4), **75**, 119

2-Methyl-2-butene: 2-Butene, 2-methyl-; (513-35-9), **77**, 34

3-Methyl-3-buten-1-ol: 3-Buten-1-ol, 3-methyl-; (763-32-6), **78**, 203

Methyl tert-butyl ether: Propane, 2-methoxy-2-methyl- (9); (1634-04-4), **75**, 3; **77**, 199

Methyl carbamates by Hofmann rearrangement, **78**, 234

    table, **78**, 238

N-METHYL-N-(4-CHLOROPHENYL)ANILINE: BENZENAMINE, 4-CHLORO-N-METHYL-N-PHENYL-; (174307-94-9), **78**, 23

2-METHYLCYCLOHEXANONE: CYCLOHEXANONE, 2-METHYL-; (583-60-8), **76**, 240

Methyl 3-cyclopentene-1-carboxylate: 3-Cyclopentene-1-carboxylic acid, methyl ester: (58101-60-3), **75**, 197

Methyl cyclopropanecarboxylate: Cyclopropanecarboxylic acid, methyl ester; (2868-37-3), **78**, 142

(1'S,2'S)-METHYL-3O,4O-(1',2'-DIMETHOXYCYCLOHEXANE-1',2'-DIYL)-α-D-MANNOPYRANOSIDE: (α-D-MANNOPYRANOSIDE, METHYL 3,4-O-(1,2-DIMETHOXY, 1,2-CYCLOHEXANEDIYL)-, [3[S(S)]]-); (163125-35-7), **75**, 170

(3R),(4E)-METHYL 3-(DIMETHYLPHENYLSILYL)-4-HEXENOATE: 4-HEXENOIC ACID, 3-(DIMETHYLPHENYLSILYL)-, METHYL ESTER, [R-(E)]-; (136174-52-2), **75**, 78

(3S),(4E)-METHYL 3-(DIMETHYLPHENYLSILYL)-4-HEXENOATE: 4-HEXENOIC ACID, 3-(DIMETHYLPHENYLSILYL)-, METHYL ESTER, [S-(E)]-; (136314-66-4), **75**, 78

4-Methyl-1,3-dioxane: 1,3-Dioxane, 4-methyl-; (1120-97-4), **75**, 177

trans-2-METHYL-2,3-DIPHENYLOXIRANE: OXIRANE, 2-METHYL-2,3-DIPHENYL-, trans-; (23355-99-9), **78**, 225

7-Methylene-8-hexadecyn-6-ol: 8-Hexadecyn-6-ol, 7-methylene-; (170233-66-6), **76**, 264

Methyl formate: Formic acid, methyl ester; (107-31-3), **75**, 171

Methyl α-D-galactopyranoside: α-D-Galactopyranoside, methyl; (3396-99-4), **77**, 213

Methyl 3-hydroxy-2-methylenebutanoate: Butanoic acid, 3-hydroxy-2-methylene-, methyl ester; (18020-65-0), **75**, 106

METHYL 3-HYDROXYMETHYL-4-METHYL-2-METHYLENE-PENTANOATE: PENTANOIC ACID, 3-HYDROXY-4-METHYL-2-METHYLENE-, METHYL ESTER; (71385-30-1), **77**, 107

Methyl iodide: Methane, iodo-; (74-88-4), **75**, 19; **78**, 4

METHYL (S)-2-ISOCYANATO-3-PHENYLPROPANOATE: BENZENEPROPANOIC ACID, α-ISOCYANATO-, METHYL ESTER, (S)-; (40203-94-9), **78**, 220

Methyl (4S)-4,5-O-isopropylidenepent-(2Z)-enoate: 2-Propenoic acid, 3-(2,2-dimethyl-1,3-dioxolan-4-yl)-, methyl ester, [S-(Z)]-; (81703-94-8), **75**, 140

Methyllithum: Lithium, methyl-; (917-54-4), **75**, 99; **76**, 193, 253

Methylmagnesium bromide: Magnesium, bromomethyl-; (75-16-1), **77**, 154; **78**, 44

Methyl α-D-mannopyranoside; (617-04-9), **75**, 171

METHYL METHANETHIOSULFONATE: METHANESULFONOTHIOIC ACID, S-METHYL ESTER: (2949-92-0) **78**, 99

METHYL N-(p-METHOXYPHENYL)CARBAMATE: CARBAMIC ACID,N-(4-METHOXYPHENYL)-, METHYL ESTER; (14803-72-6) **78**, 234

METHYL 2,3-O-(6,6'-OCTAHYDRO-6,6'-BI-2H-PYRAN-2,2'-DIYL)-α-D-GALACTOPYRANOSIDE: α-D-GALACTOPYRANOSIDE, METHYL, 2,3-O-(OCTAHYDRO[2,2'-BI-2H-PYRAN]-2,2'-DIYL-,[2(2R,2'R)]-; (144102-32-9), **77**, 212

Methyl Orange: Benzenesulfonic acid, 4-[[4-(dimethylamino)phenyl]azo]-, sodium salt; (547-58-0), **77**, 237

METHYL (R)-(+)-β-PHENYLALANATE: BENZENEPROPANOIC ACID, β-AMINO-, (R)-, METHYL ESTER; (37088-67-8), **77**, 50

(R)-2-METHYL-1-PHENYL-3-HEPTANONE: 3-HEPTANONE, 2-METHYL-1-PHENYL-, (R)-; (159213-12-4), **77**, 32

[R-(R*,S*)]-β-METHYL-α-PHENYL-1-PYRROLIDINEETHANOL: 1-PYRROLIDINEETHANOL, β-METHYL-α-PHENYL-, [R-(R*,S*)]-;(127641-25-2), **77**, 12

Methyl (S)-2-phthalimido-4-methylthiobutanoate: 2H-Isoindole-2-acetic acid, 1,3-dihydro-α-[2-(methylthio)ethyl]-1,3-dioxo-, methyl ester, (S)-; (39739-05-4), **76**, 123

METHYL (S)-2-PHTHALIMIDO-4-OXOBUTANOATE: 2H-ISOINDOLE-2-ACETIC ACID, 1,3-DIHYDRO-1,3-DIOXO-α-(2-OXOETHYL)-, METHYL ESTER, (S)-; (137278-36-5), **76**, 123

2-Methyl-2-propanol: 2-Propanol, 2-methyl-; (75-65-0), **75**, 225; **77**, 34

2-Methylpropanoyl chloride: Propanoyl chloride, 2-methyl-; (79-30-1), **75**, 118

2-METHYL-4H-PYRAN-4-ONE: 4H-PYRAN-4-ONE, 2-METHYL-; (5848-33-9), **77**, 115

Methylpyridines, selective monohalogenation, **78**, 86

　　　table, **78**, 88

4-Methyl-2-pyridyl triflate: Methanesulfonic acid, trifluoro-, 4-methyl-2-pyridinyl ester; (179260-78-7), **78**, 52

308

6-Methyl-2-pyridyl triflate: Methanesulfonic acid, trifluoro-, 6-methyl-2-pyridinyl ester; (154447-04-8), **78**, 52

1-Methyl-2-pyrrolidinone: 2-Pyrrolidinone, 1-methyl-; (872-50-4), **76**, 240; **77**, 135

Methyl serinate hydrochloride, **77**, 65

trans-α-Methylstilbene: Benzene, 1,1'-(1-methyl-1,2-ethenediyl)bis-, (E)-; (833-81-8), **78**, 225

α-Methylstyrene: Benzene, (1-methylethenyl)-; (98-83-9), **76**, 159

(S$_S$,R)-(+)-Methyl N-(p-toluenesulfinyl)-3-amino-3-phenylpropanoate: Benzenepropanoic acid, β-[[(4-methylphenyl)sulfinyl]amino]-, methyl ester, [S-(R*,S*)]-; (158009-86-0), **77**, 52

Methyl triflate: Methyl trifluoromethanesulfonate: Methanesulfonic acid, trifluoro-, methyl ester; (333-27-7), **77**, 81

5-Methyl-2-(trifluoromethanesulfonyl)oxypyridine: Methanesulfonic acid, trifluoro-, 5-methyl-2-pyridinyl ester; (154447-03-7), **78**, 52

Methyl (triphenylphosphoranylidene)acetate: Propanoic acid, 2-(triphenylphosphoranylidene)-, methyl ester; (2605-67-6), **75**, 139

Methyl vinyl ketone: 3-Buten-2-one; (78-94-4), **78**, 249

Michael reaction, catalyzed by transition metals, **78**, 251

Mineral salt bath (MSB), **76**, 77

Mitsunobu displacement, **76**, 185

4 Å Molecular sieves: Zeolites, 4 Å; (70955-01-0), **75**, 12, 189

MONOALKYLATION, REGIOSELECTIVE, OF KETONES, VIA MANGANESE ENOLATES, **76**, 239

      table, **76**, 247, 249

MONO-C-METHYLATION OF ARYLACETONITRILES AND METHYL ARYLACETATES BY DIMETHYL CARBONATE, **76**, 169

      table, **76**, 177

Mosher esters, **77**, 7, 37, 71

2-NAPHTHYLMAGNESIUM BROMIDE: MAGNESIUM, BROMO-2-NAPHTHALENYL-; (21473-01-8), **76**, 13

Negishi cross-coupling reaction, **78**, 57

Nickel-catalyzed coupling, **78**, 42

Nitric acid; (7697-37-2), **75**, 90

Nitriles, into secondary and tertiary amines, **76**, 133

    table, **76**, 140

Nitrilotriacetic acid, CANCER SUSPECT AGENT: Glycine, N-bis(carboxymethyl)-; (139-13-9), **76**, 78

Nitrobenzene: Benzene, nitro-; (98-95-3), **77**, 2

o-Nitrobenzenesulfinic acid: Benzenesulfinic acid, 2-nitro-: (13165-79-2), **76**, 185

o-Nitrobenzenesulfonyl chloride: Benzenesulfonyl chloride, 2-nitro-; (1694-92-4), **76**, 183

o-Nitrobenzenesulfonyl hydrazide: Benzenesulfonic acid, 2-nitro-, hydrazide; (5906-99-0), **76**, 180, 183

2-Nitroethanol: Ethanol, 2-nitro-; (625-48-9), **78**, 16

Nitrogen manifold, **78**, 75, 81

Nitrogen oxides, **75**, 90

(2S,3S)-2-NITRO-5-PHENYL-1,3-PENTANEDIOL; **78**, 14

3-NITROPROPANAL:: PROPANAL, 3-NITRO-; (58657-26-4), **77**, 236

3-Nitropropanal diethyl acetal: Propane, 1,1-diethoxy- 3-nitro-; (107833-73-8), **77**, 241

3-NITROPROPANAL DIMETHYL ACETAL: PROPANE, 1,1-DIMETHOXY-3-NITRO-; (72447-81-5), **77**, 236

3-NITROPROPANOL: 1-PROPANOL, 3-NITRO-; (25182-84-7), **77**, 236

NMR Shift reagents, **76**, 18

Nonacarbonyldiiron: Iron, tri-μ-carbonylhexacarbonyldi-, (Fe-Fe); (15321-51-4) **78**, 189

1-Nonyne; (3452-09-3), **76**, 264

(1R.2S)-Norephedrine: Benzeneethanol, α-(1-aminoethyl)-, [R-(R*,S*)]- ; (492-41-4), **77**, 12

2,3,7,8,12,13,17,18-Octaethylporphyrin, **76**, 291

1-Octyn-3-ol; (818-72-4), **76**, 264

Oligonucleoside phosphorothioates, syntheses of, **76**, 272

Optically active C_4 building blocks, **78**, 130

Optically active diols, synthesis of, **76**, 106

Organocopper chemistry, **76**, 257

ORGANOLITHIUMS, **76**, 228

Organometallic chemistry, **76**, xx

Osmium oxide: See Osmium tetroxide, **75**, 109

Osmium tetroxide: Osmium oxide, (T-4)-; (20816-12-0), **75**, 108

11-OXATRICYCLO[4.3.1.1^{2,5}]UNDEC-3-EN-10-ONE, (1α,2β,5β,6α)-; (42768-72-9), **78**, 212

Oxalyl chloride: Ethanedioyl dichloride; (79-37-8), **76**, 112; **77**, 37, 67

Oxalyl chloride-dimethyl sulfoxide (Swern reagent), **76**, 112

Oxidative degradation, of aromatic compounds, by *Pseudomonas*, **76**, 81

2-(3-OXOBUTYL)CYCLOPENTANONE-2-CARBOXYLIC ACID, ETHYL ESTER: CYCLOPENTANECARBOXYLIC ACID, 2-OXO-1-(3-OXOBUTYL)-, ETHYL ESTER; (61771-81-1), **78**, 249

1-OXO-2-CYCLOHEXENYL-2-CARBONITRILE: 1-CYCLOHEXENE-1-CARBONITRILE, 6-OXO-; (91624-93-0) **78**, 254

Oxone: Peroxymonosulfuric acid, monopotassium salt, mixt. with dipotassiumsulfate and potassium hydrogen sulfate; (37222-66-5), **78**, 225

Ozone; (10028-15-6) **78**, 254

Ozonolysis apparatus, **78**, 255

Palladium, 10% on carbon, **75**, 226; **77**, 135

Palladium acetate: Acetic acid, palladium (2+) salt; (3375-31-3), **75**, 53, 61; **78**, 3, 24, 36

Palladium(II) bis(benzonitrile)dichloride: Palladium, bis(benzonitrile)dichloro-; (14220-64-5), **75**, 70

Palladium (II) chloride bisacetonitrile: Palladium, bis(acetonitrile)dichloro-; (14592-56-4), **77**, 213

PALLADIUM CATALYST, LIGANDLESS, **75**, 61

20% Palladium hydroxide on carbon, **75**, 21

Paraformaldehyde; (30525-89-4), **78**, 169

Paraldehyde: 1,3,5-Trioxane, 2,4,6-trimethyl-; (123-53-7), **78**, 124

Parr shaker, **75**, 24, 226

PENTA-1,2-DIEN-4-ONE: 3,4-PENTADIEN-2-ONE; (2200-53-5), **78**, 135

Pentaerythritol: 1,3-Propanediol, 2,2-bis(hydroxymethyl)-; (115-77-5), **75**, 89

Pentaerythrityl tetrachloride: Propane, 1,3-dichloro-2,2-bis(chloromethyl)-; (3228-99-7), **75**, 89

Pentaerythrityl trichlorohydrin: 1-Propanol, 3-chloro-2,2-bis(chloromethyl)-; (813-99-0); **75**, 89

2,4-Pentanedione; (123-54-6), **77**, 2

(2-Pentanol; (6032-29-7), **76**, 255

2-PENTYL-3-METHYL-5-HEPTYLFURAN: FURAN, 5-HEPTYL-3-METHYL-2-PENTYL-; (170233-67-7), **76**, 263

Perchlorate salts, **77**, 210

cis-4a(S),8a(R)-PERHYDRO-6(2H)-ISOQUINOLINONES, **75**, 223

pH 7 Buffer, **76**, 253

Phase transfer catalysis, **76**, 273

Phenanthrenes, substituted, **76**, 296

    table, **76**, 300

1,10-Phenanthroline; (66-71-7), **76**, 91, 255

Phenol: HIGHLY TOXIC; (108-95-2), **75**, 118

Phenylacetonitrile: Benzeneacetonitrile; (140-29-4), **76**, 169

Phenylacetylene: Benzene, ethynyl-; (536-74-3), **75**, 146

(S)-Phenylalanine: L-Phenylalanine; (63-91-2), **76**, 110

L-Phenylalanine methyl ester hydrochloride: L-Phenylalanine methyl ester, hydrochloride; (7524-50-7), **78**, 220

Phenyl butanoate: Butanoic acid, phenyl ester; (4346-18-3), **75**, 119

Phenyl chlorothionoformate: Carbonochloridothioic acid, O-phenyl ester; (1005-56-7), **78**, 239

(2S,3S)-(+)-(3-PHENYLCYCLOPROPYL)METHANOL: CYCLOPROPANEMETHANOL, 2-PHENYL-, (1S-)-; (110659-58-0), **76**, 86

Phenyl decanoate: Decanoic acid, phenyl ester; (14353-75-4), **75**, 119

5-PHENYLDIPYRROMETHANE: 1H-PYRROLE, 2,2'-(PHENYLMETHYL)BIS-; (107798-98-1), **76**, 287

PHENYL ESTER ENOLATES, IN SYNTHESIS OF β-LACTONES, **75**, 116

(R)-(-)-2-Phenylglycinol: Benzeneethanol, β-amino-, (R)-; (56613-80-0), **75**, 19

6-PHENYLHEX-2-YN-5-EN-4-OL: 1-HEXEN-4-YN-3-OL, 1-PHENYL-; (63124-68-5), **76**, 214

Phenylmagnesium bromide: Magnesium, bromophenyl-; (100-58-3), **76**, 103; **78**, 5

N-[(1R)-Phenyl-(2R)-methoxyethyl]-isovaleraldehyde imine, **75**, 20

(S)-1-(PHENYLMETHOXY)-4-PENTEN-2-OL: 4-PENTEN-2-OL, 1-(PHENYLMETHOXY)-, (S)-;(88981-35-5), **75**, 12

Phenyl 2-methylpropanoate: Propanoic acid, 2-methyl-, phenyl ester; (20279-29-2), **75**, 116

3-Phenylpropanal: Benzenepropanal; (104-53-0) **78**, 16

2-PHENYLPROPIONIC ACID: BENZENEACETIC ACID, α-METHYL-; (492-37-5), **76**, 169

2-Phenylpropionitrile: Benzeneacetonitrile, α-methyl-; (1823-91-2), **76**, 171

3-Phenylpropyltriphenylphosphonium bromide: Phosphonium, triphenyl(3-phenylpropyl)-, bromide; (7484-37-9), **75**, 153

preparation, **75**, 155

3-PHENYL-2-PROPYNENITRILE: 2-PROPYNENITRILE, 3-PHENYL-; (935-02-4), **75**, 146

4-Phenylpyridine N-oxide: Pyridine, 4-phenyl-, 1-oxide; (1131-61-9), **76**, 47

2-PHENYL-1-PYRROLINE: 2H-PYRROLE, 3,4-DIHYDRO-5-PHENYL-; (700-91-4), **75**, 215

(PHENYL) [2(TRIMETHYLSILYL)PHENYL]IODONIUM TRIFLATE: IODONIUM, PHENYL-, 2-(TRIMETHYLSILYL)PHENYL-, SALT WITH TRIFLUOROMETHANESULFONIC ACID (1:1); (164594-13-2) **78**, 104

Phosphate buffer, aqueous, **77**, 67

Phosphomolybdic acid: Molybdophosphoric acid ($H_3PMo_{12}O_{40}$), hydrate; (51429-74-4), **77**, 38,101

Phosphorus oxychloride: Phosphoryl chloride; (10025-87-3), **77**, 199

Phosphorus pentoxide: Phosphorus oxide; (1314-56-3), **76**, 265

Phosphorus trichloride; (7719-12-2), **76**, 29

PHOTOINDUCED ADDITION, OF ALCOHOLS, **75**, 139

Photochemical reaction, **77**, 122, 164, 249

Photodeoxygenation, **77**, 157

Photolysis, **75**, 125

 of ergosterol, **76**, 283

Photolysis products from ergosterol: pro-, pre-, lumi-, and tachy-, **76**, 283

Phthalic anhydride: 1,3-Isobenzofurandione; (85-44-9), **76**, 123

Pinacol: 2,3-Butanediol, 2,3-dimethyl-; (76-09-5), **75**, 98; **77**, 178

Pinacol rearrangement, **76**, 296

N-Pivaloyl-o-toluidine: Propanamide, 2,2-dimethyl-N-(2-methylphenyl)-; (61495-04-3), **76**, 215

Poly(methylhydrosiloxane: PMHS: Poly[oxy(methylsilylene)]; (9004-73-3), **78**, 240

Polyphenylene, **76**, 81

Porphyrin model studies, **76**, 291

Potassium tert-butoxide: 2-Propanol, 2-methyl-, potassium salt; (865-47-4), **75**, 224; **76**, 200; **78**, 114, 144

Potassium bromate: Bromic acid, potassium salt ; (7758-01-2), **77**, 141

Potassium dihydrogen phosphate: Phosphoric acid, monopotassium salt; (7778-77-0), **76**, 78

Potassium (R)-(+)-2.3-epoxypropanoate: Oxiranecarboxylic acid, potassium salt, (R)-; (82044-23-3), **75**, 37

Potassium glycidate: See Potassium (R)-(+)-2,3-epoxypropanoate, **75**, 38

Potassium hexafluorophosphate: Phosphate(1-), hexafluoro-, potassium; (17084-13-8), **77**, 207

Potassium hydride; (7693-26-7), **75**, 20

Preculture preparation, **76**, 77

4a,9a-Propano-4H-cyclopenta[5,6]pyrano[2,3-d]-1,3-dioxin-6,12(5H)-dione, **75**, 191

Propanoic acid: See Propionic acid, **75**, 80

1-Propanol; (71-23-8), **75**, 53

Propargyl alcohol: 2-Propyn-1-ol; (107-19-7), **76**, 178, 200

2-Propargyloxytetrahydropyran: 2H-Pyran, tetrahydro-2-(2-propynyloxy)-; (6089-04-9), **76**, 178

[1.1.1]PROPELLANE: TRICYCLO[1.1.1.0^{1,3}]PENTANE; (35634-10-7), **75**, 98; **77**, 249

Propionaldehyde: Propanal; (123-38-6), **76**, 152

Propionic acid: Propanoic acid ; (79-09-4), **75**, 80

Propionic anhydride: Propanoic acid, anhydride; (123-62-6), **77**, 22

Propyl alcohol: See 1-Propanol, **75**, 53

1-PROPYNYLLITHIUM: LITHIUM, 1-PROPYNYL-; (4529-04-8), **76**, 214

      addition to electrophiles, table, **76**, 218

Protected vicinal amino alcohols, synthesis of, **77**, 78

Protection, for diequatorial vicinal diols, table, **77**, 218

      of trans-hydroxyl groups in sugars, **75**, 173

        table, **75**, 174

      of primary and secondary amines, **75**, 167, 232

      selective, of 1,3-diols, **75**, 177

        table, **75**, 182

      selective, of primary alcohols, **75**, 186

Protective groups:

      Boc-, **76**, 72

2-cyanoethyl, **77**, 193

1-ethoxyethyl, removal of, **76**, 207

in phosphoramidite syntheses, **76**, 273

phthaloyl, for amino groups, **76**, 129

Pseudoephedrine amides, **77**, 22, 29

(R,R)-(-)-Pseudoephedrine: Benzenemethanol, α-[1-(methylamino)ethyl]-, [R-(R*,R*)]-; (321-97-1), **76**, 58

recovery, **76**, 63

(1S,2S)-(+)-Pseudoephedrine: Benzenemethanol, α- [1-(methylamino)ethyl]-, (R*,S*)-(±)-; (90-82-4), **76**, 58; **77**, 22

(R,R)-(-)-Pseudoephedrine L-allylglycinamide: 4-Pentenamide, 2-amino-N-(2-hydroxy-1-methyl-2-phenylethyl)-N-methyl, [1S-[1R*(S*), 2R*]]-; (170642-23-6), **76**, 59

(R,R)-(-)-Pseudoephedrine D-allylglycinamide diacetate, **76**, 70

(R,R)-(-)-Pseudoephedrine L-allylglycinamide diacetate, **76**, 70

(R,R)-(-)-Pseudoephedrine glycinamide: Acetamide, 2-amino-N-(2-hydroxy-1-methyl-2-phenylethyl)-N-methyl-, [R-(R*,R*)]-; (170115-98-7), **76**, 57; [S-(R*,R*)]-; (170115-96-5), **76**, 71

(R,R)-(-)-Pseudoephedrine glycinamide diacetate, **76**, 70

(R,R)-(-)-Pseudoephedrine glycinamide monohydrate, **76**, 58

(1S,2S)-Pseudoephedrine-(R)-2-methylhydrocinnamamide, **77**, 23, 30, 31, 32, 42

(1S,2S)-Pseudoephedrinepropionamide, **77**, 22

Pseudomonas putida 39/D, **76**, 77

4H-Pyran-4-one; (108-97-4), **78**, 114

as a phenol precursor, **78**, 119

Pyridine; (110-86-1),**75**, 69, 89, 210, 225; **76**, 142, 102, 277; **77**, 114; **78**, 2, 27, 52, 143, 240

Pyridinium chloride: Pyridinium hydrochloride: (628-13-7), **76**, 102

Pyridinium tetrafluoroborate: Pyridine, tetrafluoroborate (1-); (505-07-7), **75**, 211

2-Pyridyl triflates, **78**, 58

γ-Pyrones, unsymmetrically substituted, **77**, 117

    table, **77**, 119

Pyrrole: 1H-Pyrrole; (109-97-7), **76**, 288

Pyrrolidine; (123-75-1), **78**, 212

1-PYRROLINES,

    2-SUBSTITUTED, **75**, 215

        table, **75**, 221

    2,3-disubstituted,

        table, **75**, 222

Quinine: Cinchonan-9-ol, 6'-methoxy-, (8a, 9R)-; (130-95-0), **75**, 224

Quininone: Cinchon-9-one, 6'-methoxy-, (8a)-; (84-31-1), **75**, 224

Quinolizidines, **76**, 36

Racemic pinitol, **76**, 81

Reagents and compounds, useful, syntheses of, **76**, xxi

Rearrangements, **76**, xix

Reductive cyclizations with Ti reagents, table, **76**, 149-150

Reductive ring expansion, **78**, 78

Resolution, OF 1,1'-BI-2-NAPHTHOL, **76**, 1

    of a dl-diamine, **76**, 26

    of trans-1,2-diaminocyclohexane, **75**, 9

Ritter reaction, **76**, 52

(Salen)metal complexes, **75**, 10

(Salen)Mn-catalyzed epoxidation reactions, **76**, 53

Salicyladehydes, 3,5-substituted, **75**, 10

Scandium(III) trifluoromethanesulfonate: Methanesulfonic acid, trifluoro-, scandium (3+) salt; (144026-79-9), **77**, 45

Secondary amines, from nitriles, **76**, 137

L- or (S)-Serine; (56-45-1), **75**, 37; **77**, 65

Shaker, benchtop, orbital incubator, **76**, 77

Silver nitrate, : Nitric acid, silver salt; (7761-88-8), **77**, 207

~ 10 wt.% on silica gel, **76**, 265

Silver(I) oxide: Silver oxide (Ag$_2$O) (9); (20667-12-3), **75**, 70

Silver tetrafluoroborate: Borate(1-), tetrafluoro-, silver (1 +); (14104-20-2), **78**, 213

Sodium; (7440-23-5), **75**, 79; **77**, 65; **78**, 5

Sodium azide (26628-22-8), **78**, 203

Sodium bis(trimethylsilyl)amide: Silanamine, 1,1,1-trimethyl-N-(trimethylsilyl)-, sodium salt; (1070-89-9), **77**, 52, 254; **78**, 153

Sodium bisulfite: Sodium hydrogen sulfite: Sulfurous acid, monosodium salt (7631-90-5), **75**, 161; **76**, 255; **77**, 250; **78**, 171

Sodium borohydride: Borate (1-), tetrahydro-, sodium-; (16940-66-2), **76**, 134; **77**, 71, 81, 237

Sodium tert-butoxide: 2-Propanol, 2-methyl-, sodium salt (865-48-5), **78**, 15, 23

Sodium chlorite, **77**, 34

Sodium fluoride; (7681-49-4), **77**, 136

Sodium hexamethyldisilazide: Silanamine, 1,1,1-trimethyl-N-(trimethylsilyl)-, sodium salt; (1070-89-9), **76**, 37

Sodium hydride; (7646-69-7), **75**, 153, 215; **77**, 92; **78**, 125, 180

Sodium hydrogen sulfite: Sulfurous acid, monosodium salt; (7631-90-5), **78**, 171

Sodium hypochlorite: Hypochlorous acid, sodium salt; (7681-52-9), **76**, 47

for scavenging sulfur compounds, **76**, 124

Sodium iodide; (7681-82-5), **76**, 255; **77**, 91

Sodium metaperiodate: Periodic acid, sodium salt; (7790-28-5), **75**, 108

Sodium methoxide: Methanol, sodium salt; (124-41-4), **78**, 64

Sodium nitrite: Nitrous acid, sodium salt; (7632-00-0), **75**, 37; **77**, 237; **78**, 52

Sodium omadine: See Sodium 2-pyridinethiol-1-oxide, **75**, 125

Sodium periodate: Periodic acid, sodium salt; (7790-28-5), **78**, 124

Sodium 2-pyridinethiol-1-oxide (sodium omadine); (3811-73-2), **75**, 125

Sodium sulfide, **76**, 95

Sodium sulfite: Sulfurous acid, disodium salt; (7757-83-7), **76**, 90

Sodium tetraborate decahydrate: Borax; (1303-96-4), **76**, 79

Sodium thiosulfate: Thiosulfuric acid, disodium salt (7772-98-7), **77**, 92, 143

Sodium β-trimethylsilylethanesulfonate: Ethanesulfonic acid, 2-(trimethylsilyl)-, sodium salt; (18143-40-3), **75**, 161

Sonication, **76**, 232

Squarates, alkyl, **76**, 193

Squaric acid: 3-Cyclobutene-1,2-dione, 3,4-dihydroxy-; (2892-51-5), **76**, 190

Stannanes, chiral α-alkoxyallylic, **77**, 103

Stereochemical purity, determination of, **76**, 52

Stereodefined synthesis, of allenes, **76**, 185

Sterile loop, **76**, 77

STILLE COUPLINGS, **77**, 135

"Streaking for isolation", **76**, 79

SULFINIMINES (THIOOXIMINE S-OXIDES), **77**, 50

Sulfonyl halides, reduction with zinc, **78**, 99

Sulfur dioxide, **75**, 90

Sulfuric acid, fuming: Sulfuric acid, mixt. with sulfur trioxide; (8014-95-7), **76**, 47

Superacidic triflic acid, **76**, 296

SUZUKI COUPLING, **75**, 69

      ACCELERATED, **75**, 61

      MODIFIED, **75**, 53

Synthetic transformations, **76**, xix

Synthons, haloalkenes, **76**, 224

TADDOLS, **76**, xvii, 14

DL-(±)-Tartaric acid: Butanedioic acid, 2,3-dihydroxy-, (R*,R*)-; (133-37-9), **76**, 25

L-Tartaric acid: Butanedioic acid, 2,3-dihydroxy-, [R-(R*,R*)]-; (87-69-4),**75**, 2; **76**, 25, 48

Tedlar bag, **75**, 131

Tertiary amines, from nitriles, **76**, 133

1,4,7,10-TETRAAZACYCLODODECANE; (294-90-6), **78**, 73

1,2,4,5-Tetrabromobenzene: Benzene, 1,2,4,5-tetrabromo-; (636-28-2), **75**, 201

Tetrabutylammonium bromide: 1-Butanaminium, N,N,N-tributyl-, bromide; (1643-19-2), **76**, 271

Tetrabutylammonium fluoride: 1-Butanaminium, N,N,N-tributyl-, fluoride; (429-41-4), **76**, 254; **78**, 106

Tetrabutylammonium fluoride trihydrate: Ammonium, tetrabutyl-, fluoride, hydrate; (22206-57-1), **77**, 79; **78**, 106

Tetrabutylammonium hydrogen sulfate: 1-Butanaminium, N,N,N-tributyl-, sulfate (1:1); (32503-27-8), **78**, 204

Tetraethylammonium bromide: Ethanaminium, N,N,N-triethyl-, bromide; (71-91-0), **76**, 264

(+)-(E,1R,3S)-Tetracarbonyl[(3-benzyloxy)-1-(phenylsulfonyl)-η^2-but-1-ene]iron(0): Iron, tetracarbonyl[[[[(2-3η)-1-methyl-3-(phenylsulfonyl)-2-propenyl]oxy]methyl] benzene]-, stereoisomer; (168431-28-5), **78**, 189

  removal of paramagnetic impurities, **78**, 193

(1R,2S,3R)-TETRACARBONYL[1-3η]-1-(PHENYLSULFONYL)-BUT-2-EN-1-YL]IRON(1+) TETRAFLUOROBORATE: IRON(1+), TETRACARBONYL[(1,2,3-η)-1-(PHENYLSULFONYL)-2-BUTENYL]-, STEREOISOMER, TETRAFLUOROBORATE(1-); (162762-06-3), **78**, 189

  as an allylic cation synthon, **78**, 194

anti-1,4,5,8-TETRAHYDROANTHRACENE-1,4:5,8-DIEPOXIDE:1,4:5,8-DIEPOXYANTHRACENE, 1,4,5,8-TETRAHYDRO-, (1α,4α,5β,8β)- ; (87207-46-3), **75**, 201 syn-1,4,5,8-TETRAHYDROANTHRACENE-1,4:5,8-DIEPOXIDE: 1,4:5,8-DIEPOXYANTHRACENE, 1,4,5,8-TETRAHYDRO-, (1α,4α,5α,8α)-; (87248-22-4), **75**, 201

1,2,3,4-TETRAHYDROCARBAZOLE: 1H-CARBAZOLE, 2,3,4,9-TETRAHYDRO-; (942-01-8), **78**, 36

Tetrahydrothiopyran-4-one: 4H-Thiopyran-4-one, tetrahydro-; (1072-72-6), **78**, 225

Tetrakis(dimethylamino)diboron: Diborane (4) tetramine, octamethyl-; (1630-79-1), **77**, 177

Tetrakis(triphenylphosphine)palladium(0); Palladium, tetrakis(triphenyl-phosphine)-, (T-4)-; (14221-01-3), **78**, 53

 alternatives to, **75**, 57

α-Tetralone: 1(2H)-Naphthalenone, 3,4-dihydro-; (529-34-0), **76**, 230

2,3,4,4-Tetramethoxy-2-cyclobuten-1-one: 2-Cyclobuten-1-one, 2,3,4,4-tetramethoxy-; (67543-98-0), **76**, 192

1,1,2,2-Tetramethoxycyclohexane: Cyclohexane, 1,1,2,2-tetramethoxy-; (163125-34-6), **75**, 170

(R,R)-(+)-N,N,N',N'-Tetramethyltartaric acid diamide: Butanediamide, 2,3-dihydroxy-, N,N,N',N'-tetramethyl-, [R-(R*,R*)]-; (26549-65-5), **76**, 89

 recovery, **76**, 90

5,10,15,20-Tetraphenylporphyrin, **76**, 291

Tetravinyltin; (1112-56-7), **76**, 193

Thiazole Aldehyde Synthesis, **77**, 85

4,4'-Thiobis[2-(2-tert-butyl-m-cresol): Phenol, 4,4'-thiobis[2-(1,1-dimethylethyl)-3-methyl-; (4120-97-2), **76**, 201, 202

Thionyl chloride; (7719-09-7), **75**, 89, 162

Thionyl chloride-dimethylformamide, **75**, 162

THIOOXIMINE S-OXIDES (SULFINIMINES), **77**, 50

Thiophenol: Benzenethiol; (108-98-5), **75**, 100; **78**, 169

Thiophosphoryl chloride, HIGHLY TOXIC; (3982-91-0), **76**, 271

Thiourea;(62-56-6), **77**, 186

Thiosulfonic S-esters, **78**, 99

table, **78** 103

L-Threitol 1,4-bismethanesulfonate: 1,2,3,4-Butanetetrol, 1,4-dimethanesulfonate, [S-(R*,R*)]-; (299-75-2), **76**, 101

Thymine: 2,4(1H,3H)-Pyrimidinedione, 5-methyl-; (65-71-4), **77**, 163

Tin(IV) chloride: Stannane, tetrachloro-; (7646-78-8), **77**, 164

Titanium(III) chloride; (7705-07-9), **76**, 143

Titanium(IV) isopropoxide: See Titanium tetraisopropoxide, **75**, 12

Titanium reagents, for reductive cyclizations, **76**, 149

Titanium tetraisopropoxide: 2-Propanol, titanium(4+) salt; (546-68-9), **75**, 12; **78**, 142

o-Tolueneboronic acid: See o-Tolylboronic acid, **75**, 61

Toluene dioxygenase, **76**, 80

(S)-(+)-p-Toluenesulfinamide: Benzenesulfinamide, 4-methyl-, (S)-; (1517-82-4), **77**, 51

p-Toluenesulfinic acid: Benzenesulfinic acid, 4-methyl-; (536-57-2), **77**, 198

p-Toluenesulfinic acid, sodiun salt: Benzenesulfinic acid, 4-methyl-, sodium salt; (824-79-3), **77**, 198

p-Toluenesulfonic acid monohydrate: Benzenesulfonic acid, 4-methyl-, monohydrate; (6192-52-5), **76**, 180, 200, 203; **77**, 115, 125, 238, 241; **78**, 64

p-Toluenesulfonic anhydride (4124-41-8), **77**. 225

p-Toluenesulfonyl chloride: Benzenesulfonyl chloride, 4-methyl-; (98-59-9), **77**, 220, 225

6^A^-O-p-TOLUENESULFONYL-β-CYCLODEXTRIN: β-CYCLODEXTRIN, 6^A^-(4-METHYLBENZENESULFONATE); (67217-55-4), **77**, 220, 225

1-(p-Toluenesulfonyl)imidazole: 1H-Imidazole, 1-[(4-methylphenyl)sulfonyl]-; (2232-08-8), **77**, 225

m-Toluoyl chloride: Benzoyl chloride, 3-methyl-; (1711-06-4), **75**, 2

o-Tolylboronic acid: o-Tolueneboronic acid ; (16419-60-6), **75**, 61

N-(α-Tosylbenzyl)formamide: Formamide, N-[[(4-methylphenyl)-sulfonyl]phenylmethyl]-; (37643-54-2), **77**, 199

α-TOSYLBENZYL ISOCYANIDE: BENZENE, 1-[(ISOCYANOPHENYL--METHYL)SULFONYL]-4-METHYL-; (36635-66-2), **77**, 198

Tosylmethyl isocyanides, substituted, table, **77**, 202

Transmetallation, **76**, 258

1,1,1-TRIACETOXY-1,1-DIHYDRO-1,2-BENZIODOXOL-3(1H)-ONE: 1,2-BENZIODOXOL-3(1H)-ONE, 1,1,1-TRIS(ACETYLOXY)-1,1-DIHYDRO-; (97413-09-0), **77**, 141

1,3,5-O-Tribenzoyl-α-D-ribofuranose: α-D-Ribofuranose, 1,3,5-tribenzoate; (22224-41-5), **77**, 163

1,3,5-O-Tribenzoyl-2-O-[(3-trifluoromethylbenzoyl]-α-D-ribofuranose: α-D-Ribofuranose, 1,3,5-tribenzoate 2-[3-(trifluoromethyl)benzoate; (145828-13-3), **77**, 163

Tri-tert-butylphosphine: Phosphine, tris(1,1-dimethylethyl)-; (13716-12-6), **75**, 81

Tributylstannane: Stannane, tributyl-; (688-73-3), (in situ) **78**, 239

2-(Tributylstannyl)thiophene: Stannane, tributyl-2-thienyl-; (54663-78-4), **77**, 135

Tributyltin hydride: Stannane, tributyl-; (688-73-3), **77**, 98

Tricarbonyl (η^6-cycloheptatriene)chromium(0): Chromium, tricarbonyl[(1,2,3,4,5,6-η)-1,3,5-cycloheptatriene]-; (12125-72-3), **77**, 121

Trichloroacetic acid: Acetic acid, trichloro-; (76-03-9), **76**, 288

Trichloroacetonitrile: Acetonitrile, trichloro-; (545-06-2), **78**, 180

Trichloromethyl chloroformate: See Diphosgene, **75**, 48

Trichlorosilane: Silane, trichloro-; (10025-78-2), **78**, 4

endo-TRICYCLO[3.2.1.0^{2,4}]OCT-6-ENE:TRICYCLO[3.2.1.0^{2,4}]OCT-6-ENE, (1α, 2α, 4α, 5α,)-, (3635-94-7), **77**, 255

TRICYCLO[1.1.1.0^{1,3}]PENTANE: See [1.1.1]PROPELLANE, **75**, 98

Triethylamine: Ethanamine, N,N-diethyl-; (121-44-8), **75**, 45, 108; **76**, 112, 123, 202, 264; **77**, 3, 65, 80, 163, 200; **78**, 4, 136, 179

Triethylamine hydrobromide: Ethanamine, N,N-diethyl-, hydrobromide; (636-70-4), **78**, 136

Triethylamine trishydrofluoride: Ethanamine, N,N-diethyl-, trishydrofluoride; (73602-61-6), **76**, 159

Triethylenetetramine: 1,2-Ethanediamine, N,N'-bis(2-aminoethyl)-; (112-24-3), **78**, 74

Triethyl orthoformate: Ethane, 1,1',1''-[methylidynetris(oxy)]tris-; (122-51-0), **77**, 241

Triethyl phosphite: Phosphorous acid, triethyl ester; (122-52-1), **78**, 170

Triethyl phosphonoacetate: Acetic acid, (diethoxyphosphinyl)-, ethyl ester; (867-13-0), **78**, 125

Triflic anhydride: Methanesulfonic acid, trifluoro-, anhydride; (358-23-6), **76**, 81

Trifluoroacetic acid: Acetic acid, trifluoro-; (76-05-1), **75**, 130; **76**, 207, 288; **77**, 30, 53

Trifluoroacetic anhydride: Acetic acid, trifluoro-, anhydride; (407-25-0), **76**, 95, 190, 193

1,1,1-TRIFLUORO-2-ETHOXY-2,3-EPOXY-5-PHENYLPENTANE: OXIRANE, 2-ETHOXY-3-(2-PHENYLETHYL)-2-(TRIFLUOROMETHYL)-, cis-(±)-; (141937-91-9), **75**, 153

(Z)-1,1,1-TRIFLUORO-2-ETHOXY-5-PHENYL-2-PENTENE: BENZENE, (4-ETHOXY-5,5,5-TRIFLUORO-3-PENTENYL)-, (Z)-; (141708-71-6), **75**, 153

Trifluoromethanesulfonic acid: Methanesulfonic acid, trifluoro-; (1493-13-6), **76**, 294; **78**, 105, 178

Trifluoromethanesulfonic anhydride: Methanesulfonic acid, trifluoro-, anhydride; (358-23-6), **78**, 2, 27, 52, 91

3-(Trifluoromethyl)benzoyl chloride: Benzoyl chloride, 3-(trifluoromethyl)-; (2251-65-2), **77**, 163

Trimethyl borate: Boric acid, trimethyl ester; (121-43-7), **76**, 87

Trimethyl orthoacetate: Ethane, 1,1,1-trimethoxy-; (1445-45-0), **75**, 80

Trimethyl orthoformate: Methane, trimethoxy-; (149-73-5), **75**, 170, 171; **76**, 190, 288; **77**, 238

A [β-(TRIMETHYLSILYL)ACRYLOYL]SILANE, **76**, 199

Trimethylsilyl chloride: Silane, chlorotrimethyl-; (75-77-4), **76**, 24, 202, 252

2-TRIMETHYLSILYLETHANESULFONYL CHLORIDE: ETHANESULFONYL CHLORIDE, 2-(TRIMETHYLSILYL)-; (106018-85-3), **75**, 161

3-(Trimethylsilyl)propanesulfonic acid, sodium salt: 1-Propanesulfonic acid, 3-(trimethylsilyl)-, sodium salt; (2039-96-5), **77**, 190

2-(Trimethylsilyl)thiazole: Thiazole, 2-trimethylsilyl-; (79265-30-8), **77**, 79

1,3,5-Trioxane; (110-88-3), **75**, 189

Triphenylarsine: Arsine, triphenyl-; (603-32-7), **75**, 70; **77**, 135

(R)-(+)-1,1,2-Triphenylethanediol: 1,2-Ethanediol, 1,1,2-triphenyl-, (R)-; (95061-46-4), **77**, 45

Triphenylphosphine: Phosphine, triphenyl-; (603-35-0), **75**, 33, 53, 155; **76**, 180; **78**, 135, 143

Triphenylphosphine oxide hydrobromide: Phosphine oxide, triphenyl-, compd. with hydrobromic acid; (13273-31-9) **78**, 136

Triphosgene: Carbonic acid, bis(trichloromethyl) ester; (32315-10-9), **75**, 45; **78**, 220

Tris(acetonitrile)chromium tricarbonyl: Chromium, tris(acetoniytrile)tricarbonyl-; (16800-46-7), **77**, 121

Tris(chloromethyl)acetic acid: Propionic acid, 3-chloro-2,2-bis(chloromethyl)-; (17831-70-8); **75**, 90

Tris(dibenzylideneacetone)dipalladium(0): Palladium, tris(1,5-diphenyl-1,4-penta-dien-3-one)di-; (52409-22-0), **78**, 23

Tris(dimethyamino)borane: Boranetriamine, hexamethyl-; (4375-83-1), **77**, 176

Tropylium tetrafluoroborate: See Cycloheptatrienylium tetrafluoroborate, **75**, 210

Tryptones (Bacteriological), See: Peptones, Bacteriological; (73049-73-7*), **76**, 80

Ullmann methoxylation, **78**, 63

    table, **78**, 71

Ultrasonic cleaning bath, **77**, 81

Undecenoic acid: 10-Undecenoic acid ; (112-38-9), **75**, 124

N-(10-Undecenoyloxy)pyridine-2-thione: 2(1H)-Pyridinethione, 1-[(1-oxo-10-undecenyl)oxy]-; (114050-28-1), **75**, 124

Vilsmeier-Haack reagent, **75**, 167

Vinyl acetate: Acetic acid ethenyl ester; (108-05-4), **75**, 79

N-Vinylation, **75**, 45

Vinyllithium: Lithium, ethenyl-; (917-57-7), **76**, 190, 193

N-Vinyl-2-pyrrolin-2-one: 2-Pyrrolidinone, 1-ethenyl-; (88-12-0), **75**, 215

Vinyltrimethylsilane: Silane, ethenyltrimethyl-; (754-05-2), **75**, 161

VITAMIN D$_2$: 9,10-SECOERGOSTA-5,7,10(19),22-TETRAEN-3-OL, (3β)-; (50-14-6), **76**, 275

Vitamin D$_2$ 3,5-dinitrobenzoate: Ergocalciferol, 3,5-dinitrobenzoate; (4712-11-2), **76**, 276

Vortex mixer, **75**, 2

WITTIG OLEFINATION, OF PERFLUORO ALKYL CARBOXYLIC ESTERS, **75**, 153

Zeolites, 4 Å: See 4 Å Molecular sieves, **75**, 189

Zinc; (7440-66-6), **76**, 24, 143, 252; **78**, 99

Zinc chloride; (7646-85-7), **75**, 177; **78**, 53